ASTROPHYSICAL SOURCES FOR GROUND-BASED GRAVITATIONAL WAVE DETECTORS

Related Titles from AIP Conference Proceedings

565   Young Supernova Remnants: Eleventh Astrophysics Conference
Edited by Stephen S. Holt and Una Hwang, May 2001, 0-7354-0001-6

556   Explosive Phenomena in Astrophysical Compact Objects: First KIAS
Astrophysics Workshop
Edited by Heon-Young Chang, Chang-Hwan Lee, Mannque Rho, and Insu Yi,
March 2001, 1-56396-987-4

537   Waves in Dusty, Solar, and Space Plasmas
Edited by F. Verheest, M. Goossens, M. A. Hellberg, and R. Bharuthram, October 2000,
1-56396-962-9

526   Gamma-Ray Bursts: 5th Huntsville Symposium
Edited by R. Marc Kippen, Robert S. Mallozzi, and Gerald J. Fishman, June 2000,
CD-ROM included, 1-56396-947-5

523   Gravitational Waves: Third Edoardo Amaldi Conference
Edited by Sydney Meshkov, June 2000, 1-56396-944-0

516   26th International Cosmic Ray Conference: ICRC XXVI, Invited, Rapporteur,
and Highlight Papers
Edited by Brenda L. Dingus, David B. Kieda, and Michael H. Salamon
May 2000, 1-56396-939-4

515   GeV-TeV Gamma Ray Astrophysics Workshop: Towards a Major
Atmospheric Cherenkov Detector VI
Edited by Brenda L. Dingus, Michael H. Salamon, and David B. Kieda,
May 2000, 1-56396-938-6

493   General Relativity and Relativistic Astrophysics: Eighth Canadian Conference
Edited by C. P. Burgess and R. C. Myers, November 1999, 1-56396-905-X

456   Laser Interferometer Space Antenna: Second International LISA Symposium
on the Detection and Observation of Gravitational Waves in Space
Edited by William M. Folkner, December 1998, 1-56396-848-7

To learn more about these titles, or the AIP Conference Proceedings Series, please visit the webpage
http://www.aip.org/catalog/aboutconf.html

ASTROPHYSICAL SOURCES FOR GROUND-BASED GRAVITATIONAL WAVE DETECTORS

Philadelphia, Pennsylvania
30 October–1 November 2000

EDITOR
Joan M. Centrella
Drexel University, Philadelphia, Pennsylvania
and
NASA/Goddard Space Flight Center
Greenbelt, Maryland

AMERICAN INSTITUTE OF PHYSICS

Melville, New York, 2001
AIP CONFERENCE PROCEEDINGS ■ VOLUME 575

Editor:

Joan M. Centrella
NASA/Goddard Space Flight Center
Laboratory for High Energy Astrophysics
Code 661
Greenbelt, MD 20771
USA
E-mail: jcentrel@lheamail.gsfc.nasa.gov

L.C. Catalog Card No. 2001091141
ISBN 0-7354-0014-8
ISSN 0094-243X
Printed in the United States of America

CONTENTS

Preface

As the 21st century begins, gravitational wave astronomy is poised for unprecedented expansion and discovery. First-generation gravitational wave interferometers such as LIGO, VIRGO, GEO and TAMA will soon embark on their first scientific data-taking runs. In addition, LIGO II and other advanced interferometers are being planned, and new technology for bars is being developed worldwide. At the same time, new initiatives in astronomy and astrophysics are going forward with an impressive array of new telescopes and other detectors such as the planned space-based gravitational wave interferometer LISA. In concert with these developments, theoretical and numerical modeling of gravitational wave sources continues to provide a better understanding of possible sources and a growing ability to calculate waveforms from simulations.

Examining the expected gravitational wave frequencies and other characteristics of astrophysical sources is essential to stimulate and influence detector development. To this end, the *Workshop on Astrophysical Sources for Ground-Based Gravitational Wave Detectors* was held at Drexel University on October 30 - November 1, 2000. The workshop brought together gravitational wave physicists, astronomers, astrophysicists, and numerical relativists, with a focus on the capabilities of the instruments involved and the new directions that are being proposed. This book comprises the proceedings of the workshop.

The workshop and the preparation of this book were made possible by the support of the National Science Foundation, and the Department of Physics and the College of Arts and Sciences at Drexel University. Special thanks go to Lisa Lowe, who served as secretary for both the conference and the preparation of these proceedings. She is largely responsible for the smooth flow of the meeting, and for the timely appearance of these proceedings. Dept. Head Michel Vallieres, Jacqueline Sampson, Janice Murray, Dan Brennan, Wolf Nadler, Ernest Mamikonyan, and Bradley Kenney of the Physics Dept. are also gratefully acknowledged for their efforts in insuring that the workshop ran smoothly. And the workshop speakers and authors are thanked for their essential contributions towards the success of both the meeting and these proceedings.

One final note of appreciation goes to Richard Isaacson of the National Science Foundation. It was primarily through his efforts that Gravitational Physics became a separate office in the Division of Physics at the NSF, and that experimental, theoretical, and computational research in gravitation became a key part of the NSF Physics program. Richard has served as Program Officer for Gravitational Physics at the NSF throughout the efforts to conceive, build, and fund the LIGO detectors. His insightful and dedicated service on behalf of experimental, theoretical, and numerical relativity was celebrated at the conference banquet through a talk given by Charles Misner and the testimonials of other colleagues.

Joan M. Centrella
Conference Chair
Philadelphia
March 2001

Scientific Organizing Committee

Joan Centrella - chair *Drexel University*
Barry C. Barish *Caltech - LIGO*
L. Samuel Finn *The Pennsylvania State University*
Jay Gallagher *University of Wisconsin-Madison*
William Hamilton *Louisiana State University*
Steve McMillan *Drexel University*
David Spergel *Institute for Advanced Study & Princeton University*
Kip Thorne *Caltech*
Michael Vogeley *Drexel University*
Rainer Weiss *MIT - LIGO*

Ground-Based Gravitational Wave Detectors

First Generation Interferometers

Barry C. Barish

California Institute of Technology
Pasadena, CA 91125

Abstract. The status and plans for the first generation long baseline suspended mass interferometers TAMA, GEO, LIGO and Virgo are presented, as well as the expected performances.

INTRODUCTION

The effect of the propagating gravitational wave is to deform space in a quadrupolar form. The characteristics of the deformation are indicated in Fig. 1.

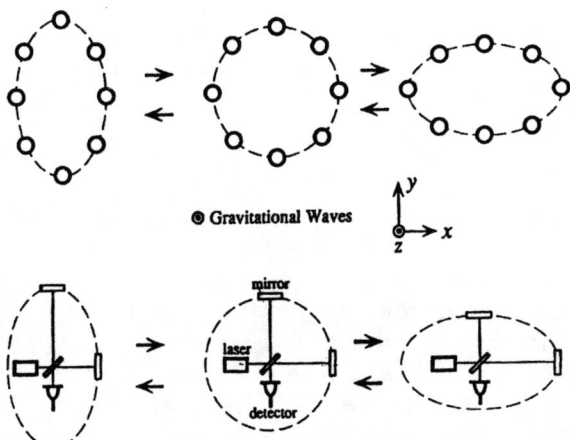

FIGURE 1. The effect of gravitational waves for one polarization is shown at the top on a ring of free particles. The circle alternately elongates vertically while squashing horizontally and vice versa with the frequency of the gravitational wave. The detection technique of interferometry being employed in the new generation of detectors is indicated in the lower figure. The interferometer measures the difference in distance in two perpendicular directions, which if sensitive enough could detect the passage of a gravitational wave

For an astrophysical source, one can estimate the frequency of the emitted gravitational wave. An upper limit on the gravitational wave source frequency can be estimated from the Schwarzshild radius $2GM/c^2$ of the radiated object. We do not

CP575, *Astrophysical Sources for Ground-Based Gravitational Wave Detectors*, edited by J. M. Centrella
© 2001 American Institute of Physics 0-7354-0014-8/01/$18.00

expect strong emission for periods shorter than the light travel time $4\pi GM/c^3$ around its circumference. From this we can estimate the maximum frequency as about $10^4\ Hz$ for a solar mass object. Of course, the frequency can be much lower as illustrated by the 8 hour period of PSR1916+13, which is emitting gravitational radiation. Frequencies in the higher frequency range $1Hz < f < 10^4\ Hz$ are potentially reachable using detectors on the earth's surface, while the lower frequencies require putting an instrument into space. The physics goals of the terrestrial detectors and the LISA space mission are complementary, much like different frequency bands are used in observational astronomy for electromagnetic radiation

The strength of a gravitational wave signal depends crucially on the quadrupole moment of the source. We can roughly estimate how large the effect could be from astrophysical sources. If we denote the quadrupole moment of the mass distribution of a source by Q, a dimensional argument, along with the assumption that gravitational radiation couples to the quadrupole moment yields:

$$h \sim \frac{G\ddot{Q}}{c^4 r} \sim \frac{G(E_{kin}^{non-symm.}/c^2)}{c^2 r} \tag{1}$$

where G is the gravitational constant and $E_{kin}^{non-symm.}$ is the non-symmetrical part of the kinetic energy.

For the purpose of estimation, let us consider the case where one solar mass is in the form of non-symmetric kinetic energy. Then, at a distance of the Virgo cluster we estimate a strain of $h \sim 10^{-21}$. This is a good guide to the largest signals that might be observed. At larger distances or for sources with a smaller quadrupole component the signal will be weaker

LONG BASELINE INTERFEROMETRY

A Michelson interferometer operating between *freely suspended* masses is ideally suited to detect the antisymmetric (compression along one dimension and expansion along an orthogonal one) distortions of space induced by the gravitational waves as was illustrated in figure 1. Other optical configurations or interferometer schemes, like a Sagnac, might also be used and could have advantages, but the present generation of interferometers discussed here are of the Michelson type.

The simplest configuration, a white light (equal arm) Michelson interferometer is instructive in visualizing many of the concepts. In such a system the two interferometer arms are identical in length and in the light storage time. Light brought to the beam splitter is divided evenly between the two arms of the interferometer. The light is transmitted through the splitter to reach one arm and reflected by the splitter to reach the other arm. The light traverses the arms and is returned to the splitter by the distant arm mirrors. The roles of reflection and transmission are interchanged on this return and, furthermore, due to the Fresnel laws of E & M the return reflection is accompanied by a sign reversal of the optical electric field. When the optical electric fields that have come from the two arms are recombined at the beam splitter, the beams that were treated to a reflection (transmission) followed by a transmission

4

(reflection) emerge at the antisymmetric port of the beam splitter while those that have been treated to successive reflections (transmissions) will emerge at the symmetric port.

In a simple Michelson configuration the detector is placed at the antisymmetric port and the light source at the symmetric port. If the beam geometry is such as to have a single phase over the propagating wavefront (an idealized uniphase plane wave has this property as does the Gaussian wavefront in the lowest order spatial mode of a laser), then, providing the arms are equal in length (or their difference in length is a multiple of 1/2 the light wavelength), the entire field at the antisymmetric port will be dark. The destructive interference over the entire beam wavefront is complete and all the light will constructively recombine at the symmetric port. The interferometer acts like a light valve sending light to the antisymmetric or symmetric port depending on the path length difference in the arms.

If the system is balanced so that no light appears at the antisymmetric port, the gravitational wave passing though the interferometer will disturb the balance and cause light to fall on the photodetector at the dark port. This is the basis of the detection of gravitational waves in a suspended mass interferometer. In order to obtain the required sensitivity, the arms of the interferometer must be long.

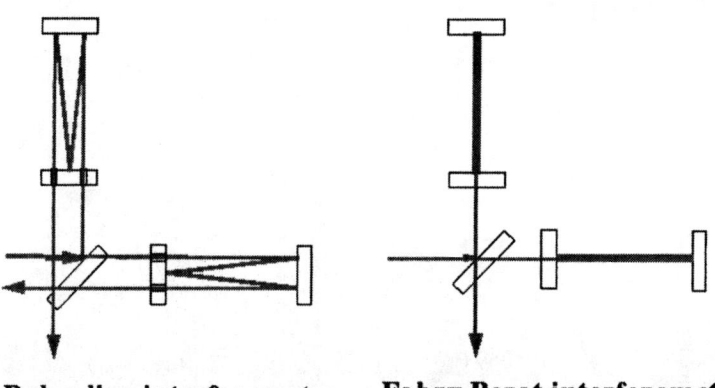

Delay line interferometer   **Fabry Perot interferometer**

FIGURE 2. Folded optical configurations for interferometer. The arrangement on the left is called a delay line interferometer and the one on the right using a resonant cavity is a Fabry Perot interferometer. The GEO600 interferometer is a delay line interferometer, while the all the other long baseline interferometers use Fabry Perot resonant cavities.

The amount of motion of the arms to produce an intensity change at the photodetector depends on the optical length of the arm; the longer the arm the greater is the change in length up to a length that is equal to 1/2 the gravitational wave wavelength. Equivalently the longer the interaction of the light with the gravitational wave, up to 1/2 the period of the gravitational wave, the larger is the optical phase shift due to the gravitational wave and thereby the larger is the intensity change at the photodetector. The initial long baseline interferometers, besides having long arms also will fold the optical beams in the arms in optical cavities or delay lines to gain further increase in the path length or equivalently in the interaction time of the light with the gravitational wave (Fig. 2). The initial LIGO interferometers will store the light about 50 times longer than the beam transit time in an arm. (A light storage time of about 1 millisecond.)

Another feature employed in these interferometers is to increase the change in intensity due to a phase change at the antisymmetric port by making the entire interferometer into a resonant optical storage cavity. The fact that the interferometer is operated with no light emerging at the antisymmetric port and all the light that is not lost in the mirrors or scattered out of the beam returns toward the light source via the symmetric port, makes it possible to gain a significant factor by placing another mirror between the laser and the symmetric port and 'reuse the light'. This technique is general referred to as power recycling. By choosing this mirror's position properly and by making the transmission of this mirror equal to the optical losses inside the interferometer, one can "match" the losses in the interferometer to the laser so that no light is reflected back to the laser. As a consequence, the light circulating in the interferometer is increased by the reciprocal of the losses in the interferometer. This is equivalent to increasing the laser power and does not affect the frequency response of the interferometer to a gravitational wave. The power gain achieved by this scheme can be a factor of 10 or even 100.

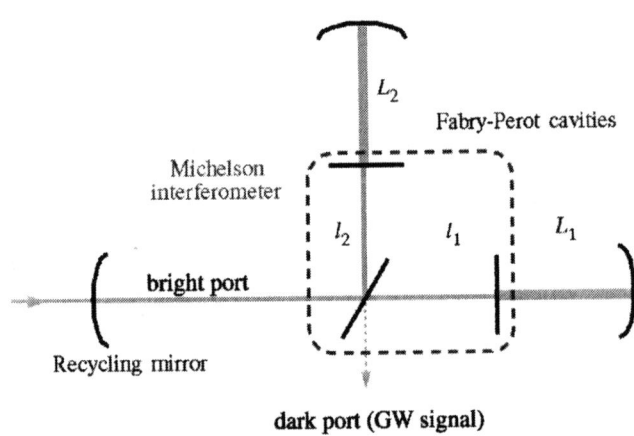

FIGURE 3. The optical configuration of a Michelson interferometer with Fabry-Perot arms is shown. A relative change in length of the two arms causes a phase shift destroying the destructive interference and the dark port detects a signal

The system just described is called a power recycled Fabry-Perot Michelson interferometer and it is this type of configuration that will be used in the initial interferometers (Fig. 3). There are many other possible types of interferometer configurations, such as narrow band interferometers with the advantage of increased sensitivity in a narrow frequency range. This can be accomplished by adding yet another mirror in the output port and is generally called signal recycling. Such interferometers are planned for future versions of the various interferometers facilities and the GEO600 interferometer has already incorporated this capability.

The interferometer parameters of the new set of detectors just being brought online have been chosen such that the initial sensitivity will be consistent both with the dimensional arguments given above and with estimates needed for possible detection of known sources. Although the rate for these sources are very uncertain, large increases in sensitivity are anticipated as the detectors are improved. This is because the planned incremental improvements in sensitivity correspond to the cube of that improvement in the event rate, which scales as the volume searched.

THE INTERFEROMETER NOISE FLOOR

The success of the detector ultimately will depend on how well we one can to control the noise in the measurement of these small strains. Noise is broadly but also usefully categorized in terms of those phenomena which limit the ability to sense and register the small motions (sensing noise limits) and those that perturb the masses by causing small motions (random force noise). Eventually one reaches the ultimate limiting noise, the quantum limit, which combines the sensing noise with a random force limit. This orderly and intellectually satisfying categorization presumes that one is careful enough as experimenters in the execution of the experiment that one has not produced less fundamental, albeit, real noise sources that are caused by faulty design or poor implementation. These might be referred to as technical noise sources and in real life these have often been the impediments to progress and mask the limiting noise sources of the interferometer. The primary noise sources for the initial detectors are illustrated in Fig. 4, where the estimated levels of the various noise sources are shown for LIGO. The other interferometers have similar curves with some difference in detail due to the different trade-offs that have been made.

In order to control the technical noise sources, extensive use is made of two concepts. The first is the technique of modulating the signal to be detected at frequencies far above the $1/f$ noise due to the drift and gain instabilities experienced in all instruments. For example, the optical phase measurement to determine the motion of the fringe is carried out at radio frequency rather than near DC. Thereby, the low frequency amplitude noise in the laser light will not directly perturb the measurement of the fringe position. (The low frequency noise still will cause radiation pressure fluctuations on the mirrors through the asymmetries in the interferometer arms.) A second concept is to apply feedback to physical variables in the experiment to control the large excursions at low frequencies and to provide damping. The variable is measured through the control signal required to hold it stationary. Here a good example is the position of the interferometer mirrors at low frequency. The interferometer fringe is maintained at a fixed phase by holding the mirrors at fixed positions at low frequencies. Feedback forces to the mirrors effectively hold the mirrors "rigidly". In the initial LIGO interferometers the forces are provided by permanent magnet/coil combinations. The mirror motion that would have occurred is then read in the control signal required to hold the mirror.

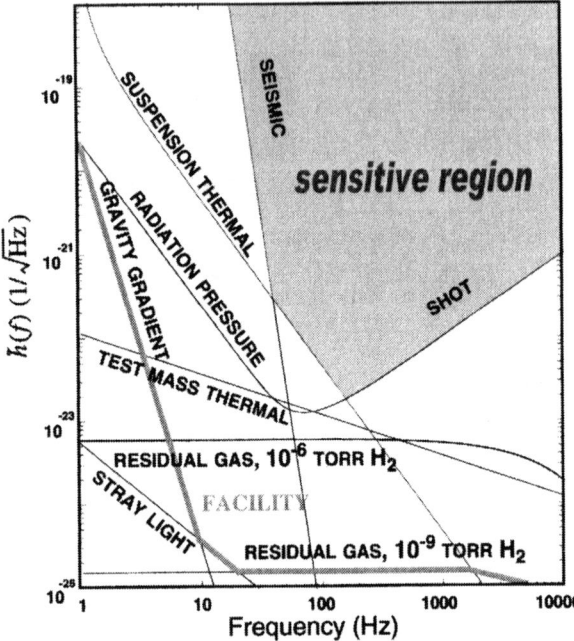

Figure 4. Limiting noise sources for the initial LIGO detectors. Note that the interferometer is limited by different sources at low frequency (e.g. seismic), middle frequencies by suspension thermal noise, and at high frequencies by shot noise (or photo statistics). Lurking below are many other potential noise sources

Great care must be taken to control the technical noise sources. In order to test and understand the sensitivity and limiting noise, extensive tests have been performed with a 40 meter LIGO prototype interferometer on the Caltech campus. This interferometer essentially has all the pieces and the optical configuration used in LIGO, so represents a good place to understand noise and performance before the full-scale LIGO interferometers are in operation. The 40 m prototype device has achieved a displacement sensitivity of $h \sim 10^{-19}$ m/Hz$^{1/2}$, which is close to the displacement sensitivity that is required in the 4 km LIGO interferometers. Fig. 5 shows the measured noise curve in this prototype instrument. The modeled noise sources are shown and for the most part are a good representation of the observed performance, including the most of the line structures that are due to wire resonances and other such sources. The same model has been used to determine the expected sensitivity curve for LIGO.

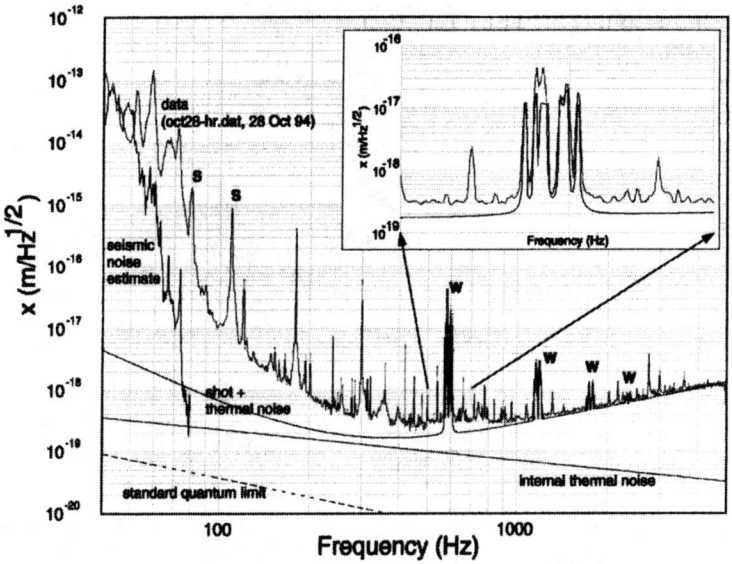

Figure 5. The displacement noise measured in the 40m suspended mass interferometer LIGO prototype on the Caltech campus. The general shape and level are well simulated by our understanding of the limiting noise sources - seismic noise at the lowest frequencies, suspension thermal noise at the intermediate frequencies, and shot noise at the highest frequencies. Also, the primary line features are understood as various resonances in the suspension system.

STATUS OF THE INTERFEROMETER PROJECTS

TAMA300

The first of the new generation of interferometers (TAMA300) has begun initial operation over the past year. TAMA300 is a Fabry-Perot Michelson interferometer with arm lengths of 300m. The site is at the National Astronomical Observatory of Japan (Tokyo, Mitaka). The interferometer uses a 10W injection-locked LD-pumped Nd:YAG laser and employs a 10m-length ring-type cavity mode cleaner to condition the input beam. The interferometer is designed for power recycling with gain of x10, which has not yet been employed. The design sensitivity is $h_{RMS} = 3 \times 10^{-21}$. The best sensitivity achieved to date is a displacement noise of about 4×10^{-18} m/Hz$^{1/2}$, corresponding to a strain sensitivity of approximately $h \sim 1 \times 10^{-20}$/Hz$^{1/2}$.

The improvement of the noise curve as the interferometer has been commissioned over the past year is shown in Fig. 6. The figure shows the improvements in the noise curve during the first eight months of operation, and Fig. 7 shows a break down of the main contributors to the curve from various noise sources.

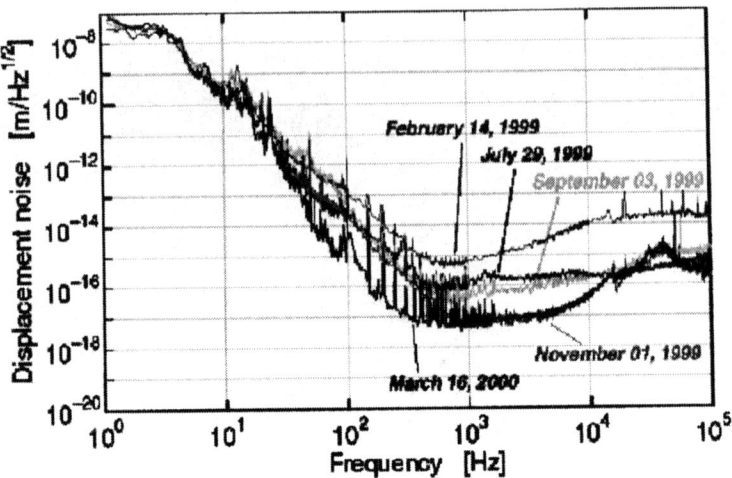

FIGURE 6. The performance of the TAMA300 interferometer, the first of the new generation of interferometers to begin operation is shown. The top curve shows the dramatic improvement from Sept 1999 when the interferometer established its first noise curves until March 16, 2000 when the noise curve had been improved by about 4 orders of magnitude.

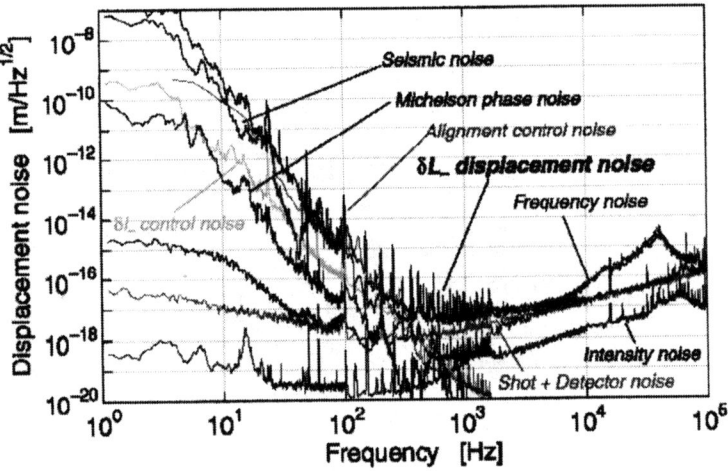

Figure 7. For the March run the noise in TAMA has been broken up into its various components. Improvements are planned in the near future to improve the seismic and other dominant noise sources

GEO600

The GEO600 interferometer facility, a U.K.- German collaboration is located in Hanover, Germany. It has 600m long arms and incorporates signal recycling and an advanced suspension system. The triple suspension system (fig 8) consists of an upper mass/cantilever spring and intermediate mass and a test mass / reaction mass combination. The controls are applied to the upper masses to reduce noise. 4 x 180mm silica wires are welded to the silica test mass. GEO construction is complete and commissioning is underway. The vacuum system is operational, central optics are installed and a long arm test of the interferometer is in progress.

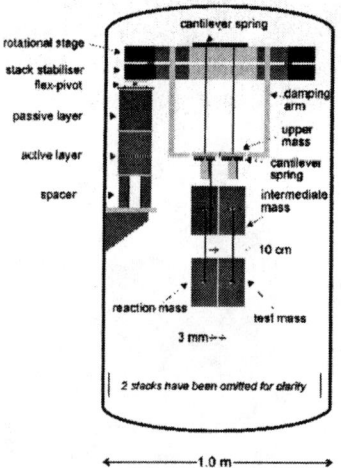

Figure 8. On the left is a schematic of the GEO600 triple suspension system, which employs drawn silica wires bonded to the test masses. The controls are applied only to the upper masses to reduce noise.

LIGO

LIGO at Hanford has both a half-length (2km) and a full-length (4km) interferometer installed in the same vacuum chamber. The extra constraint of requiring a half-size signal in the shorter interferometer will be used to eliminate common noise and lower the singles rate in the coincidence between the sites. Over the past year, the long 2 km arm cavities were locked one at a time for typical few hour times, at which point lock was lost due to tidal effects. The electronics to compensate for tidal effects had not yet been installed

Following the locking of the long arms early in 2000, attention was shifted to the power-recycled Michelson part of the interferometer, formed by the input mirrors, as well as the beam splitter and the recycling mirror. To make LIGO as sensitive as possible, one wants as much light as possible returning to the beam splitter. This occurs when the recycling mirror is placed at the correct distance from the beam splitter, 'trapping' the reflected light in the Michelson interferometer. This causes the laser light in the interferometer to build up to a high level. For the full system, recycling will cause the light to build up by about a factor of thirty. When this build up occurs, the power-recycled Michelson interferometer is resonating. Such a resonant state was achieved with the short arm system during the summer.

The final step was to lock the full interferometer, achieving 'first lock' in the LIGO interferometers. This was achieved this fall for short lock periods as shown in Fig. 9. Robust locking is now achieved in the recombined configuration and a very successful one-week engineering test run in his configuration was carried out this month. A variety of electronic dynamic range and gain issues are presently limiting the stability of the lock for the power recycled configuration, which will be the focus of the activities in the coming months, as well as the commissioning of the interferometer at Livingston. The first coincidence test running between the two sites is planned for summer 2001.

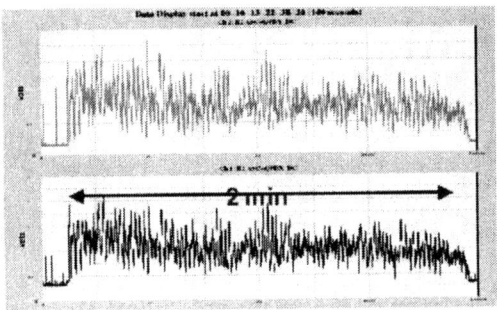 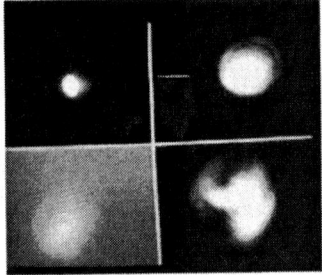

FIGURE 9. Data recording the first lock in LIGO of the full power recycled Michelson Fabry-Perot interferometer. The traces on the left show a 2 minute stretch where lock was held stable, while the figure on the right shows images of the locked beam at the recycling mirror (bottom left), beam splitter (bottom right); and at the end test masses of each 2km arm (top images). The large contrast in the top photos is an indication of the intensity of the resonating beam in the arms. The difference in sizes of the images in the two arms is an artifact of the magnification of the cameras.

Virgo

The Virgo Experiment is a French-Italian collaboration and is located at Cascina, Italy not far from Pisa. The infrastructure housing the central interferometer is complete and the sophisticated seismic-suspension system is being implemented. This Virgo seismic-suspension system, which is designed to give the best sensitivity of the new generation of detectors at low frequencies, is illustrated in Fig. 10, as well as the performance from a full-scale prototype. All four long suspensions for the entire central interferometer are scheduled for completion by the end of 2000.

The laser, input optics and mode cleaner have been installed and testing of this system, which will lead to a commissioned short arm version of Virgo are planned while the long arms are under construction. The first long arm is scheduled to be completed by 2002, as well as the large optics. The full long baseline interferometer is scheduled to begin operations by the end of 2002.

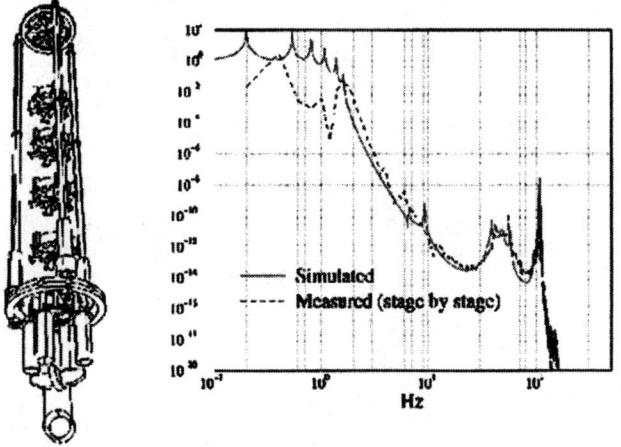

Figure 10. The Virgo long suspension with five intermediate stages is shown on the left. The figure on the right shows the suspension vertical transfer function measured and simulated (prototype).

AIGO

Initial steps toward a long baseline interferometer facility have been initiated near Perth, Australia. A collaboration of Australian Universities (ACIGA) is developing the facility, as well as actively collaborating with the other interferometer groups. The consortium is doing research on high power lasers, new optical configurations, suspensions, test masses, etc. The central facility for AIGO has been constructed and work is underway to build the central optics and a short arm interferometer as the first step toward the hope of eventually extending the arms to a long baseline interferometer to be used in conjunction with those being developed in the northern hemisphere. In addition, an R & D program to test optics at the high powers envisioned for second-generation interferometers is being planned.

CONCLUSIONS

The first generation interferometers having sensitivities that should allow searches with $h \sim 10^{-21}$ are becoming operational. These will be the first experimental searches that approach the sensitivity where detection of gravitational waves is plausible. The design sensitivities of the major interferometer projects are shown in Fig. 11. In addition to the anticipated 'stand-alone' searches using each of these detectors, discussions and preparations are underway to bring the data together from the different detectors to do coincidence work to both improve overall sensitivity and confidence in detection. Significant improvements to these detectors are anticipated to be initiated in about 5 years that should improve the sensitivity to $h \sim 10^{-22}$.

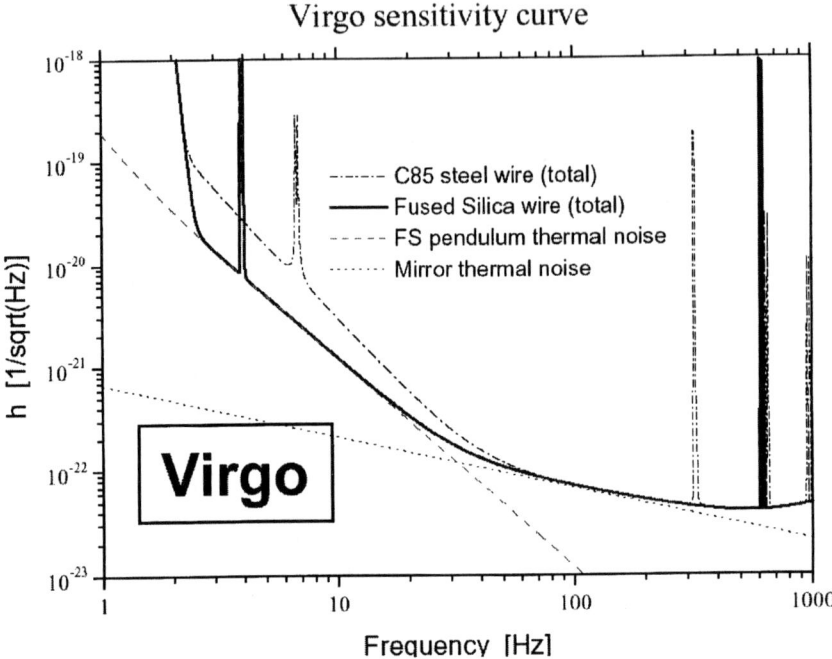

FIGURE 11. The design sensitivities of the Virgo interferometer is shown. This is illustrative of what will be obtained in the next few year from the new generation of detectors.

REFERENCES

1. LIGO http://www.ligo.caltech.edu/
2. Virgo: http://www.virgo.infn.it/
3. GEO600 http://www.geo600.uni-hannover.de/
4. TAMA http://tamago.mtk.nao.ac.jp/
5. AIGO http://www.gravity.uwa.edu.au/AIGO/AIGO.html

The Second Generation LIGO Interferometers

Peter Fritschel

LIGO Project, Massachusetts Institute of Technology, 175 Albany St., NW17-161, Cambridge, MA 02139

Abstract. The interferometers being planned for second generation LIGO promise an order of magnitude increase in broadband strain sensitivity–with the corresponding cubic increase in detection volume–and an extension of the observation band to lower frequencies. In addition, one of the interferometers may be designed for narrowband performance, giving further improved sensitivity over roughly an octave band above a few hundred Hertz. This article discusses the physics and technology of these new interferometer designs, and presents their projected sensitivity spectra.

The initial LIGO interferometers were designed–as much as possible–using proven concepts and technologies; the idea was to get into the business of collecting and analyzing gravitational wave data as quickly, yet to build into the LIGO facilities the capability of supporting much more sensitive instruments in the future. Following the collection of one integrated year of science data with the initial detectors, we would continue carrying out this strategy by implementing an advanced set of interferometers (LIGO II). The strain sensitivity goals for this next generation of LIGO interferometers advance the detection likelihood from being plausible to probable. The designs being developed promise an improvement over the initial LIGO strain sensitivity by a factor of 15 in the spectral region of maximum sensitivity, 100 Hz < f < 200 Hz, and also reduce the lower end of the sensitive band from 40 Hz to 10 Hz. The impressive effect of these improvements is that the first few hours of LIGO II operation will surpass the space-time volume probed by the entire initial LIGO science run.

To achieve these sensitivity advances, virtually every aspect of the initial LIGO interferometer design must be upgraded (see Fig. 1). A new seismic isolation system is needed to push the seismic wall down to 10 Hz; new suspensions and test masses are needed to dramatically reduce thermal noise; much higher laser power and a more efficient interferometer configuration are needed to push down sensing noise; more massive test masses are needed to combat the increased radiation pressure. As Fig. 2 shows, the sensitivity is brought very close to the limit imposed by gravity gradients at 10 Hz, but otherwise it is still far from the facility limitations, leaving that challenge to yet later generation instruments.

Table 1 highlights the performance and design differences between the LIGO I and LIGO II interferometers. The configuration is a power-recycled and signal-recycled Michelson interferometer, with Fabry-Perot cavities in the arms–thus the basic LIGO I configuration is kept, and a signal recycling mirror is added at the output[1]. This extra mirror provides capability to tailor the interferometer response to strain: we can choose the frequency of maximum response and the bandwidth about that point, and use these parameters to optimize the detection sensitivity to a particular source or source class. Figure 3 illustrates the types of possible response curves. The peak response frequency

CP575, *Astrophysical Sources for Ground-Based Gravitational Wave Detectors,* edited by J. M. Centrella
© 2001 American Institute of Physics 0-7354-0014-8/01/$18.00

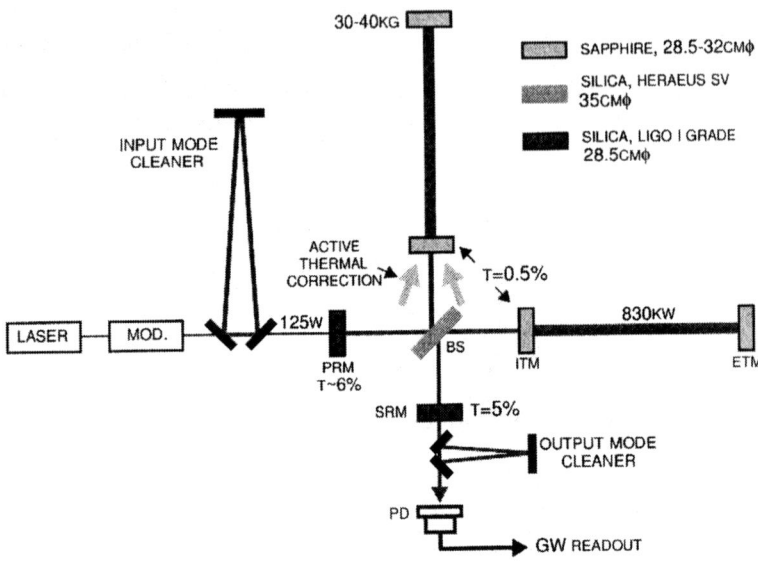

FIGURE 1. Schematic of a LIGO II interferometer, with mirror reflectivities optimized for neutron star binary inspiral detection. Several new features compared to initial LIGO are shown: more massive, sapphire test masses; 20× higher input laser power; signal recycling; active correction of thermal lensing; an output mode cleaner. This is a snapshot of an evolving design.

can be selected by varying the microscopic position of the signal recycling mirror, although for the maximum response the mirror reflectivity must also be changed. For LIGO II, we optimize the response for detection of neutron star binary inspirals; this fixes the signal recycling mirror reflectivity, leaving some limited in-situ freedom to affect the response via the signal mirror's microscopic position. Compared to an optimized interferometer without signal recycling, the additional mirror increases the inspiral detection range by 25-30%.

The diagram in Fig. 3 helps conceptualize the effect of signal recycling. The signal sidebands produced in the arm cavities by gravitational-wave (or other) strains 'see' a compound output coupler formed by the arm input mirrors and the signal recycling mirror. The equivalent, frequency-dependent reflectivity of this compound mirror may be higher or lower than the reflectivity of the arm input mirror, depending on the individual mirror reflectivities and the signal frequency. The main laser field is on a Michelson dark fringe and thus does not see the signal recycling mirror; the arm buildup of the laser field is determined only by the optical properties of the arm cavities, the power recycling mirror and the beam splitter. At one end of the design space, the arm cavity bandwidth is made very narrow (high finesse), but is effectively increased for the signal field by the signal recycling mirror, limited by losses in the signal cavity (formed by the signal recycling mirror, the beam splitter, and the arm input mirrors); in this case the laser power in the power cavity (formed by the power recycling mirror, the beam split-

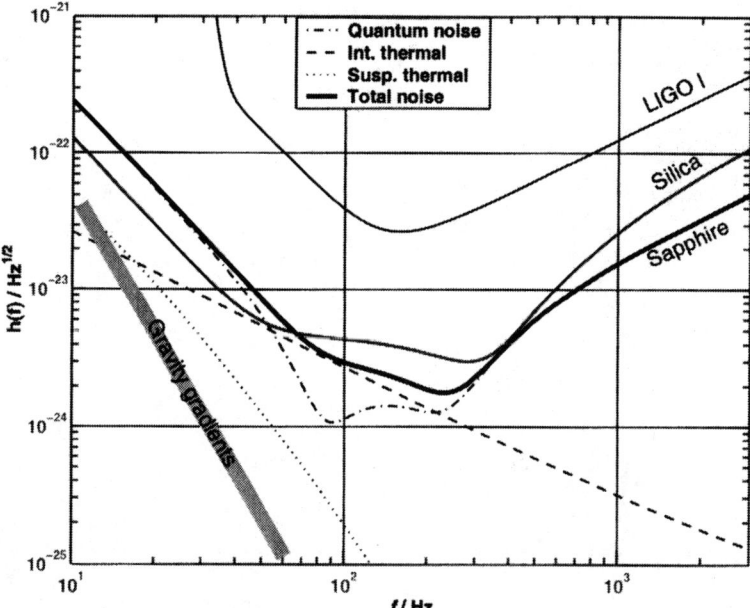

FIGURE 2. Strain sensitivity projections for LIGO II, showing the total noise curve using sapphire test masses (thick black line), and fused silica test masses (thick grey line), as well as the quantum and thermal noise components for the sapphire design. Also shown is the estimate for gravity gradient noise during quiescent times, from reference 2.

ter, and the arm input mirrors) is relatively low. At the other end of the design space, the arm cavity bandwidth is made very wide (low finesse), but is effectively narrowed by the signal recycling mirror for the signal field; in this case there is a relatively high buildup in the power cavity, limited again by optical losses. The choice of mirror reflectivities is made considering the impact of the different losses: nonlinear thermal absorption losses argue for lower power recycling gain, and thus higher arm finesse; whereas losses in the signal recycling cavity limit how high the arm finesse can be made and still effectively lowered for the signal field through their coupling into this cavity. Given our estimates of these losses, the optimization between these two effects appears to occur around an arm finesse of 1000.

Until recently, sensing noise was calculated simply by multiplying the inverse of the response function (given in radians/strain) by the shot noise implied by the power on the beam splitter, given in rad/$\sqrt{Hz}$. Quantum radiation pressure was then calculated by enforcing the photon number-phase uncertainty relation. Recent work has shown that this oversimplifies the situation (see K. Thorne's article in this proceedings). The full quantum mechanical approach shows that the addition of the signal recycling mirror results in a correlation between quantum shot noise and quantum radiation pressure [3]. At high and low frequencies the noise may still be said to be due to photon counting statistics and radiation pressure, respectively, but at intermediate frequencies–in fact in the

17

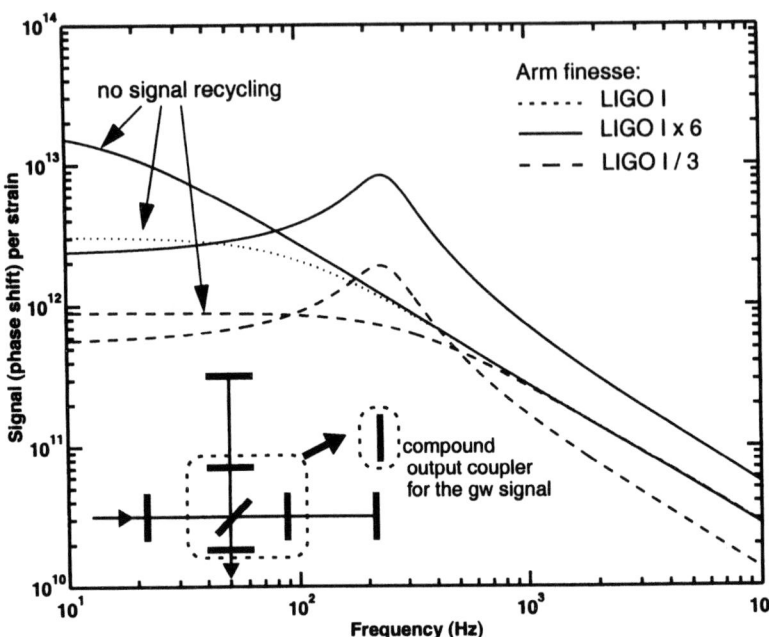

FIGURE 3. Interferometer gravitational-wave response curves with and without signal recycling. Shown is the optical phase shift per strain for three values of arm finesse: same as LIGO I; 6x the LIGO I finesse; 1/3 of the LIGO I finesse. For the latter two arm finesses, a response curve with signal recycling is shown; the high arm finesse signal recycled curve corresponds to the nominal LIGO II design, and the other curve illustrates that the same response shape can be obtained with a different arm finesse, by adjusting the signal mirror parameters. The low arm finesse signal recycled case has a overall smaller response, but this is compensated by the fact that the beamsplitter power would be higher, in theory yielding the same sensitivity for these two cases. In practice, nonlinear thermal lensing losses favor the design with higher arm finesse and lower beamsplitter power.

most sensitive band–the two effects becomes indistinguishable. Thus in plots of strain sensitivity, where once there were two curves for shot noise and radiation pressure, there is now a single 'quantum noise' curve.

The LIGO II design comes close to being a completely quantum noise limited interferometer, with the exception of the 50-250 Hz region, where internal thermal noise of the test masses is dominant. Improved internal thermal noise performance relative to LIGO I comes from using sapphire for the test mass material. Sapphire has a much higher quality factor than the initial fused silica optics–2×10^8 versus $2 - 3 \times 10^6$ for a typical LIGO I test mass. Unfortunately, this advantage is offset somewhat by sapphire's higher thermo-elastic damping[4]. In this noise mechanism, thermodynamical temperature fluctuations act through the material's coefficient of thermal expansion to produce fluctuations of the test mass surface. Sapphire suffers relatively strongly from this because of its high thermal expansion and thermal conductivity coefficients. The effect can be countered to some degree by increasing the beam size on the mirrors, thereby reducing the temperature gradients in the material and the corresponding heat loss. As a result we have increased the beam size from 3.7 cm and 4.3 cm radius (on the input and end mirrors, respectively, in the initial interferometers) to 5.5 cm radius on all

Parameter	LIGO I	LIGO II
Equivalent strain noise, minimum	$3\times10^{-23}/\sqrt{\text{Hz}}$	$2\times10^{-24}/\sqrt{\text{Hz}}$
Neutron star binary inspiral detection range[a]	19 Mpc	285 Mpc
Stochastic background sensitivity[b]	3×10^{-6}	1.5–8×10^{-9}
Interferometer configuration	Power-recycled Michelson w/ FP arm cavities	LIGO I, plus signal recycling
Input laser power	6 W	125 W
Test masses	fused silica, 11 kg	sapphire, 30-40 kg
Suspension system	single pendulum, steel wires	quad pendulum, silica fibers/ribbons
Seismic isolation system, type	passive, 4-stage	active, 2-stage
Seismic wall frequency	40 Hz	10 Hz

Table 1: Comparison of interferometer design and performance for LIGO I and LIGO II.

a. Range for all three LIGO interferometers, assuming for LIGO II that two are optimized for binary inspirals and the third is a narrowband instrument.
b. The number is $h_{100}^{2} \cdot \Omega_{gw}$, where h_{100} is the Hubble constant in units of 100 km/s/Mpc and Ω_{gw} is the GW energy density per unit logarithmic frequency interval in units of the closure density.

mirrors in LIGO II. Somewhat larger beams may be possible, though competing issues of diffraction loss and cavity mode stability would have to be carefully addressed. Also at issue is the current uncertainty in the values of the thermal expansion and conductivity coefficients for the sapphire material of choice; for some published values the thermo-elastic noise estimate is high enough that sapphire loses its performance advantage over silica. Finally, sapphire material development is not as mature as fused silica. Though producing large pieces appears to be possible, good uniformity may be difficult to achieve; sapphire also displays much higher optical absorption than silica–less harmful in theory given sapphire's high thermal conductivity, but potentially more problematic if it is spatially nonuniform.

Research and development of sapphire is underway in industry and within the LIGO Science Collaboration to address these issues, but given the uncertainties with sapphire we are also maintaining a design based on fused silica test masses. Silica offers room for improvement compared to initial LIGO. When lossy materials such as magnets (used in initial LIGO to control the mirror position) are kept off the surface, the intrinsic Q of fused silica is seen to be much higher, with $Q \approx 5\times10^{7}$ recently measured[5]. Increasing the beam size helps here as well, though not as quickly as with thermo-elas-

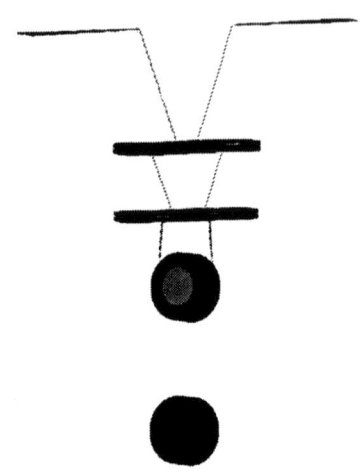

FIGURE 4. Quadruple pendulum for the LIGO II test masses. The bottom mass is the test mass optic, and is suspended from the penultimate mass by fused silica fibers (either circular or ribbon geometry). The upper stages use cantilevered blade springs for high vertical compliance. All local damping is applied at the upper-most suspended stage.

tic damping (the latter being insignificant in silica, since it is much less thermally conductive and expansive than sapphire).

For either material, the test mass size is increased significantly for LIGO II. Simply increasing the mass from 10 kg to 30-40 kg is important to keep the low-frequency regime of quantum noise low (otherwise known as radiation pressure). Larger diameter is needed to support the increased beam size in the cavities. We are aiming to use the largest test masses obtainable with the requisite quality, though we are still far from the 1-ton test mass often touted for the ultimate interferometer design. Both test mass materials may also suffer from mechanical loss in the multi-layer dielectric coatings which are deposited to give the required reflectivities; their effects are currently being investigated experimentally, but the concern is that they may be more lossy than the underlying substrate and lead to increased thermal noise.

Thermal noise in the mirrors' suspension systems has also been greatly reduced in the LIGO II design. This is primarily the result of using fused silica fibers to suspend the test mass, which exhibit approximately $10^4 \times$ lower intrinsic loss than the steel wires used in initial LIGO[6]. The fiber ends must be attached to the test mass and the stage above carefully–the mechanical joints must not produce additional loss. This is accomplished through the new technique of hydroxy-catalysis bonding. The suspension thermal noise can be further reduced by using a ribbon geometry in the fiber. For the same level of stress in the fiber, an appropriately oriented ribbon (width much greater than the thickness) will be more compliant along the suspended mass's sensing direction, further diluting the intrinsic loss of the fiber. These advances have made suspension thermal noise–one of the dominant noise sources for initial LIGO–essentially insignificant for

LIGO II. Providing sufficient damping of the rigid body modes of such a low-noise pendulum, without introducing excess noise, is a major design challenge. The solution is to use a multiple pendulum, and damp all modes from the top stage so that the noise in the damping controls can be filtered by the lower stages. A four-stage suspension is needed for sufficient filtering (Fig. 4); it is an extension of the triple pendulums used in the GEO 600 interferometer. In addition to the optic axis isolation provided by the four stages, much greater vertical isolation is achieved through the use of cantilevered blade springs for high vertical compliance.

The four suspension stages naturally provide a great deal of seismic noise attenuation as well–a factor of 10^7 at 10 Hz. Additional attenuation is given by the seismic isolation system, which is a two-stage active isolation system [7], a large departure from initial LIGO's 4-stage passive system. This system uses a collection of high-sensitivity seismometers and geophones to sense and stabilize via feedback all rigid body degrees-of-freedom of its two stages. The isolation it provides is complementary to the suspension isolation, giving an attenuation of 10^7 from 1-10 Hz, with significant attenuation down to 0.1 Hz. Isolation at frequencies below the gravitational wave band is crucial for controlling various technical noise sources, such as laser amplitude noise and noise in the interferometer's global control system.

The heavy filtering of seismic noise provided by the suspensions and seismic isolation create a 'seismic wall' in frequency space–at the test masses, seismic noise falls roughly as $1/f^9$ around 10 Hz. This creates a seismic cutoff frequency, f_c, where seismic noise moves the test masses by the same amount as the predominant fundamental noise source, either quantum noise (radiation pressure) or thermal noise. The seismic cutoff frequency goal in LIGO II is 10 Hz, pushed down from 40 Hz for initial LIGO. This goal is based partly on consideration of specific astrophysical source detection, and partly on technical feasibility. The estimated neutron star binary inspiral range does not in fact change significantly for f_c below ~20 Hz, though detection of more speculative black hole-black hole mergers could benefit from $f_c < 20$ Hz [8]. Technically, a cutoff at approximately 10 Hz appears feasible without undo risks in the design, so we choose to preserve strain sensitivity down to about 10 Hz.

The 20-fold increase in laser power for LIGO II is the result of continued incremental progress in laser technology. It will continue to be a diode-pumped Nd:YAG laser as in LIGO I, and the wavelength will remain at 1064 nm. The design may be a master oscillator-power amplifier type, as in LIGO I, or a high-power oscillator injection locked to a stable, low-power master. The higher power raises many technical issues regarding power dissipation in the system. This is most troublesome in the beam splitter and arm cavity input mirrors, where bulk or surface absorption leads to thermal lensing and elastic deformation losses. Since these losses scale with the laser power, the effect on interferometer performance is nonlinear, such that for given absorption coefficients there is a relatively hard upper limit to the power sustainable in the system. The LIGO I interferometer is already operating very near this upper limit. High power is of course needed in the arm cavities–where some fraction may be absorbed by the mirror surfaces–to achieve low quantum sensing noise, but the signal recycling configuration allows us to lower the beam splitter power (and thus the power in the input mirror sub-

21

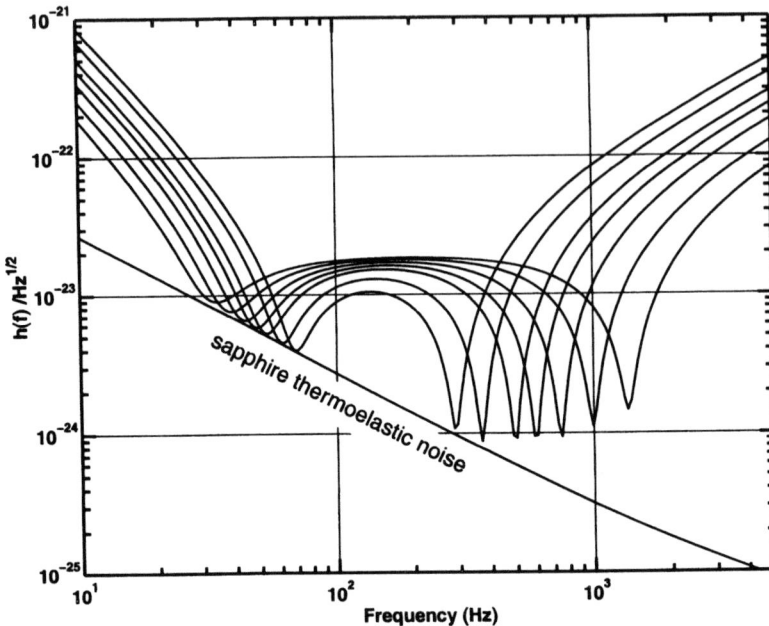

FIGURE 5. Strain sensitivity curves for a narrowband interferometer. With a single signal recycling mirror chosen to give optimum performance around 700 Hz, good performance between ~500–1000 Hz can be achieved by tuning the signal mirror position microscopically; the set of curves shown span a mirror motion of about 10^{-2} wavelength. At the lower end of the octave, sapphire's thermoelastic noise limits the performance; at higher frequencies, above ~500 Hz, sapphire has a clear advantage over fused silica for narrowband performance.

strates) to mitigate bulk absorption. We also plan on providing active compensation of the laser beam absorption effects. The idea is to reduce the thermal gradients induced via the main beam heating by adding heat preferentially around the outer volume of the optic, either through a circular radiative heating element positioned close to the optic, or through an external laser beam than is scanned across the optic surface. Such a technique should reduce the optical path distortions by an order of magnitude or more.

As Fig. 2 shows, the interferometer optimized for neutron star binary inspiral detection displays good broadband performance as well[1], simply because the inspiral signal is relatively broadband. While such an interferometer could be tuned to higher frequency by adjusting the signal mirror position, it would not have very good performance without changing the signal mirror reflectivity. To provide better high frequency sensitivity, an option being considered is to design the third interferometer for narrowband performance. This is achieved with a higher signal mirror reflectivity; as Fig. 5

---

1. If we define the bandwidth as delimited by the frequencies where the strain noise is 10× the minimum noise level, the bandwidth for the sapphire design is 30 Hz – 1.5 kHz.

shows, an appropriate reflectivity value can still yield good narrowband performance over roughly an octave band, e.g. 500-1000 Hz.

Thus the current strategy is to upgrade all three LIGO interferometers, optimizing two for NS-NS inspiral detection (one at each site), with the third interferometer– increased in length from 2km to 4km–made a tunable, narrowband instrument. The plan is to begin the upgrades at the start of 2006, implementing first one complete interferometer before starting the other two in parallel. I have tried to present here a description (with uncertainties!) of a design which will surely evolve with time, but will continue to be motivated both by what is technically feasible and by astrophysical benchmarks.

ACKNOWLEDGEMENTS

The second generation LIGO interferometer designs derive from the work of the entire LIGO Laboratory and the LIGO Science Collaboration, all of whom I thank for their contributions. The author is supported by National Science Foundation grant PHY-9210038.

REFERENCES

1. B.J. Meers, Phys. Rev. D **38**, 2317-2326 (1988).
2. S.A. Hughes and K.S. Thorne, Phys. Rev. D **58** (1998).
3. A. Buonanno and Y. Chen, gr-qc/00010001.
4. V.B. Braginsky, M.L. Gorodetscky, and S.P. Vyatchanin, Phys. Lett. A **264**, 1 (1999).
5. S.D. Penn, G.M. Harry, A.M. Gretarsson, S.E. Kittelberger, P.R. Saulson, J.J. Schiller, J.R. Smith, S.O. Swords, gr-qc/0009035.
6. A.M. Gretarsson, S. Rowan, G. Cagnoli, G.M. Harry, J. Hough, S.D. Penn, P.R. Saulson, W.J. Startin, Phys. Lett. A **A270**, 108-114 (2000).
7. S.J. Richman, J.A. Giaime, D.B. Newell, R.T. Stebbins, P.L. Bender, J.E. Faller, Rev. Sci. Inst. **69:6**, 2531 (1998).
8. K. Thorne, private communication.

Resonant Detectors of Gravitational Radiation

William O. Hamilton

Department of Physics and Astronomy
Louisiana State University
Baton Rouge, LA 70893

Abstract. The operation of resonant detectors is described along with the international network of resonant gravitational wave detectors. Searches have been performed for burst and cw sources and these are briefly described. Upper limits on gravitational bursts have been established by the network. Limitations of the existing detectors is detailed along with plans for network and individual detector improvement.

INTRODUCTION

It is very important to acknowledge the great contributions of Joseph Weber to the research that we now call gravitational physics. The detectors that I will be describing are called Weber bars for a good reason. Joe was the first to use them and many of the ideas that we now use to design our supports and transducers were first tried in Weber's laboratory. As a matter of fact, we should also acknowledge that the interferometric detector was also first tried by Bob Forward who was one of Weber's first students.

This, however, is not an article about the history of the field. In what follows I will briefly describe the principle of operation of the resonant detector and the noise sources limiting their performance. I will then describe what we have learned from the operation of resonant detectors and show some examples of how sensitive they are at present. I will conclude with a description of the International Gravitational Event Collaboration (IGEC) and what results it has obtained to date.

HOW RESONANT DETECTORS WORK

We start by noting that a gravitational wave acts as a tidal force; that is to say, it does not exert a force on an isolated point mass but, because it is a time dependent change in the curvature of the local space-time, it acts like a time dependent force gradient or tidal force on an extended body. This causes a quadrupolar distortion. Thus it follows that a tidal force will couple to those normal modes of an extended

CP575, *Astrophysical Sources for Ground-Based Gravitational Wave Detectors,* edited by J. M. Centrella
© 2001 American Institute of Physics 0-7354-0014-8/01/$18.00

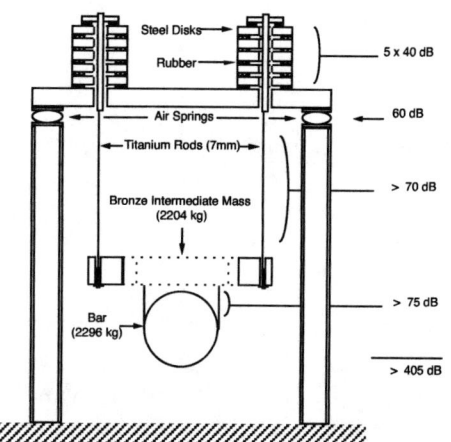

Estimated Attenuation

Steel Disks → 5 x 40 dB
Rubber
Air Springs → 60 dB
Titanium Rods (7mm)
Bronze Intermediate Mass
(2204 kg) → > 70 dB
Bar
(2296 kg) → > 75 dB
> 405 dB

FIGURE 1. The Allegro detector photographed during assembly. The bar can be seen below the helium reservoir and the supporting mass.

FIGURE 2. Allegro support system design and isolation. Only support components are shown. The containers for the steel and rubber isolators are barely visible at the top of Figure 1.

body that have a quadrupole moment. The lowest frequency such mode in a bar is the fundamental longitudinal mode: that mode which results in oscillations in the length of the bar while the diameter of the bar changes in the opposite phase to keep the density of the material constant. Since the fundamental longitudinal mode of Allegro is at a frequency of 912 Hz, the bar's principal region of sensitivity will be in the neighborhood of that frequency. The measurement and isolation strategy then is directed toward measuring changes in the length of the resonant bar while isolating the bar and measuring apparatus against noise which might affect that length measurement. For instance, the bar is supported by being balanced on a single cable at its center. In that way any vibration that gets to the bar will only, to first order, excite bending modes and not longitudinal modes. The figures illustrate the general design philosophy.

A bar antenna has different sensitivity to radiation coming from different directions. It turns out that the bar has an antenna pattern similar to that of a simple dipole antenna. That is, the bar is most sensitive to radiation with a propagation vector perpendicular to the axis of the bar. That is, the direction of propagation that will result in the maximum length change and hence that will cause the maximum change of energy in the bar's fundamental longitudinal mode.

The transducer, which converts the mechanical motion to an electrical signal, is mounted on the end of the bar. In all of the antennas now being used the

$$h_c \approx 10^{-20} \left[\frac{10 \text{ kpc}}{d} \right] \qquad\qquad \text{Pessimistic} \qquad\qquad \text{[Müller, '82]}$$

$$h_c \approx 5 \times 10^{-19} \left[\frac{10 \text{ kpc}}{d} \right] \qquad\qquad \text{Optimistic} \qquad\qquad \text{[Ipser and Monagan, '84]}$$

$$h_c \approx 4 \times 10^{-18} \left[\frac{10 \text{ kpc}}{d} \right] \qquad\qquad \text{Very Optimistic} \qquad\qquad \text{[Eardley, '83]}$$

For collapse to a black hole:

$$f_c \approx 13 \text{ kHz} \left[\frac{M_\odot}{M_{\text{hole}}} \right] \qquad\qquad \text{Frequency of the radiation}$$

$$h_c \approx 10^{-17} \left[\frac{10 \text{ kpc}}{d} \right] \left[\frac{1 \text{ kHz}}{f_c} \right] \left[\frac{\epsilon}{.01} \right]^{\frac{1}{2}} \qquad\qquad \text{[Stark and Piran, '86]}$$

transducer is some type of tuned accelerometer. The accelerometer is tuned to the bar frequency and so we have a case of two coupled oscillators, the bar and the transducer. This results in the detector output consisting of the amplitude of the modes at the sum and difference frequencies of the two coupled oscillators. A discussion of transducers is a fascinating subject that could occupy a whole volume but it is beyond the scope of this paper and will not be discussed here. Details can be found in the papers of Solomonson [1].

We also need to look at the size of the gravitationally induced strain that we are trying to detect. The table, borrowed from Kip Thorne's article in Hawking and Israel's book, gives the best theoretical estimates that were available at the time that we were first putting the experiment together [2]. Since we haven't seen anything, many of the current best guesses have crept toward even smaller strains, mostly by postulating the source to be outside our galaxy. Note that the table has scaled the strains to a source at the center of the galaxy.

The numbers in the table indicate the difficulty that all of the experiments face. h_c is the gravitational wave signal and is the size of the strain that is induced in the bar antenna. Thus to detect a strain of 10^{-18} in a 3 meter bar we need to be able to detect the results of a wave that would change the length of a bar by 3×10^{-18} meters: a fraction of the diameter of a nucleus. The important details that make such measurements possible are such things as adequate vibration isolation (the estimated isolation of various stages are shown in Figure 2), electrical isolation, and elimination of electrical noise and back-action from the measurement system. Figure 3 shows the measured strain sensitivity of the Allegro detector. One must be cautious in interpreting this picture. To measure a pulse of gravitational radiation one must make a measurement that encompasses a wider bandwidth than that shown. The measurement band shown in the figure is about 50 Hz. To measure a 1 millisecond pulse we would need approximately a 1 kHz bandwidth. Looking at the most sensitive region of the figure we may estimate that the antenna's effective bandwidth is about 1 Hz. Therefore we might expect the effective sensitivity to be

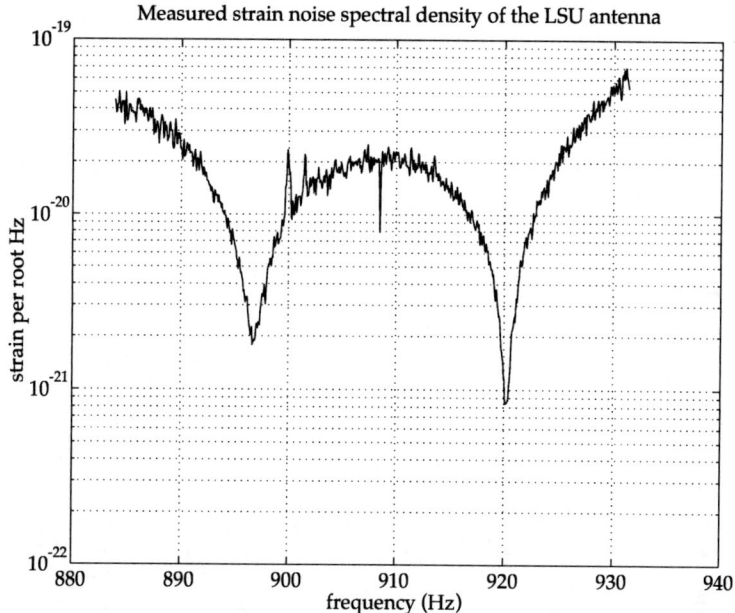

FIGURE 3. The measured strain sensitivity of the Allegro detector.

about a factor of 30 less than the peak value shown. As a matter of fact, the rms antenna noise level would correspond to a random gravitational signal of 5×10^{-19} which is just about what we would guess from the figure. All of the resonant detectors in the world are roughly comparable to Allegro in sensitivity.

The world's active resonant antennas are:

Detector	Frequencies		Location
Allegro	895 Hz	920 Hz	Baton Rouge, Louisiana
Auriga	912 Hz	930 Hz	Padova, Italy
Explorer	905 Hz	921 Hz	CERN
Nautilus	908 Hz	924 Hz	Frascati, Italy
Niobé	694 Hz	713 Hz	Perth, Western Australia

WHAT HAVE RESONANT DETECTORS DONE?

The first thing to note is that resonant detectors have *not* detected gravitational radiation yet. However those of us who operate them have begun to set upper limits on rates and strengths of gravitational wave bursts and we have looked for cw gravitational radiation in the regions where our detectors are most sensitive. Most importantly, we have demonstrated that a single detector can not do it alone. We now realize that the anticipated signals are so small and so infrequent that it is virtually impossible to make an unambiguous detection with a single detector. Thus, over the past ten years, what were previously competing groups have begun to cooperate to form a worldwide network of resonant detectors.

Burst Sources

The upper limits for bursts set by the International Gravitational Event Collaboration (IGEC) are shown in Figure 4 [3]. The figure shows upper limit of the rate of bursts above the threshold shown on the horizontal axis. The quantity used for the threshold is the amplitude of the fourier transform of the burst at the bar frequency. Notice how the upper limit becomes more stringent when there are more detectors operating. We will come back to this point later.

The original sources envisaged for resonant detectors were supernovae. It was expected that a supernove would give a burst of gravitational radiation and that that burst would have significant energy content in the frequency range in which the bar detectors are sensitive. The detection strategy was to look for changes in the amplitude and/or phase of the fundamental longitudinal mode of the antenna. If a change were measured which was of a size to be statistically unlikely then that change would be a possible candidate for a gravitational wave burst. If two separate detectors measured such a change simultaneously then that would be evidence for a gravitational wave signal.

An important thing to note from the above remarks is that, to detect a gravita-

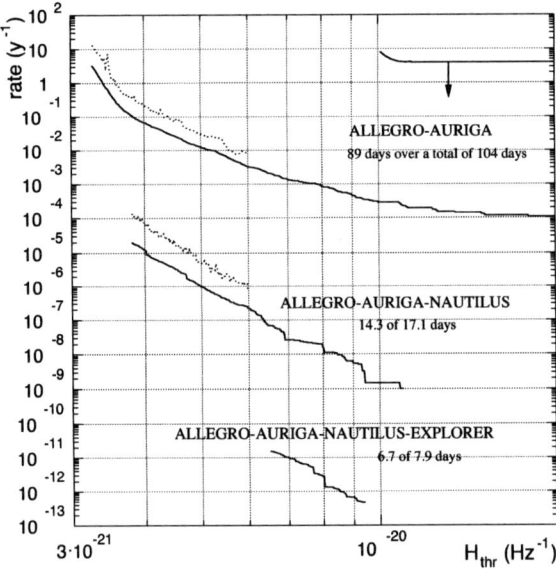

FIGURE 4. Upper limits for gravitational bursts for half of 1998.

28

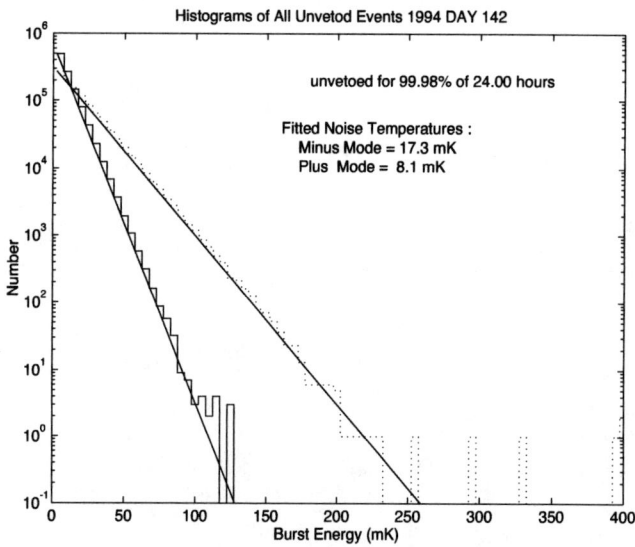

FIGURE 5. Histogram of the filtered events for one day in 1994.

tional wave burst, it is not enough to have a sensitive detector. It is also crucial that the noise of the detector be well characterized so that an unlikely change will stand out. We have come to refer extra noise as "non-gaussian noise". It is also sometimes referred to by noting whether the noise of a given detector is stationary. By that we mean whether or not we have well characterized gaussian processes exciting the various modes of the system. This point is illustrated in Figure 5.

The change in the amplitude of the bar's normal modes is sampled periodically (125 samples per second for Allegro). Figure 5 shows a histogram of all samples for a day. The energy units are milliKelvin and represent the change in the energy of the mode from the previous sample. Notice that the histograms are linear on the semi-log plot. The exponential distribution is what is theoretically expected if the modes are excited by thermal noise. If there were non-gaussian noise then there would have been a large number of points which did not fit the stationary distribution. Any single gravitational wave event would have to result in an energy significantly above the shown distributions to be considered significant. There are only 4 possible samples which could be considered to be candidates for a gravitational wave event shown on the graph. The energy change in the bar for any given sample is the weighted average of the energy change in each mode.

Figure 6 shows 10 days of data from Allegro. All non-vetoed events which exceed an arbitrarily chosen threshold are shown. There are 9 obvious candidates for comparison with other detectors.

Each individual detector group then makes a list of possible events and compares with a list from other antennas. To do this intelligently there must be a good

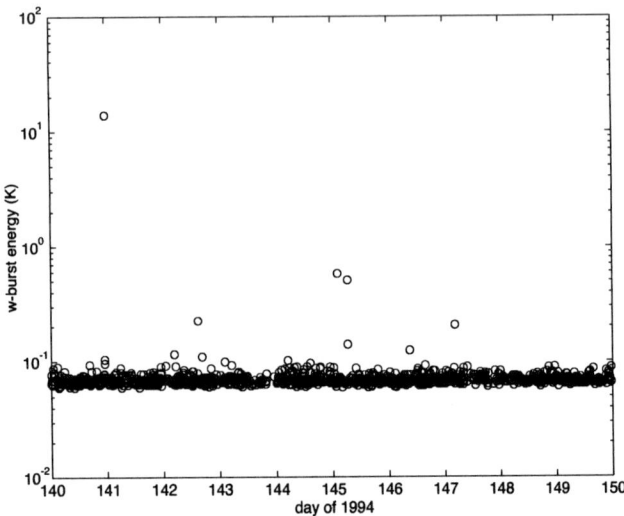

FIGURE 6. Ten days of data in 1994 for Allegro. All events above threshold are shown.

calibration of each antenna so that the amplitude and time response are well known. This is shown for Allegro in Figure 7.

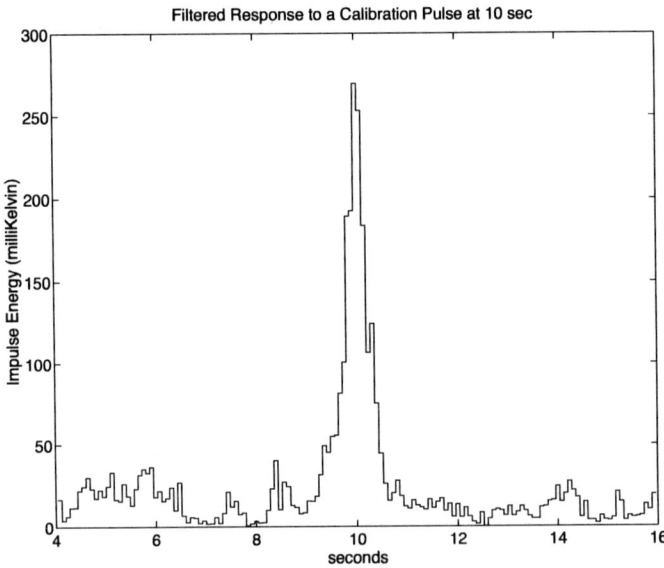

FIGURE 7. Allegro's response to a 1 millisecond calibration pulse.

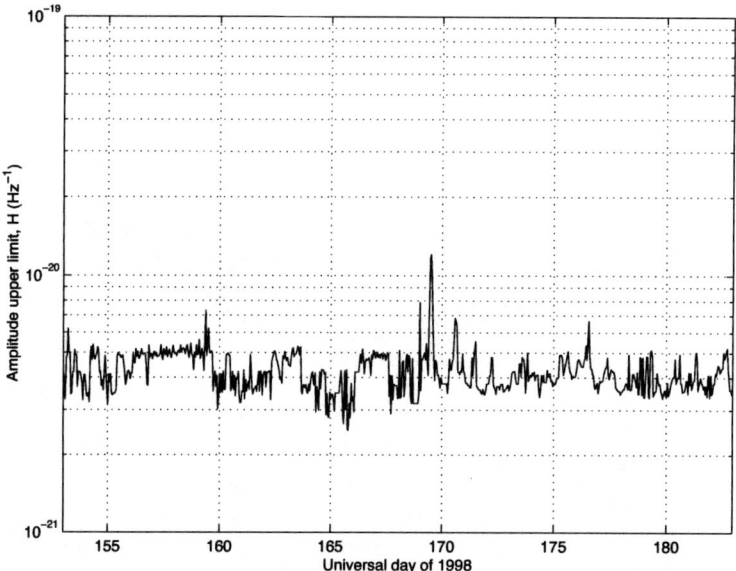

FIGURE 8. The upper limit, calculated hour by hour, for gravitational wave burst events. The vertical axis is the fourier transform of the burst amplitude at the bar resonant frequency.

Comparing event lists then requires knowing the time precisely and in deciding on a suitable coincidence window. The ideal coincidence window would be as narrow as possible, so as to reduce the probability of accidental coincidences. Looking at Figure 7 one sees however that the response of these detectors, at least in the way they were being operated in 1994, will not allow a window much narrower than ±0.5 seconds. In most of the coincidence comparisons done to date a window of ±1 second was used.

We know that the occurrence of events in a single detector is a random process. Thus the occurrence of coincidences is a Poisson process and we know that the rate with which accidentals occur is

$$R = R_A R_B \Delta t$$

where R_A is the rate at which threshold is exceeded in antenna A, etc., and Δt is the width of the time window for coincidences. Thus it is necessary to set a threshold for the number of candidate events to be considered from each antenna in order to make the probability of getting an accidental coincidence sufficiently low. This probability will be governed by the best guess as to the probability of receiving a real gravitational burst signal.

Clearly the more detectors which are comparing event lists the lower the probability of getting an accidental event at the same time on all detectors. Thus, if the sensitivity of a given type of detector is good enough to see a signal, it makes sense to have as many of those detectors as possible operating. In this way the threshold

31

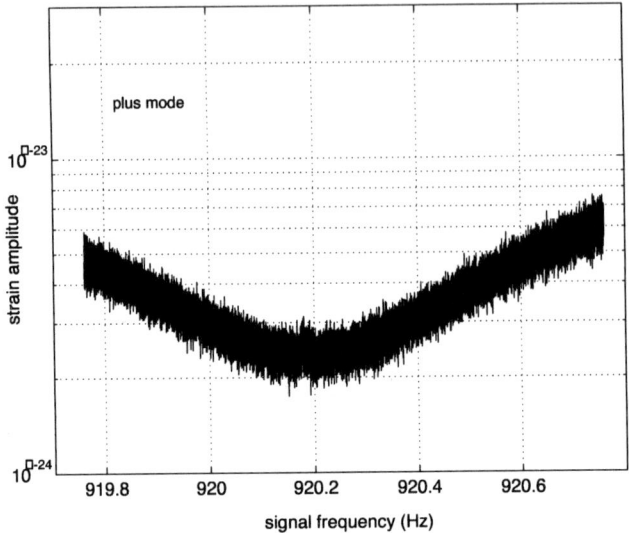

FIGURE 9. Upper limit for c.w. signals from the galactic center.

on individual detectors could be lowered and yet the false alarm probability of the network could still kept suitably low. This is the point illustrated previously in Figure 4 though in that case the thresholds on the individual detectors was kept the same, resulting in an extremely low false alarm probability.

The upper limit need not be calculated for an extended period. Figure 8 illustrates the upper limit for the amplitude of gravitational wave bursts during one month of 1998. The points were calculated for one hour intervals. The variation occurs due to one detector or another becoming noisy or going off line, resulting in a higher limit for a given false alarm probability [3].

Continuous Waves

Several searches have been made for c.w. signals such as might be emitted from a millisecond pulsar. Such searches are computationally very expensive since one must adjust for the motion of the antenna due to both rotation of the earth and movement of the earth around the sun. Moreover because of the earth's rotation the antenna pattern of the antenna rotates around the celestial sphere. Typically one picks a source location and performs the calculations for that. One of the first such searches was performed by Mauceli [4]. Figure 9 shows the results from his thesis for the upper limit for signals coming from the galactic center. These results gain from averaging 29 long sections of data, thereby enabling a very narrow effective bandwidth. Similar searches using different calculational techniques have been performed by the Rome group [5].

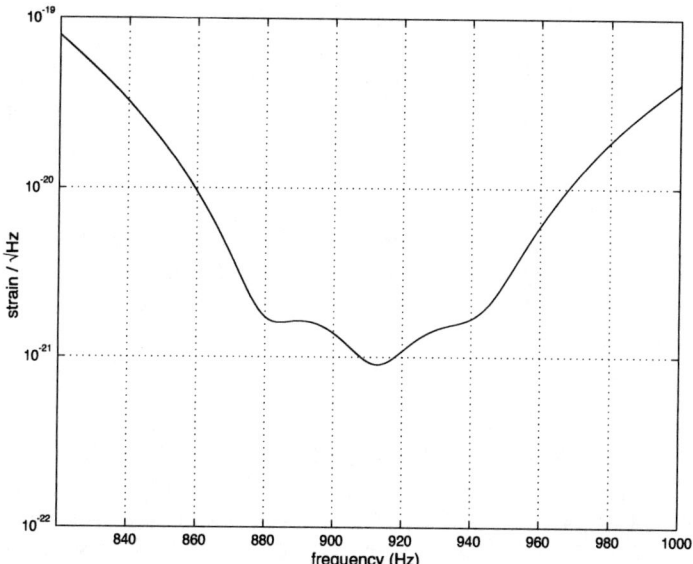

FIGURE 10. Anticipated Allegro performance with a two mode transducer and improved SQUID.

THE FUTURE OF RESONANT DETECTORS

The principal noise sources in the existing resonant detectors are external vibrations getting past the vibration isolation and thermal and electronic noise in the amplifiers (generally a SQUID amplifier). Vibration isolation is simply a matter of better engineering. Thermal noise can be reduced by cooling the antennas to lower temperature. This part of the future is already here, in that both the Nautilus and Auriga detectors in Frascati and Legnaro have operated for extensive periods at a temperature of 100 milliKelvin, a factor of 40 lower than the Allegro detector. Thus we should expect that they should have better energy sensitivity by a factor of at least 40. Better engineering should improve that even more. Preliminary data from the Rome group indicates that Nautilus now is considerably more sensitive than the first data given in [3].

We also can improve on the design of the transducer and use a better SQUID. Better SQUIDs are been worked on by all of the groups and the Rome group is already using one. Figure 10 shows how the existing Allegro detector might be expected to behave if it were operated with a modified transducer the amplifier of which is replaced by a new SQUID already tested by Paik and Wellstood's Maryland group.

Perhaps the most interesting development for the future is the testing and development of the spherical detector. The idea of using the quadrupole modes of a spherical mass as a detector of gravitational waves had been noticed by Weber and first worked out by Ashby and Dreitlein [6] and Wagoner and Paik [7]. The invention by Johnson of the TIGA configuration [8] and the proof of its performance by Merkowitz [9,10] renewed interest in spherical detectors. The SFERA project in Rome and the Graviton project in Brazil are the first major experiments to test and build an omnidirectional detector. These detectors will be equally sensitive to gravitational radiation of any polarization and incident from any direction. Their integrated sensitivity improvement should be approximately a factor of 54 over the equivalent bar antenna. They will provide a complementary technology to the interferometers and when equipped with the type of transducer used to generate Figure 10 could work with interferometers to reduce the false alarm probability for higher frequency signal detection.

THE INTERNATIONAL GRAVITATIONAL EVENT COLLABORATION (IGEC)

Clearly gravitational wave detection is a collaborative effort. No single detector can make a discovery because the signals are just too weak. The early experiments tended to compete for a first detection and that competition probably held the field back several years. The historical result was that all of the detectors took data in different ways and there was no standardized format for sharing and comparing data. There were early collaborative efforts involving Explorer and Allegro [11,12] but these tended to be *ad hoc* and did not create a framework that was easily extended to other experiments.

This led to the creation of the IGEC in 1997 with a protocol for data exchange and an agreement for publication policy. The orgainization also addresses the need to coordinate times for operation or repair of the different detectors. The charter members of the IGEC were the principals from the gravity wave groups at Rome, Legnaro, Perth, and Baton Rouge. The IGEC was also formed with the idea that it could be expanded and adapted to the inclusion of other types of data, such as that generated by interferometers. The IGEC agreed to make its data public and set up a web site (http://igec.lnl.infn.it) to make that data available. All of the documents and protocols are available on the web site. The first paper from the IGEC was reference [3] and the publishing of that paper has led to a renewed effort for more collaborative publication.

The IGEC has also issued an invitation to the other gravitational wave detectors to join and submit any data that might be of joint interest. For instance, TAMA made its first runs at a time that several of the IGEC antennas were also operating. There might be joint benefit in working out the way to compare data.

ACKNOWLEDGEMENTS

The Allegro detector has been supported throughout its history by grants from the National Science Foundation. Substantial support has also been provided by

LSU. Many individuals have contributed to its progress over the years: faculty members at a number of institutions, postdoctoral associates and graduate students. Siong Heng has provided invaluable help for this paper. The current membership of the IGEC is working extremely hard to bring gravitational wave astronomy nearer to realization. We must also acknowledge the contributions of the early pioneers who are no longer with us: Eduardo Amaldi, Bill Fairbank, I. Hirakawa and Joe Weber.

REFERENCES

1. Solomonson, N. et al. *Rev. Sci. Instrum.* **65**, 174 (1994)
2. Thorne, K.S., "Gravitational Radiation", in *300 Years of Gravitation*, edited by Stephen W. Hawking and Werner Israel, Cambridge: Cambridge University Press, 1987.
3. Allen, Z. et al., *Phys. Rev. Lett.* **85**, 5046-5050 (2000).
4. Mauceli,E. , Ph.D. thesis, Louisiana State University, 1997.
5. Astone, P. et al., "Search for periodic g.w. sources with the Explorer detector", gr-qc/0011072 (November 2000).
6. Ashby, N. and Dreitlein, J., *Phys. Rev. D*, **12**, 336 (1975).
7. Wagoner, R.V. and Paik, H.J., *Gravitazione Sperimentale*, Accademia Nazionale dei Lincei, Roma (1977).
8. Johnson, W.W. and Merkowitz, S.M., *Phys. Rev. Lett.*, **70**, 2367 (1993).
9. Merkowitz, S.M., Ph.D. thesis, Louisiana State University, 1994.
10. Merkowitz, S.M. and Johnson, W.W., *Phys. Rev. D*, **53**, 5377 (1996)
11. Astone, P. et al. *Phys. Rev. D*, **59**, 122001 (1999)
12. Mauceli, E. et al. *Phys. Rev. D*, **56**, 6081-6084 (1997)

Astronomy and Astrophysics in
the 21st Century

Particle, Nuclear and Gravitational Wave Astrophysics in the Decadal Survey

Thomas K. Gaisser

Bartol Research Institute, University of Delaware
Newark, DE 19716

Abstract. The Panel on Particle, Nuclear and Gravitational Wave Astrophysics (PNGW) was one of seven science panels that feeds into the current decadal survey, *Astronomy and Astrophysics in the New Millennium*. The theme of this panel is multi-messenger astronomy, including high energy cosmic rays, TeV gamma-rays, neutrinos and gravitational waves. This paper is a short summary of my talk for the Workshop on Astrophysical Sources for Ground-Based Gravitational Wave-Detectors about the science of the PNGW panel and its report.

I INTRODUCTION

The decadal survey, *Astronomy and Astrophysics in the New Millennium* was released on May 18, 2000. It is available from the National Academy Press and on the web [1]. The main report summarizes and prioritizes science initiatives surveyed and recommended by seven science panels. The panel reports were also reviewed and are published as an appendix to the main report in a separate volume [2]. This note is a brief summary of my talk about the report of the PNGW panel presented during this workshop. While I have attempted to give an accurate summary of the report, the amount of text here is not necessarily proportional to the priority or importance of each topic, but reflects somewhat my own areas of interest. The reader is urged to consult the authoritative AASC report [1] and its appendix containing the panel reports [2] for a complete and balanced picture.

II MULTI-MESSENGER ASTRONOMY

This theme is appropriate because the PNGW panel deals with astrophysical observations of charged particles (cosmic rays), neutrinos, and gravitational waves, as well as gamma-rays of very high energy and with searches for dark matter. All

CP575, *Astrophysical Sources for Ground-Based Gravitational Wave Detectors*, edited by J. M. Centrella
© 2001 American Institute of Physics 0-7354-0014-8/01/$18.00

other science panels deal only with observations of photons in various wave-length bands, from radio to X-ray and gamma-ray. The PNGW panel report covers a broad range of science characterized by connections between astronomy and several other fields, including cosmology, relativity, and particle and nuclear physics.

Gravitational waves provide information on bulk motions of matter in the most energetic events in nature, such as coalescence of black holes. High energy γ-rays trace populations of accelerated particles. Cosmic-ray protons and nuclei carry information about the cosmic accelerators in which they originate. Neutrinos emerge from deep inside regions opaque to photons.

Cosmic-rays, neutrinos and high-energy γ-rays are all manifestations of Nature's cosmic particle accelerators. Protons and electrons are accelerated in sources such as supernova blast waves, relativistic jets in active galactic nuclei and gamma-ray burst sources. Electrons produce photons by processes including bremsstrahlung and inverse Compton scattering. Protons produce neutrinos as well as γ-rays whenever they interact (with ambient gas or radiation fields) to produce pions. The neutrinos come from decay of charged pions and the photons from π^0-decay.

III   SCIENCE RECOMMENDATIONS

A   LISA

The highest priority recommendation of the PNGW panel is the Laser Interferometer Space Antenna (LISA) [3], the goals of which include detection of gravitational waves from coalescence of massive black holes in colliding Galaxies in the cosmos and from white dwarf binaries in our Milky Way galaxy. This proposed mission is described in another talk at this Workshop [4].

B   VERITAS

The second recommendation in overall priority of PNGW is the Very Energetic Radiation Imaging Imaging Telescope Array System (VERITAS) [5]. This ground-based observatory will be complementary to the Gamma Large Area Space Telescope (GLAST) [6], a priority recommendation of the Panel on High Energy Astrophysics from Space. VERITAS will have significantly improved sensitivity over existing ground-based gamma-ray telescopes (such as the Whipple telescope [7], which is its ancestor) and will also have a lower energy threshold to provide overlap with GLAST. Together, these detectors will map the γ-ray sky. A question of particular interest is to understand the workings of gamma-ray blazars, a new class of objects discovered by EGRET [8]. Presently there is a gap in energy coverage between detectors in space (e.g. EGRET) and ground-based telescopes (e.g. Whipple). Full coverage of energies from sub-GeV to the TeV range will not only help understand how AGN blazars work, but will also probe the infra-red background radiation by measuring how the energy-cutoff of these sources depends on their redshift. The idea is that the more distant sources will appear to be cutoff at lower energy because of increased absorption of their γ-rays by infra-red photons.

The field of ground-based very high energy gamma-ray astronomy is in a state of active development internationally, with many new experiments, some already operating and some under construction or proposed. There is a wealth of information in recent reviews [9–11] and workshop proceedings [12,13].

C   Program in particle astrophysics

PNGW also recommends a balanced program in particle and nuclear astrophysics. This includes searches for dark matter, measurements of high energy cosmic rays, neutrino astronomy, and solar neutrinos.

1   Dark matter searches

Deciphering the nature of dark matter remains one of the most important goals of astrophysics. With the Axion [14], CDMS II [15] and other experiments, cosmologically important regions of parameter space are now being explored. It is likely that new experiments will be needed within the decade, the nature of which will be determined by results of the ongoing experiments. Indirect searches for neutrino-induced muons from WIMP annihilation in the Sun or the core of the Earth [16] are also important in the high mass range, complementary to the direct detection experiments.

2   Galactic cosmic rays

To discriminate among models of galactic cosmic rays, the Advanced Cosmic-ray Composition Experiment for Space Station (ACCESS) [17] aims to extend direct measurements of cosmic ray nuclei to approaching 10^{15} eV. It will also be able to determine the source spectra produced by cosmic accelerators which are presently obscured by energy-dependent effects of propagation. By measuring the ratio of secondary to primary nuclei in the $\sim$TeV range, the propagation effects can be separated and the source spectrum determined.

3   Ultra-high energy cosmic rays

Among the objects that are possible sources of the highest energy cosmic rays with energies $> 10^{19}$eV are relativistic jets of active galactic nuclei and gamma-ray burst sources, both of which are also at the top of the list of targets for high-energy gamma-ray astronomy. There are, however, problems with both of these models, and the sources are essentially unknown at present. Other, even more exotic sources are also possible, as summarized in a recent review [18]. The experimental challenge is to overcome the extremely low intensity of the highest energy particles by constructing detectors with effective areas of thousands of square kilometers. The new Hi-Res Fly's Eye [19] is now collecting data, and construction of the Auger detector [20] in Argentina is underway There are proposals under discussion for an expanded detector complex on the ground in the Northern hemisphere. The ultimate air shower detector would be a Fly's Eye in space [21,22] that could monitor a million square kilometers for giant cosmic ray cascades in the atmosphere. An up-to-date overview of all aspects of ultra-high energy cosmic rays is given in Refs. [23,24].

4   Neutrino astronomy

Closely related to studies of ultra-high energy cosmic rays is high energy neutrino astronomy. Neutrinos are expected from energetic cosmic accelerators whenever there is pion production in or near the sources. Neutrinos in principle distinguish models of high energy gamma-ray sources that involve acceleration of hadrons from those that involve primarily electrons; in addition, neutrinos can emerge from deep inside regions opaque to high energy photons. The challenge is to build large enough detectors to detect the weakly interacting neutrinos. Following on the successful deployment and operation of detectors in Lake Baikal [25] and at the South Pole [26], there are active proposals and development projects to construct kilometer-scale neutrino detectors, either in the deep ocean [27,28] or in the Antarctic ice sheet [29],

An important aspect of neutrino astronomy is to maintain a supernova watch with a network of neutrino detectors around the world [30]. Super-Kamiokande [31] will play a central role in this effort with its ability to reconstruct low energy neutrino interactions with better sensitivity than Kamioka [32] and IMB [33] had when they detected neutrinos from SN1987A. While the high energy neutrino detectors cannot reconstruct low energy events, a supernova neutrino burst could be detected as a simultaneous rise in counting rate of many individual modules. Gravitational wave detectors would also be part of the supernova watch.

5   Solar neutrinos

In light of several solar neutrino experiments sensitive to various parts of the solar neutrino spectrum, together with data from helioseismology, the solar neutrino deficit is now understood to reflect novel physical properties of neutrinos rather than any misunderstanding of the solar model [34]. A second generation of solar neutrino experiments is now in progress [35–38]. By comparing the total rate of interactions of all types of neutrinos with the interaction rate of electron neutrinos, it will be possible to check directly whether electron neutrinos produced in the Sun are changing identity on the way to Earth [36]. A third generation of solar neutrino experiments will be required to measure the energy spectrum of the dominant low energy neutrinos from the pp fusion reaction chain and thus to complete the program of measuring both the energy spectrum and flux of solar neutrinos. Various approaches are under development [39–42].

IV POLICY ISSUES

The PNGW report concludes with comments about facilities and recommendations for funding agencies. A common feature of these projects is that they are carried out at remote sites, in some cases deep underground. This raises questions of infra-structure and involvement of existing laboratories. International collaboration is also an important aspect in many instances. Another common feature is the interdisciplinary nature of the scientific goals. Implications for cross-disciplinary and inter-agency cooperation and coordination by funding agencies and for advisory committee structures are noted.

As a postscript, I note that the NRC Committee on Physics of the Universe is developing a report that addresses science (and science policy) at the intersection of physics and astronomy [43].

REFERENCES

1. *Astronomy and Astrophysics in the New Millennium*, National Academy Press, 2001 (http://books.nap.edu/catalog/9839.html).
2. *Astronomy and Astrophysics in the New Millennium: Panel Reports*, National Academy Press (http://www.nap.edu/catalog/9840.html).
3. LISA web site: http://lisa.jpl.nasa.gov/documents.html.
4. Thomas Prince, these Proceedings.
5. F. Krennrich *et al.* in *GeV-TeV Gamma Ray Astrophysics Workshop* (A.I.P. Conf. Proc. #515, ed. B.L. Dingus, M.H. Salamon & D.B. Kieda, 1999) p. 515. VERITAS web site: http://veritas.sao.arizona.edu/veritas/.
6. GLAST web site: http://www-glast.sonoma.edu/.
7. T.C. Weekes *et al.*, Ap.J. 342 (1989) 379 and J.P. Finley *et al.* in *GeV-TeV Gamma Ray Astrophysics Workshop*, op. cit., p. 301.
8. R. Mukherjee *et al.*, Ap.J. 490 (1997) 116.
9. René A. Ong, Physics Reports 305 (1998) 93.
10. C.M. Hoffman, C. Sinnis, P. Fleury & M. Punch, Revs. Mod. Phys. 71 (1999) 897.
11. T.C. Weekes in *GeV-TeV Gamma Ray Astrophysics Workshop*, op. cit., p. 3.
12. *TeV Astrophysics of Extragalactic Sources* (ed. M. Catanese & T.C. Weekes) Astroparticle Physics 11 (1999).
13. *International Symposium on High Energy Gamma-Ray Astrophysics* (Heidelberg, June, 2000, to be published in A.I.P. Conference Proceedings).
14. L.J. Rosenberg & K. van Bibber, Physics Reports 325 (2000) 1.
15. CDMS web page: http://cdms.berkeley.edu/.

16. L. Bergstrom, T. Damour, J. Edsjo, L.M. Krauss & P. Ullio, Journal of High Energy Physics 08 (1999) 010.
17. ACCESS web page: http://lheawww.gsfc.nasa.gov/ACCESS/.
18. P. Bhattacharjee & G. Sigl, Physics Reports 327 (2000) 109.
19. C.C.H. Jui in *26th International Cosmic Ray Conference* (ed. B.L. Dingus, D.B. Kieda & M.H. Salamon, A.I.P. Conf. Proc. #516 (1999) p. 370. HiRes web site: http://www.cosmic-ray.org/.
20. Auger Project web site: http://www.auger.org/.
21. Extreme Universe Space Observatory web site: http://www.ifcai.pa.cnr.it/Ifcai/euso.html
22. Orbiting Wide-angle Light Collectors web site: http://owl.gsfc.nasa.gov/
23. A.A. Watson, Physics Reports 333-334 (2000) 309.
24. M. Nagano & A.A. Watson, Revs. Mod. Phys. 72 (2000) 689.
25. I.A. Belolaptikov *et al.*, Astropart. Physics 7 (1997) 263. Baikal web site: http://www.ifh.de/baikal/baikalhome.html.
26. E. Andres *et al.*, Astroparticle Physics 13 (2000) 1. AMANDA web site: http://amanda.berkeley.edu/.
27. Antares web site: http://antares.in2p3.fr/.
28. Nestor web site: http://www.nestor.org.gr/.
29. IceCube web site: http://phenom.physics.wisc.edu/IceCube/.
30. SNEWS (SuperNova Early Warning System): http://hep.bu.edu/ snnet/
31. Super-Kamiokande web site: http://www-sk.icrr.u-tokyo.ac.jp/doc/sk/index.html.
32. K.S. Hirata *et al.* Phys. Rev. Lett. 58 (1987) 1490.
33. R.M. Bionta *et al.* Phys. Rev. Lett. 58 (1987) 1494.
34. J.N. Bahcall, M.H. Pinsonneault & S. Basu, astro-ph/0010346.
35. Super-Kamiokande Collaboration, Phys. Rev. Letters 82 (1999) 1810 and 2430.
36. Sudbury Neutrino Observatory web site: http://www.sno.phy.queensu.ca/.
37. Borexino web site: http://almime.mi.infn.it/.
38. KamLAND web site: http://www.awa.tohoku.ac.jp/html/KamLAND/.
39. HERON web site: http://www.physics.brown.edu/research/heron/.
40. LENS web site: http://cdfinfo.in2p3.fr/Experiences/Lens/.
41. HELLAZ web site: http://sg1.hep.fsu.edu/hellaz/.
42. ICARUS web site: http://www.aquila.infn.it/icarus/.
43. Committee on Physics of the Universe: http://www.nationalacademies.org/bpa/projects/cpu/.

The National Virtual Observatory

Robert J. Hanisch

Space Telescope Science Institute
Computing and Information Services Division
3700 San Martin Drive
Baltimore, Maryland 21218
hanisch@stsci.edu

Abstract. The National Virtual Observatory is a distributed computational facility that will provide access to the "virtual sky"—the federation of astronomical data archives, object catalogs, and associated information services. The NVO's "virtual telescope" is a common framework for requesting, retrieving, and manipulating information from diverse, distributed resources. The NVO will make it possible to seamlessly integrate data from the new all-sky surveys, enabling cross-correlations between multi-Terabyte catalogs and providing transparent access to the underlying image or spectral data. Success requires high performance computational systems, high bandwidth network services, agreed upon standards for the exchange of metadata, and collaboration among astronomers, astronomical data and information service providers, information technology specialists, funding agencies, and industry. International cooperation at the onset will help to assure that the NVO simultaneously becomes a global facility.

I INTRODUCTION

The National—or more properly International or Global—Virtual Observatory has been highly ranked in the National Academy of Science Astronomy and Astrophysics Survey. The NVO will provide public access to a wealth of on-line data archives and catalogs, as well as the software tools needed to access and utilize them. The NVO is "virtual", however, in that it does not reside at a single brick-and-mortar facility, but rather is a distributed network of data and computational services. Through the NVO astronomers and the general public will have access to a digital sky, with data spanning the electromagnetic spectrum. The NVO will also provide a framework for enabling new, more cost-effective surveys.

In the past decade astronomy has been the benefactor of advances in computation and imaging technology. Computer processor speeds have increased by two orders of magnitude, and data storage capacities have grown enormously with associated costs decreasing. Data is being collected at unprecedented rates owing to the development of large format, highly efficient detectors. The CCD detectors on today's large telescopes collectively represent approximately $10y$ pixels, meaning

CP575, *Astrophysical Sources for Ground-Based Gravitational Wave Detectors,* edited by J. M. Centrella
© 2001 American Institute of Physics 0-7354-0014-8/01/$18.00

that many hundreds of Gbytes of data can now be collected every twenty-four hours. Significant portions of these data are acquired in dedicated surveys, which yield object catalogs with billions of entries (and with each entry having tens to hundreds of measured attributes. This plethora of data is impossible for any individual astronomer to make use of, yet the most interesting new discoveries are likely to depend on the ability to easily intercompare data from multiple instruments, in multiple bandpasses, in a statistically robust manner. This is the primary goal—and challenge—of the NVO.

The NVO will enable the discovery of new classes of rare or unusual objects through its ability to cross-correlate properties among different catalogs and surveys. We will be able to obtain large, statistically complete samples of brown dwarfs, high redshift galaxies and quasars, ultra-luminous infrared galaxies, and distant clusters. With homogeneous samples of 100 million or more objects it will be possible to make well-defined and statistically robust samples, with limitations governed by systematics rather than shot noise. We thus enter a new era of high precision astrophysics in which classes of objects can be studied as a function of mass, luminosity, morphology, chemical composition, time variability, and overall spectral energy distribution.

II WHAT IS THE NVO?

The NVO *is not* a new building or institution—rather, it is a federation of data providers linked together by an information technology infrastructure of communications and interoperability standards. In the broadest terms the NVO comprises the "virtual sky" (the collection of digital data sets and object catalogs), a "virtual telescope" (the computer middleware that manages queries, collects data, and translates metadata), and the "virtual instruments" (computer programs that analyze, cross-correlate, and visualize data). The NVO is supported by data providers (archives, catalogs), information providers (bibliographic databases), and service providers (such as supercomputer centers).

The NVO's "instruments" will include advanced visualization tools which enable the astronomer to browse through millions of objects in a multi-parameter space. When existing catalogs do not contain attributes the astronomer wishes to explore, it will be possible to easily explore the underlying image or spectral data. Through the NVO's interoperability layers and interfaces it will be possible to easily compare observations with numerical simulations. The analysis programs will be designed not only for portability, but as "agents" deployed to where the data resides.

Clearly the NVO will require advanced computer hardware and communications. One of the biggest challenges is in accommodating the enormous quantity of distributed data. Locally systems will require database engines with I/O rates of Gbytes/second, and data providers will need to be linked together with high speed networks capable of at least 10 Gbit/second transfer rates. Data analysis, data mining, and visualization tools will require a scalable computing environment in which many hundreds of processors can be used together to work on an intensive problem. These same processors, however, should be able to support many simultaneous small queries. Collectively the NVO will have to support Petabytes of mass data storage.

III INITIATING THE NVO—PRESENT CONTEXT

NASA's Office of Space Science has embraced the systematic archiving of data from space astrophysics missions for nearly two decades. This farsighted vision for comprehensive data management has culminated in an astrophysics data system composed of primary science archive research centers (SARCs): a high-energy SARC, an optical/UV SARC, and an infrared SARC. The High Energy Astrophysics SARC (HEASARC) is located at NASA/Goddard Space Flight Center, the optical/UV SARC, known as the Multimission Archive at Space Telescope (MAST), is located at the Space Telescope Science Institute in Baltimore, and the infrared SARC, known as the Infrared Science Archive (IRSA), is located at the Infrared Processing and Analysis Center (IPAC) at Caltech. The National Space Science Data Center (NSSDC) provides permanent archive services to the three SARCs and provides direct archival services for the COBE and IRAS mission data sets.

Active missions such as SIRTF, HST, and Chandra process and manage their data directly at the associated science operations center (SIRTF Science Center, Space Telescope Science Institute, Chandra X-Ray Center). Active missions coordinate with the relevant SARC to assure access to their data during the mission lifetime and arrange for long-term access after the mission is over. In some cases active missions contract directly with one of the SARCs to provide archive and data distribution services (e.g., the FUSE mission data are processed at Johns Hopkins University and archived at MAST/STScI).

NASA has also supported the development and operations of complementary astronomical information services, such as the Astrophysics Data System (ADS) abstract and bibliographic database at SAO, the Astronomical Data Center catalog collection (a component of the Astrophysics Data Facility at GSFC), and the NASA Extragalactic Database (NED) bibliographic and data service at IPAC/Caltech. NASA has also supported U.S. access to the SIMBAD database at the Centre Données astronomiques de Strasbourg (CDS) in France. These services provide many cross-links to the data in the SARCs; ADS, for example, is including direct links from its bibliographic database to the underlying data.

The relationship between the permanent archive, the active archive centers or SARCs, the active missions, and the ancillary information services in astrophysics are shown in Fig. 1.

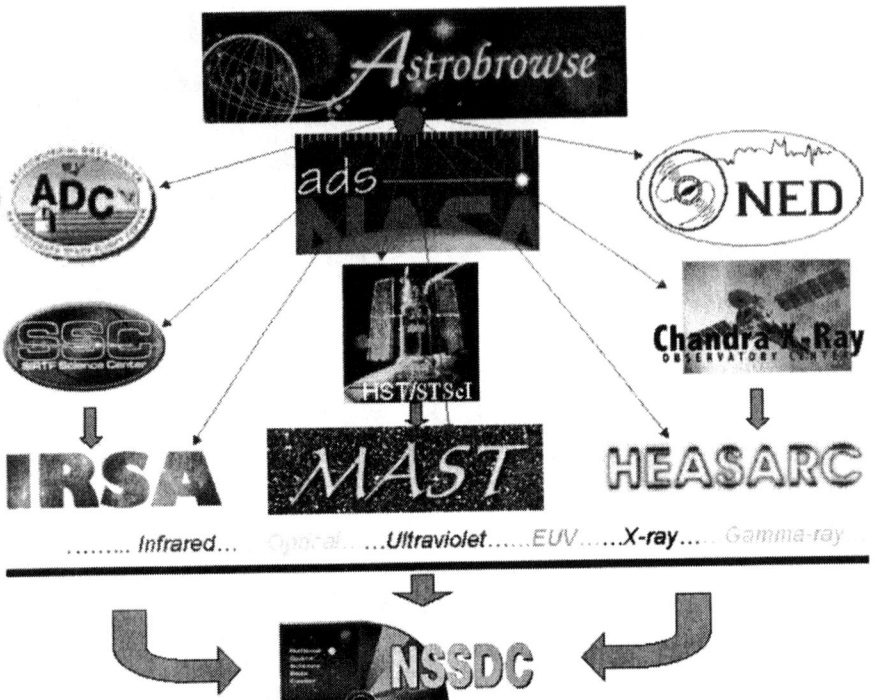

FIGURE 1. NASA's astrophysics data management structure. NASA's astrophysics data centers, catalog services, and bibliographic and thematic databases are elements in a complementary and comprehensive data and information service. Data from active missions (SIRTF, HST, Chandra) are archived and distributed from the associated science center, then either migrates or are linked to the relevant archive center (IRSA, MAST, HEASARC). The NSSDC provides permanent archive services. ADC, ADS, and NED provide catalog and bibliographic services. Astrobrowse allows users to search all information services for data of interest.

IV NASA'S SCIENCE ARCHIVE RESEARCH CENTERS

Infrared Science Archive

The Infrared Science Archive (IRSA) is located at the Infrared Processing and Analysis Center (IPAC) at Caltech. IRSA and IPAC are notable for their experience in managing the data from survey instruments, beginning with IRAS and continuing now with 2MASS. Both the underlying scan or imaging data are available as well as the extracted source catalogs. IRSA staff has been working with Caltech's "Digital Sky" project and are developing experience in indexing and cross-correlating very large databases (e.g., 2MASS vs. DPOSS; the 2MASS archive already exceeds 12 TB in size). IRSA is the U.S. portal for data from the ISO mission and will host the SOFIA and SIRTF data archives.

Multimission Archive at Space Telescope

The Multimission Archive at Space Telescope (MAST) [9] is located at the Space Telescope Science Institute in Baltimore. MAST was established as the IUE mission closed down, recognizing the close scientific relationship between HST and IUE data. MAST is the repository for optical, UV, and near-IR data, including, in addition to HST and IUE: FUSE, ASTRO (HUT, UIT, WUPPE), ORPHEUS (BEPS, IMAPS), and Copernicus. In partnership with the HEASARC, MAST provides access to the EUVE data that are physically stored at HEASARC. MAST also is the primary provider of the STScI Digitized Sky Survey and Guide Star Catalogs and is one of several sites hosting the VLA FIRST (Faint Images of the Radio Sky at Twenty centimeters) survey data. In the future MAST will host the GALEX and CHIPS mission public archives.

High Energy Astrophysics Science Archive Research Center

The High Energy Astrophysics Science Archive Research Center (HEASARC) [1] is located at the Goddard Space Flight Center. HEASARC hosts data from space missions covering the spectral range from the far UV to gamma rays. HEASARC holdings include: Ariel 5, ASCA, Beppo-SAX, CGRO, Einstein, EXOSAT, EUVE, ROSAT, XMM, and RXTE. Data from the Chandra mission will be archived at the Chandra X-ray Center during the active mission, but an access portal will be incorporated into HEASARC. HEASARC staff has led in the development of the Astrobrowse data location service and of Skyview, a first generation data integration service [12].

V CATALOG, BIBLIOGRAPHIC, AND THEMATIC INFORMATION SERVICES

Astronomical Data Center

The Astronomical Data Center (ADC) is a part of the Astrophysics Data Facility at Goddard Space Flight Center. The primary role of the ADC is to act as custodian of the many hundreds of standard catalogs that astronomers use in support of their research. ADC has all catalogs on-line and available for search and cross-correlation, either through ADC-provided user interfaces (Impress, CatsEye, ADC Data Viewer) or through the ADC External Query (AEQ) service, which allows HTTP-formatted queries against ADC catalogs to be received from other sites. Staff at the ADC has also taken on a leadership role concerning the development of XML standards for the astronomy community.

Astrophysics Data System

The Astrophysics Data System Abstract Service, located at the Smithsonian Astrophysical Observatory in Cambridge, has become a fundamental tool in astronomical research [10]. ADS indexes all of the primary astronomical research literature, including both the peer reviewed journals and the Los Alamos astro-ph preprint archive [5]. Direct links are provided from the abstracts to the full text of articles, either via the on-line journals or via scanned bit-maps of the historical literature. Increasingly the ADS is providing links to the data underlying the literature.

NASA Extragalactic Database

The NASA Extragalactic Database (NED), hosted by IPAC/Caltech, is a combined bibliographic and database service. It provides a thematic view of extragalactic astronomy, bringing together at one site extragalactic object catalogs, a name resolution service, links to important data resources, complete bibliographic information, and a suite of on-line tools (coordinate conversions, extinction calculations, velocity calculations). NED has recently introduced the "Level 5" service in which collections of the most significant publications related to key topics in extragalactic astronomy have been assembled and enhanced by adding current links to associated data or catalogs.

International Partners

A notable partner in astrophysics information services is the Centre Données de astronomiques de Strasbourg (CDS) [4], whose catalog and bibliographic services are used widely in the NASA astrophysics community. The SIMBAD database of astronomical objects and their bibliographic references [15], Vizier catalog browser [13], and Aladin correlation tool [2], are important elements of the international astrophysics information system.

VI INTEROPERABILITY INITIATIVES

The NASA archive and information services operate in an informal confederation known as the Astrophysics Data Centers Coordinating Council (ADCCC). The goals of the ADCCC are to maximize interoperability among the data centers and services, minimize redundancy of services and development efforts, and provide excellent information services to the entire astronomical community.

An example of improved interoperability is illustrated by access to resources such as data from the EUVE and ROSAT missions. The EUVE data are physically stored at the HEASARC. EUVE data are of interest to astronomers working with X-ray data and UV data, and thus interfaces to the EUVE observation catalog have been implemented at both HEASARC and MAST. These interfaces are designed to be similar to the interfaces for other data sets at each facility, and thus there is a minimal learning curve for users of either facility to gain access to the EUVE data. MAST users can retrieve EUVE data transparently, that is, without ever being aware that the data are physically stored somewhere other than in MAST. Similarly, MAST has recently added a direct link to ROSAT data, also at HEASARC, to enable easy cross-correlation between UV/optical and X-ray data. The HST observation catalog is routinely updated and provided to ADC and CDS to enable local catalog cross-correlation, but with direct links to the HST archive for preview images and making data requests. The Digital Sky project, managed through IRSA, has constructed the first portable services for positional cross-comparison of large catalogs.

Astrobrowse [7] [11] is a front-end search facility that is aware of over 1500 astronomical resources at some 80 sites worldwide. It is a data location service that can format queries for information on a given object in selected bandpasses or in selected data types. Users can located sources of potentially interesting data with one query and need not visit each of hundreds of web sites manually. Development of Astrobrowse has been fostered through ADCCC collaborations.

Another example of interconnectivity is found in NED, which returns not only basic data on individual galaxies held locally but also provides extensive sets of direct Web links for each object to catalog entries and original image databases held at NASA mission centers, and to a variety of ground-based archives around the world. NED users have transparent access to IRAS and 2MASS images via interfaces provided by IRSA.

Members of the ADCCC have worked to improve links between the scientific literature and the underlying data. The ADC includes in its catalog collections many key tables from the literature, and takes on a curatorial role: reviewing tables for errors, making corrections, and then providing access via its catalog browsing and comparison tools. The ADS has pioneered efforts to interlink distributed services (electronic journals, the SIMBAD database, the ADC catalogs, and NED) and is now including links from its abstracts to archival data, enabling researchers to quickly browse—and retrieve for detailed analysis as desired—the data on which a published paper is based. ADS also links to the NED and CDS databases for many abstracts. The NED and SIMBAD name resolvers are now used throughout the community, integrated into virtually all astronomical on-line services.

The ADCCC has been fostering new technological developments aimed at enabling interoperability among astronomy and space science data centers. For example, the ADC at NASA/GSFC has led community efforts in developing XML standards for astronomical catalogs, complementing efforts at IRSA, STScI, and CDS. ADCCC data centers have joined in the ISAIA (Integrated System for Archival Information Access) initiative, whose goal is to define protocols and standards for queries and responses from space science data services [6].

VII MIDDLEWARE FOR INTEROPERABILITY

The Virtual Observatory will require an interoperability layer to provide transparent access to distributed archives and information services. This layer operates primarily with metadata, translating user queries into a generic transport protocol that can be received by a data service and mapped onto site or service-specific queries. Similarly, query responses are remapped into the transport protocol and returned to an integrator, a program that receives query responses from distributed services and presents them in a coherent, consistent manner to the end-user.

A key requirement for a middleware layer is that it not impose constraints on how information providers manage their data internally, and the threshold for participation must be kept low. It must be simple—a matter of hours—to implement data dictionaries or interfaces compliant with the middleware layer. Finally, the middleware layer must retain the identity of the participating services so that users understand the provenance of the data they are using and know where to find assistance in using the data.

VIII METADATA STANDARDS AND PROFILES

The key to interoperability—to locating data of interest in distributed archives and information services and then being able to retrieve and integrate that data—is having standard metadata. Metadata is the information that describes the data, the most common example of which in astronomy is the information in a FITS file header. However, the FITS data format standard primarily defines the syntax rules for header information and the associated data structures. There is considerable room for creativity in the definitions of header keywords, such that many similar or identical quantities are described by any number of different terms. Moreover, extracted object catalogs are of critical interest to the virtual observatory, but here again the same measured quantities (a visual magnitude, for example) can have many different labels [14].

In the ISAIA the concept of a *profile* has been developed to implement metadata standards and to provide a mechanism for the exchange of metadata between data and information services and the integrator. The profile defines standard metadata labels for the data descriptors needed for locating and retrieving information. ISAIA profiles have three components: the resource profile, the query profile, and the returned information profile. The resource profile includes the metadata describing the data and information resources themselves. It characterizes the each service's data holdings and is used to determine which of the hundreds of archives and catalogs should actually be sent a query for certain types of data. For example, if a user is looking for X-ray data sets the resource profile would determine that STScI/MAST would not have data holdings of interest and the query would not be sent to that service. An example showing the elements of a resource profile is shown in Fig. 2.

The query and returned information profiles characterize the actual archival data sets and catalogs and form the basis for sending data selection queries to distributed services, and then collecting the results for uniform presentation. To take an obvious example, right ascension and declination are fundamental pieces of metadata describing an astronomical object or observation. However, we know that different astronomical catalogs use different labels, different formats, and different epochs for storing this information. One catalog might use labels "RA" and "DEC", another "right ascen" and "declin", and yet another "CRVAL1" and "CRVAL2" (following the FITS world coordinate system model). The values may be given in hexagesimal format (with variants), decimal degrees, or radians. The query and returned information profiles accommodate this by defining a standard term for right ascension and a format for exchange, say "RA" and decimal degrees.

IX   PROFILE IMPLEMENTATION AND EXCHANGE PROTOCOLS

The profiles themselves are insufficient for the meaningful exchange of information. There are two approaches to implementing the profiles and the necessary translations from the standard metadata to site and service specific terms and formats. The translations can be implemented for each service, by each service, through a thin interface layer. That is, a data service receives queries, maps the standard metadata into its local terminology, responds to the query, and then constructs an response in which the metadata is mapped back into the standard label and format. An alternative approach is to have each site define its mappings of the profile via a data dictionary, and to aggregate the data dictionaries for use by a query/response server. The query/response server sends queries to each service preformatted for that service, having translated a generic query into a site-specific one.

Data Attributes:

FACILITY	name of observatory, mission, program, etc.
DISCIPLINE	astronomy, space physics, planetary science, solar physics
INSTRUMENT HOST	name of telescope (HST, IUE, COBE, ...)
INSTRUMENT NAME	name of instrument (WFPC, NICMOS, FIRAS, ...)
INSTRUMENT TYPE	magnetometer, spectrometer, imager, photometer, ...
OBSERVED PHYS. QUANTITY	photon, electron, proton, ion, atom, molecule, magnetic field, electric field, pressure, temperature, ...
SAMPLING MODE	time series, image, aperture, spectrum, visibility, scan, ...
DATA CLASS	pointed observation, survey observation, derived (catalog), simulation, model fit, ephemeris, software, literature
DATA FORMAT	FITS, CDF, PDS, HDF, ASCII, ...
TIME SPAN	range of times covered by resource
PRINCIPLE INVESTIGATOR	name of PI for INSTRUMENT NAME
OBJECT NAME	astronomical object name, planet name, region of space
ENERGY RANGE	optical, UV, IR, 2-10keV, ...

Resource Attributes:

RESOURCE NAME	name of the data resource described by this profile
DATA SERVICE SITE	name of the data center providing the data
CURATOR	name of person or organization responsible for knowledge of the data
RESOURCE URL	URL for accessing the resource for end-users
SUB-PROFILES	identifiers for sub-profiles supported by this resource

FIGURE 2. Sample resource profile for space sciences. Terms in the profile at this level are intended to pertain to space science data sets in general. Additional terms are introduced in subprofiles for specific disciplines, allowing users both broad and deep access to space science resources. *Data attributes* are fields which, when a query is forwarded to a data provider, can result in a selection or filtering of data from the provider. *Resource attributes* are fields that describe the resource but which do not result in selection or filtering at the provider's site.

The server listens for responses, and when they are received translates them back into the standard terms and formats. The latter approach can be implemented in a straightforward fashion through use of the CDS *GLU* (Générateur des Liens Uniformes, [3], http://simbad.u-strasbg.fr/glu/glu.htx). This puts a minimal burden on participating sites: all they need to do is maintain a simple data dictionary. The former approach requires a greater level of effort on behalf of participating sites, but also allows for much higher levels of functionality (e.g., server-side preprocessing of data).

ISAIA also has the concept of hierarchical profiles. A profile can be defined at a very general level, describing, for example, the primary characteristics of astronomical images. Subprofiles could be developed for different classes of images having unique characteristics (CCD images vs. radio interferometry data cubes, for example).

In the NVO the profile definitions will most likely be expressed in XML, and queries and responses can be exchanged via HTTP. This approach has been prototyped for the Planetary Data System and Palomar Testbed Interferometer [8]. Data service providers are increasingly using Java for user interfaces given that it has much greater capabilities than HTTP/CGI. Assuming that this trend continues, the NVO middleware layer will also need to be able to access distributed services through JDBC or other mechanisms.

X SUMMARY

The NASA astrophysics data centers will be cornerstones in the emerging Virtual Observatory. We have already begun to establish the sort of inter-center partnerships and distributed data services that are requisite to the success of the NVO. The work of the ADCCC has demonstrated the feasibility of the NVO, through support for new technology and establishment of standards. We must now take the next steps in realizing the vision of the Virtual Observatory.

Acknowledgments The author thanks his colleagues at each of the data centers described for their collaboration in the ADCCC and their contributions to this paper. The ISAIA project team includes Tom McGlynn and Nick White (NASA/GSFC HEASARC), Joe King (NASA/GSFC NSSDC), Cynthia Cheung (NASA/GSFC ADC), Ray Plante and Bob McGrath (NCSA), Joe Mazzarella (Caltech/IPAC), Arnold Rots (SAO/CXC), Steve Hughes (JPL), Mike A'Hearn (UMd), Reeta Beebe (NMSU), Françoise Genova (CDS), and Paolo Giommi (BSDC). This work has been partially supported by NASA's Applied Information Research Program under grants to the Space Telescope Science Institute, the National Center for Supercomputing Applications/University of Illinois, and the Goddard Space Flight Center.

URLS

ADC	http://adc.gsfc.nasa.gov/
AEQ	http://tarantella.gsfc.nasa.gov/viewer/AEQdoc.html
Aladin	http://aladin.u-strasbg.fr/aladin.gml
Astrobrowse	http://heasarc.gsfc.nasa.gov/ab/
CDS	http://cdsweb.u-strasbg.fr/CDS.html
HEASARC	http://heasarc.gsfc.nasa.gov/
IMPReSS	http://tarantella.gsfc.nasa.gov/impress/
ISAIA	http://heasarc.gsfc.nasa.gov/isaia/
IRSA	http://irsa.ipac.caltech.edu/
MAST	http://archive.stsci.edu/mast.html
NSSDC	http://nssdc.gsfc.nasa.gov/
NVO	http://www.voforum.org/
SkyView	http://skyview.gsfc.nasa.gov/

REFERENCES

1. Angelini, L., Breedon, L., Garcia, L., Hilton, G., Stollberg, M., & White, N. 1999, AAS Meeting, 194, #83.01
2. Bonnarel, F., Fernique, P., Bienaymé, Egret, D., Genova, F., Louys, M., Ochsenbein, F., Wenger, M., & Bartlett, J. 2000, A&A Suppl., 143, 33
3. Fernique, P., Ochsenbein, F., & Wenger, M. 1998, in ASP Conf. Ser., 145, Astronomical Data Analysis Software and Systems VII, R. Albrecht, R. N. Hook, & H. A. Bushouse, eds., 466-469
4. Genova, F., Egret, D., Bienaymé, O., Bonnarel, F., Dubois, P., Fernique, P., Jasniewicz, G., Lesteven, S., Monier, R., Ochsenbein, F., & Wenger, M. 2000, A&A Suppl., 143, 1
5. Grant, C. S., Accomazzi, A., Eichhorn, G., Kurtz, M. J., & Murray, S. S. 2000, 2000, A&A Suppl., 143, 111
6. Hanisch, R. J. 2000, Computer Physics Communications, 127, 177
7. Heikkila, C. W., McGlynn, T. A., & White, N. E. 1999, in ASP Conf. Ser., Vol. 172, Astronomical Data Analysis Software and Systems VIII, ed. David M. Mehringer, Raymond L. Plante, & Douglas A. Roberts (San Francisco: ASP), 221
8. Hughes, S. 2000, in ASP Conf. Ser., Virtual Observatories of the Future, ed. R. J. Brunner, S. G. Djorkgovski, & A. Szalay (San Francisco: ASP), in press
9. Imhoff, C., Abney, F., Christian, D., Donahue, M., Hanisch, R., Kimball, T., Levay, K., Padovani, P., Postman, M., Smith, M., & Thompson, R. 1999, AAS Meeting, 194, #83.02
10. Kurtz, M. J., Eichhorn, G., Accomazzi, A., Grant, C. S., Murray, S. S., & Watson, J. M. 2000, A&A Suppl., 143, 41
11. McGlynn, T., & White, N. 1998, in ASP Conf. Ser., Vol. 145, Astronomical Data Analysis Software and Systems VII, ed. R. Albrecht, R. N. Hook, & H. A. Bushouse (San Francisco: ASP), 481

12. McGlynn, T., Scollick, K., & White, N. 1998, in IAU Symp. 179, New Horizons from Multi-Wavelength Sky Surveys, ed. B. J. McLean, D. A. Golombek, J. J. E. Hayes, & H. E. Payne (Dordrecht: Kluwer), 465
13. Ochsenbein, F., Bauer, P., & Marcout, J. 2000, A&A Suppl., 143, 23
14. Ortiz, P. F. 2000, Computer Physics Communications, 127, 188
15. Wenger, M., Ochsenbein, F., Egret, D., Dubois, P., Bonnarel, F., Borde, S., Genova, F., Jasniewicz, G., Laloë, S., Lesteven, S., & Monier, R., 2000, A&A Suppl., 143, 9

High Energy Astrophysics Missions

Nicholas E. White

Laboratory for High Energy Astrophysics
NASA's Goddard Space Flight Center,
Greenbelt, MD 20771

Abstract. NASA's Structure and Evolution of the Universe (SEU) program uses X-ray and Gamma ray observations to observe the extremes of gravity throughout the universe. This program will probe the nature of black holes, ultimately obtaining a direct image of the event horizon. It will investigate the large scale structure of the Universe to constrain the location and nature of dark matter. Finally it will search for and study the highest energy processes, that approach those found in the early universe.

INTRODUCTION

Energetic processes either in the vicinity of black holes and neutron stars, or during the creation of these compact objects, give rise to high energy (X-ray , and Gamma ray) emission. These are the same objects that are the prime candidates for gravitational radiation. Many of these X-ray and Gamma ray events may be accompanied by strong gravitational radiation signatures. Thus the fields should be highly synergistic. In this review I will highlight some of the future high energy astrophysics missions that are planned by NASA for the coming two decades.

CONSTELLATION-X

Constellation-X is a key element in NASA's Structure and Evolution of the Universe (SEU) theme aimed at understanding the great mysteries of space, time, and energy. When observations commence towards the end of the next decade, Constellation-X will address many pressing questions concerning the extremes of gravity and the evolution of the Universe. X-ray observations of broadened iron emission lines in Active Galactic Nuclei will measure black hole masses and spins and will test General Relativity in the strong gravity limit. Constellation-X will show us how black holes evolve with cosmic time and, as accretion energy may be a dominant component, will provide critical information on the total energy output of the Universe. It will be a critical step to investigating the black hole environment, to ensure the conditions are favorable for the more challenging mission (MAXIM) to follow, that will image the black hole event horizon.

The Constellation-X mission is a large collecting area X-ray facility, emphasizing observations at high spectral resolution (DE/E of 300 to 3000) while covering a broad

CP575, *Astrophysical Sources for Ground-Based Gravitational Wave Detectors*, edited by J. M. Centrella
2001 American Institute of Physics 0-7354-0014-8

energy band (0.25 to 40 keV). By increasing the telescope aperture and utilizing efficient spectrometers the mission will achieve a factor of 100 increased sensitivity over current high resolution X-ray spectroscopy missions. Constellation-X will make high resolution spectroscopy in the X-ray band routine for faint X-ray source populations. It will be to X-ray astronomy, what the Keck Observatory has come to mean for optical astronomy.

The 0.25 to 10 keV X-ray band contains the K-shell lines for all of the abundant metals (carbon through zinc), and the L-shell lines of many. The detailed X-ray line spectra are rich in plasma diagnostics which also provide unambiguous constraints on physical conditions in the sources. A spectral resolving power of at least 300 is required to separate the He-like density sensitive triplet. In the region near the iron K complex a resolving power exceeding 3000 is necessary to distinguish the lithium-like satellite lines from the overlapping helium-like transitions. Spectra of high statistical quality in an average observing time of 100,000 s for large populations of sources at the flux levels typically reached in the Chandra and XMM-Newton surveys require an effective area 20 to 100 more than provided by the spectrometers on Chandra, XMM, and Astro-E. This corresponds to 15,000 cm^2 at 1 keV and 6,000 cm^2 at 6 keV (including the spectrometer efficiency). Using a typical AGN spectrum an effective area of 1,500 cm^2 at 40 keV is required to match the sensitivity in the 0.25 to 10 keV band.

The Constellation-X design recognizes that several smaller spacecraft each carrying one *science unit* can cost less than one very large spacecraft. The program is then robust in that risks are distributed over several launches and spacecraft with no single failure leading to loss of mission. The current baseline is four satellites launched two at a time on either Delta IV or Atlas V vehicles. The mission will be placed into a L2 orbit to facilitate high observing efficiency, provide an environment optimal for cyrogenic cooling, and simplify the spacecraft design.

To cover the entire 0.25 to 40 keV band Constellation-X will utilize a matched set of high throughput focussing telescope systems. A spectroscopy X-ray telescope (SXT) covers the 0.25 to 10 keV band, and is optimized to maintain a spectral resolving power of 300 to 3000 across the bandpass. To maximize the collecting area per unit mass at minimum cost, an angular resolution requirement of 15 arc sec half power diameter (HPD) has been selected, with a goal of 5 arc sec. This exceeds the confusion limit for the limiting flux of the sources to be studied.

The SXT uses two complementary spectrometer systems to achieve the desired energy resolution: an array of high efficiency quantum micro-calorimeters with energy resolution of 2eV, and a set of reflection gratings. The gratings deflect part of the telescope beam away from the calorimeter array, with the remainder falling onto the high spectral resolution quantum calorimeter. The field of view of the calorimeter is 2.5 arc min square, with 5 arc second pixels to adequately sample the telescope point spread function. The two spectrometers are complementary, with the grating optimal for high resolution spectroscopy at low energies and the calorimeter at higher energies. The gratings also provide coverage in the 0.3 to 0.5 keV band where the calorimeter thermal and light-blocking filters cause a loss of response. This capability is particularly important for high redshift objects, where line rich regions will be moved down into this energy band.

The Constellation-X hard X-ray telescope (HXT) uses multi-layers to provide the first focusing optics system to operate in the 10 to 40 keV band. The improvement in the signal to noise results in a factor of 100 or more increased sensitivity over non-focussing methods used in this band. The HXT has a resolution of 1' HPD. It has already been demonstrated several groups that multilayers can be deposited on X-ray optics with the required performance. The GSFC-Nagoya Infocus and Caltech-Columbia HEFTE balloon payloads will prove this approach in the next couple of years. The fine position resolution, along with the required efficiency and energy resolution (R >10) can be met by a cadmium zinc telluride (CZT) detector system.

The Constellation-X mission is jointly managed by NASA & SAO. It is led by Dr Nicholas White at GSFC and Dr harvey Tananbaum at SAO. The program is the formulation phase and is targeting a first launch in 2008, with a second in 2009.

MAXIM

The Micro Arcsecond X-ray Imaging Mission (MAXIM) is designed to make a ten million fold leap in X-ray image clarity of celestial objects. The prime science goal of the MAXIM is to resolve the event horizon of the super-massive black holes found at the center of nearby galaxies. This will not only be the ultimate proof that black holes exist, but it will also allow us to directly observe the distortions of space and time caused by strong gravity, predicted by Einstein's theory of General Relativity.

X-ray telescopes in general are difficult to build because X rays only reflect at a very shallow angle (1 degree or less), referred to as grazing incidence. To obtain a true focus, they must reflect twice from very precisely figured hyperbolic and parabolic surfaces. These surfaces are, in effect, nested cylinders that are very expensive to shape to the required precision. The recently launched Chandra X-ray Observatory is the state of the art in X-ray imaging, a telescope that cost several hundred million dollars to build. Chandra achieves an impressive resolution of about 0.5 arcseconds, yet it is still far from the diffraction limit.

Micro-arcsecond resolution in the X-ray band will require the development of interferometry. Cash et al (Nature Sept 14, 2000) have recently demonstrated in the laboratory that practical interferometers can be built for the x-ray, and avoids the potential difficulties of building diffraction limited x-ray optics. Webster Cash takes what at first seems to be a disadvantage, mirrors operating at shallow angles, and turns it to an advantage. Instead of using expensive, precisely figured optics to focus the X rays, he instead utilizes two sets of readily made flat mirrors to steer an incoming X-ray beam together to create interference fringes. A two-dimensional image is created by simultaneously combining many sets of flats at different rotation angles. Because the X rays are reflected at shallow angles, the tolerances required to hold the flats aligned are about 100 times less than for a traditional (normal incidence) mirror operating at the same wavelength. To magnify 1nm waves to 100 micron fringes requires a cross angle of about 2 arcseconds, which means that the detector telescope separation will be large, of order 500 km. To simultaneous cover the uv plane, 32 flat mirrors are placed in a ring.

There are still some technological hurdles. Even at the very short X-ray wavelength, a telescope separation of 100-1,000 meters is needed to achieve the required angular resolution. This entails a fleet of up to 33 optics spacecraft flying in formation with a precision of 20 nano-meter, plus a detector spacecraft 500 kilometers behind the mirror. This is challenging, but probably no more so than missions under consideration by NASA to image extra-solar Earth-size planets. A MAXIM pathfinder mission, with a one-meter separation between the telescopes so the X-ray optics are all on one spacecraft, is a reasonable first step. This is under study by NASA for flight during the middle of the next decade. The MAXIM Pathfinder would also be powerful in its own right, providing an impressive 1,000-time improvement over the Chandra Observatory. The MAXIM program is led by Prof Webster Cash (University of Colorado), Dr Nicholas White (GSFC) and Dr Marshall Joy (MSFC)

SWIFT

The discovery by BeppoSAX of afterglow from gamma ray bursts (GRBs) has revolutionized our understanding of these enigmatic events. We now know that they are cosmological with redshift > 1 and involve the most powerful and relativistic explosions known. These explosions are thought to create a super-relativistic blast wave resulting in an afterglow that cascades down from gamma rays to radio. A panchromatic approach is now essential for the next phase of discovery. The Swift Gamma Ray Burst MIDEX will be a multiwavelength observatory that exploits the newly discovered afterglow characteristics to make a comprehensive study of ~1000 bursts. Swift will determine the origin of GRBs, tell us how the blast wave evolves and interacts with its surroundings, and identify different classes of bursts and their associated physical processes. In addition, Swift will allow GRBs to be used as probes of the early Universe.

The Swift Gamma Ray Burst Explorer has a complement of three coaligned instruments: the Burst Alert Telescope (BAT), an X-ray Telescope (XRT), and a UV optical Telescope (UVOT). BAT is a wide Field-Of-View (FOV) coded-aperture gamma ray imager that will produce arcminute GRB positions onboard within 10 seconds. The spacecraft will execute a rapid autonomous slew that will point the focusing telescopes at the BAT position in typically ~ 50s. The XRT and UVOT are an X-ray and a UV/optical focusing telescope which will produce arcsecond positions and multiwavelength lightcurves for gamma ray Burst (GRB) afterglows. Broad band afterglow spectroscopy will produce redshifts for the majority of GRBs.

The positions and images derived by the various instruments will be sent as soon as they are available from the spacecraft via the TDRSS system to the Gamma ray coordination network (GCN). The GCN broadcasts the results to the world via the Internet for rapid response by the world astronomy community for follow up observations by other ground and space based telescopes. At the next satellite pass over Malindi the more detailed data will be sent to the data center where it will be processed for public access within 30 minutes of the pass.

Swift is an international mission with participation from the UK and Italy. The PI is Dr Neil Gehrels from GSFC, with a large involvement from Penn State University led by Dr John Nousek. It is scheduled for launch in the fall of 2003.

GLAST

The energy range 20 MeV-300 GeV that the Gamma-ray Large Area Space Telescope (GLAST) will observe represents one of the last poorly-measured region of the celestial electromagnetic spectrum. After several successful exploratory missions, the Energetic Gamma Ray Experiment Telescope (EGRET) instrument on the Compton Gamma Ray Observatory, launched in 1991, made the first complete survey of the sky above 30 MeV. EGRET showed the high-energy gamma-ray sky to be surprisingly dynamic and diverse, with sources as varied as the sun and black holes at large redshifts. Most of the gamma-ray sources detected by EGRET remain unidentified. EGRET uncovered the tip of the iceberg, raising many questions, and it is in the light of EGRET's results that the great potential of the next generation gamma-ray telescope can be appreciated.

GLAST will be an imaging gamma-ray telescope vastly more capable than instruments flown previously. GLAST's area, angular resolution, field of view, and energy resolution together provide an unprecedented advance in sensitivity, a factor of 30 or more, making GLAST a very flexible tool for investigating the great range of astrophysical phenomena best studied in high-energy gamma rays. A Gamma ray burst context instrument is provided by MSFC.

The universe is largely transparent to gamma rays in the energy range of GLAST. Energetic sources near the edge of the visible universe can be detected by the light of their gamma rays. There is good reason to expect that GLAST will see known classes of sources to redshifts of 4, or even greater if the sources existed at earlier times. The small interaction cross sections for gamma rays also means that they offer a direct view into nature's highest-energy acceleration processes. EGRET discovered that blazar-class active galactic nuclei (AGNs) are bright, variable sources of high-energy gamma rays, with most of their luminosity emitted in GLAST's energy range. The emission is believed to be powered by accretion onto supermassive black holes found at the cores of distant galaxies. GLAST will increase the number of known AGN gamma-ray sources from about 60 to thousands. GLAST will probe the mechanisms of particle acceleration and gamma-ray production in AGNs. The large area and low instrumental background of GLAST will also facilitate searches for decays of exotic particles in the early universe and for annihilations of postulated weakly-interacting massive particles (WIMPs) in the halo of the Milky Way.

The primary interaction of photons in the GLAST energy range with matter is pair conversion. This process forms the basis for the underlying measurement principle by providing a unique signature for gamma rays, which distinguishes them from charged cosmic rays whose flux is as much as 10^5 times larger, and allowing a determination of the incident photon directions via the reconstruction of the trajectories of the resulting e^+e^- pairs.

the GLAST Large Area Telescope (LAT) is modular, consisting of a 4x4 array of identical towers. Each 40x40 cm^2 tower comprises a tracker, calorimeter and data acquisition module. The tracking detector consists of 18 xy layers of silicon strip detectors. This detector technology has a long and successful history of application in accelerator-based high-energy physics, and is well-matched to the requirements of high detection efficiency (>99%), excellent position resolution (<60 μm), large signal:noise (>20:1), negligible cross-talk, and ease of trigger and readout with no consumables. The calorimeter in each tower consists of 8 layers of 12 CsI bars in a hodoscopic arrangement, read out by photodiodes, for a total thickness of 10 radiation lengths. The anticoincidence detector, which covers the array of towers, employs segmented tiles of scintillator, read out by wavelength-shifting fibers and miniature phototubes.

The LAT instrument design is based on years of detailed computer simulations. A complete software model of the instrument, including gaps, structural material, noise, inefficiencies and other real-world effects, was constructed. The computer model was used to generate simulated instrument data, and then to develop realistic reconstruction algorithms. The simulations were used to (1) demonstrate the necessary background rejection performance of the instrument, (2) produce realistic triggering and readout schemes, and (3) evaluate and optimize the performance of the instrument (effective area, angular resolution, etc.) after all background rejection cuts and instrumental effects have been taken into account. These simulations have been validated as part of a vigorous R&D program, which included a beam tests at the Stanford Linear Accelerator Center (SLAC).

The GLAST Burst Monitor (GBM) augments the capabilities of the LAT for gamma-ray bursts in two ways. First, the GBM extends the energy range of burst observations from the bottom of the LAT energy range down to a few keV. This allows bursts observed with GLAST to be placed into the context of the large database of previous burst observations. Second, the GBM will provide an on-board trigger and approximate location for bursts that lie outside the field of view of the LAT. The spacecraft can then be repointed to observe delayed high-energy emission.

GLAST is a collaborative effort between NASA and DoE, with the PI Dr Peter Michelson for the main instrument, the Large Area Telescope (LAT), from SLAC. The PI for the GBM is Dr Chip Meegan from MSFC. NASA plans to launch GLAST in 2005.

EXIST

A critically important region of the astrophysical spectrum is the hard X-ray band (10 to 600 keV), which connects the predominantly thermal (x-ray) and non-thermal (gamma-ray) universe. The Energetic X-ray Imaging Survey Telescope (EXIST) mission will conduct the first high sensitivity all-sky imaging survey in the hard X-ray band (5 to 600keV) with a wide-field coded aperture telescope array. The hard x-ray band to be surveyed by EXIST is key for study of the obscured universe, particularly the heavily absorbed active galactic nuclei (AGN) of galaxies. This survey will be used to find targets for the hard X-ray telescope on Constellation-X.

EXIST consists of 8 x $1m^2$ coded aperture telescopes with combined 40 deg x 160 deg field of view. The detector plane is made up a large array of CZT pixel detectors that will provide an image of the mask pattern. The all sky survey nature of EXIST and its large mass makes it an ideal use of the International Space Station (ISS). EXIST mounted on the International Space Station (ISS) would provide full-sky imaging each orbit and achieve the ultimate hard x-ray survey sensitivity: a factor of ~1000 better than the only previous (HEAO-A4; 1979) all sky survey. EXIST is led by Dr Josh Grindlay from Harvard and is a candidate for a new start in the SEU roadmap and may be launched late in the decade.

HSIM & ACT

The low energy gamma ray spectral band contains nuclear line emission from the decay of radioactive isotopes. In a cosmic setting these are produced during supernova explosions and observations of the decay of these nuclear lines can provide critical insight into the nature of the supernova and the origin of the elements. Gamma rays are highly penetrating radiation and allow observations to be made within the enshrouding gas of the supernova explosion, as well as probe parts of the galaxy hidden by dust. The INTEGRAL mission to be launched by ESA in 2002 will follow on from the Comptel experiment on Compton GRO, to provide the most detailed view yet, employing for the first time high resolution spectroscopy. But CGRO and INTEGRAL will only uncover the tip of the expected MeV line emission science.

An enduring goal has been to study the prompt nuclear line emission (e.g. ^{57}Co) from Type 1a supernovae in the Virgo cluster at 20 Mpc. This will require a sensitivity of ~10^{-6} photons cm^{-2} s^{-1}. The High Resolution Spectral Imager Mission (HSIM) proposed by Dr Fiona Harrison at CalTech will take advantage of the fact that many important nucleosynthetic lines lie in the hard X_ray range, including ^{44}Ti (68, 78 keV), ^{57}Co (122 keV) and ^{56}Ni (158 keV). By employing multi-layers on traditional X-ray optics, it will be possible to achieve the sensitivities required to study supernovase in Virgo. The HSIM mission is a candidate for a new start in the 2008-2013 timeframe.

The Advanced Compton Telescope (ACT) is a very large spectrometer that will employ solid state technology and Ge detectors to make sensitivie surveys down to the majic ~10^{-6} photons cm^{-2} s^{-1} for the broadened lines expected from supernovae, combined with a large field of view of 10-60 degrees. The technology for this vision mission is challenging and will require considerable investment over the coming decade. The ACT study is led by Dr Jim Kurfess from NRL.

ACKNOWLEDGEMENTS

The author thanks R. Blandford, A. Bunner, W. Cash, N. Gehrels, J. Grady, J. Grindlay, F. Harrison, S. Kahn, P. Michelson, J. Ormes, S. Ritz, and H. Tananbaum for their contributions (direct and indirect) to this paper.

LIGO Data and Data Analysis

Data from the LIGO I Science Run

Albert Lazzarini

LIGO Laboratory
California Institute of Technology, Pasadena, CA 91125
LIGO-P010002-00-E

Abstract. The LIGO[1] I Science Run is planned to begin in mid-2002. The characteristics of the data stream, data volumes, data products, and data availability are discussed. The data analysis activities will be undertaken by the LIGO Scientific Collaboration (LSC[2]). These activities include operating dedicated on-site pipelines at the LIGO observatories. In addition, a dedicated off-site facility for will be dedicated to melding data from different interferometer datastreams (both LIGO and eventually those of other international projects as part of a network-wide analysis effort). Exploratory university-based research on LIGO data will likely be supported in part by the nascent US computing grid. LIGO Laboratory and the LSC are working on grid computing efforts within the GriPhyN (Grid Physics Network) collaboration research activities.

LIGO DATA CHARACTERISTICS

LIGO interferometers produce audio-band time-series data digitized at a number of frequencies that are powers of two (refer to **TABLE 1**). The data are acquired as 2-byte integers from the acquisition system. They also include computed data, which are stored as 4-byte real numbers. In total, the three LIGO interferometers will generate between 7 – 9 MB/s of data as time series from 2000+ parallel channels. Because the data rate is quite high and because the data are all time series, LIGO, along with its international partner projects, has developed a format[3] for acquiring and archiving the raw data. This so-called Frame Format captures all the data from a single interferometer that were collected during the same epoch. The frame duration is arbitrary and is typically between 1s and 32s in length. Figure 1 shows a schematic of the frame format and its organization. All data are stamped with the GPS (Global Positioning System) time, which is accurate to <100 ns, at the time of acquisition. Data from multiple detectors will be merged into composite frames spanning the same GPS epoch for coincidence analysis.

LIGO Data Volume

The datastream comprises less than 1% (3 channels at 32 kB/s each) of strain or so-called "science-channel" data that will contain the astrophysical signatures for discovery. The rest of the data stream constitutes ancillary or auxiliary channels. These channels are grouped in two broad classes. The first group (all the other interferometer channels and health/status) is used to monitor the behavior of the many electrical-mechanical-optical servos that are required to maintain the very complex

CP575, *Astrophysical Sources for Ground-Based Gravitational Wave Detectors,* edited by J. M. Centrella
© 2001 American Institute of Physics 0-7354-0014-8/01/$18.00

LIGO interferometers in a state of maximum sensitivity. This group corresponds to roughly 80% of the acquisition bandwidth. When the instrument is operating properly, these channels should show no anomalous behavior: a gravitational wave strain signal should not show up in these other channels. On the other hand, an anomalous instrumental artifact would be present in both the auxiliary channel data and the strain channel, allowing instrumental "glitches" to be detected and vetoed by suitably processing the auxiliary ("health & status") interferometer channels.

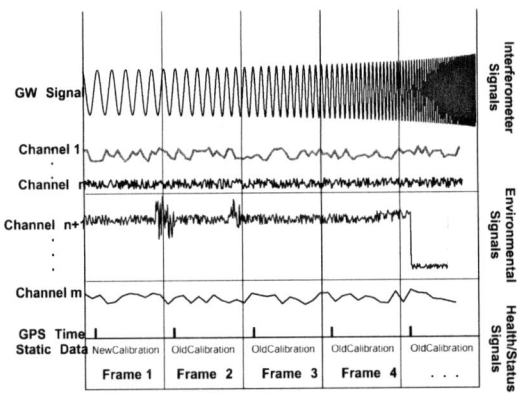

Figure 1. The LIGO datastream constitutes very many channels acquired in the time domain with sample rates up to 16384 samples/s. The data are organized in C-structures called frames which capture all the data collected by the LIGO instruments during a particular epoch. This encapsulation is intended to ensure a self-consistent set of raw data for archival.

The second class of channels is dedicated to monitoring the local terrestrial environments at the observatories. This group corresponds to roughly 20% of the acquisition bandwidth. These channels come from a variety of non-interferometric sensors that will be used to monitor at high sensitivity a large variety of geophysical and local phenomena of non-astrophysical. This is critical to the operation of LIGO because many geophysical and anthropogenic will likely be detectable by interferometers operating at sensitivities of $\sim 10^{-18}$ m RMS. The detection of the physical disturbances by other means will be used to establish vetoes for these types of real signals that are not of astrophysical significance.

TABLE 1: LIGO I Data Channel Count by Acquisition
Rate for Hanford Observatory (for 2 interferometers)

Acquisition Rate, samples/second (16 bit)	Number of Channels
16834	124
2048	532
256	109
64	205
16	208
Total No. of Channels:	1178

70

Additional LIGO datasets will be generated from the raw data. In all, it is expected that three levels of frame data will be created (Levels 1, 2,3) and distributed for data analysis. These levels, their sizes, and their expected uses are shown in **TABLE 2**. Access to Level 1 data should be infrequent and will serve primarily to verify or veto putative detections by enabling scientists to look broadly across all channels during an epoch of detection. The higher levels of data constitute progressively more refined datasets that will likely be of interest to the science teams who are analyzing the data for astrophysical signatures.

TABLE 2: LIGO I Data Products and Volumes

Mode	Raw and Derived Data for On-line Diagnostics	Level 1 Full (100%) frame data for archiving	Level 2 Strain and data summary, QA channels	Level 3 Strain best estimate
Uncompressed Rate (MB/s)	LHO: 9.479 LLO: 4.676 Total: 14.155	LHO: 4.698 LLO: 2.278 Total: 6.975	**Total: 0.300**	**Total: 0.006**
w / 50% Hardware Compression, MB/s onto tape media	-	LHO: 2.349 LLO: 1.139 Total:3.488	Total: 0.150	-
Data growth rate, per year of integrated running, TB/yr.	-	LHO: 74 LLO: 36 Total:110	Total:9.5	**Total: 0.200**
Total including redundant 100% backup, TB/yr.	-	LHO: 148 LLO: 72 **Total:220**	**Total:19**	-
Purpose	For on-line monitoring of interferometers	Deep permanent archive	Science analysis, data exchange	Science analysis, data exchange
On-site look-back time	Must use real-time control and monitoring system (CDS) disk caches	LHO Disk cache: 28 d LLO Disk cache: 28 d	-	-
Off-site look-back time	-	As long as required	In perpetuity	In perpetuity

Level 2 data correspond to the strain channel and those channels which may have sufficiently strong cross-correlations with the strain channel to enable regression and removal of the cross-correlations to be performed. In addition, a data-quality (QA) channel will be useful to quickly exclude from further analysis those data segments that have exceedingly poor sensitivity. The Level 2 datasets will be of interest to both astrophysicists who are looking at the data in detail and to interferometer scientists who are trying to understand interferometer behavior and terrestrial correlations.

71

Level 3 data correspond to a smaller subset, consisting of the strain channel, data-quality and possibly a few other measures of overall interferometer (or environmental) characteristics. Level 3 data will be the sets that are exchanged and merged among different detector projects for network analysis. It is likely that idiosyncratic instrumental channels specific to a particular instrument will not be of much utility to a network analysis team composed of members from many projects.

LIGO Databases

In addition to the raw time-series frame data, LIGO has developed a capability to store important metadata into a relational database archive. The relational database is composed of a number of inter-related database tables[4] that are used for a number of functions. Figure 2 present a schematic showing the database tables and their inter-relationships. The database provides a tabulation of those instrumental performance

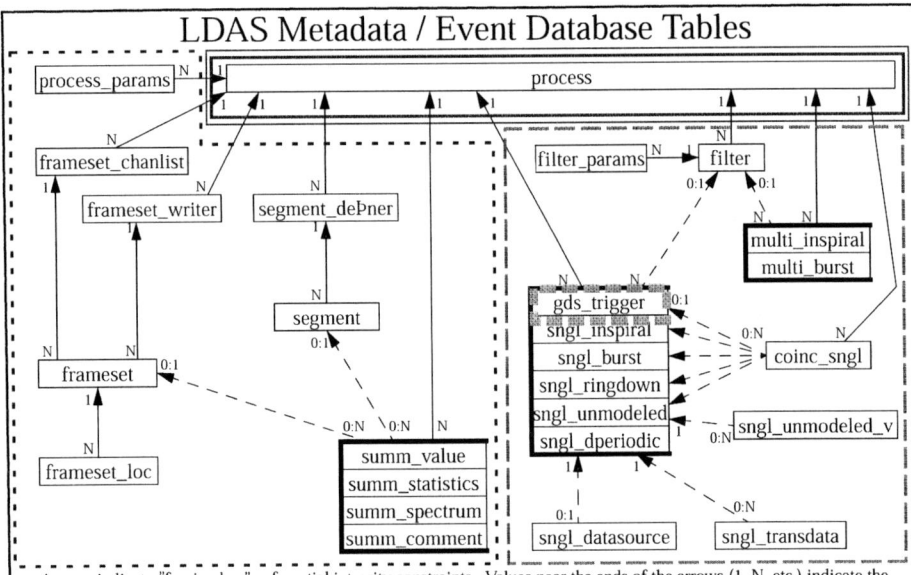

Arrows indicate "foreign key" referential integrity constraints. Values near the ends of the arrows (1, N, etc.) indicate the possible multiplicities. Dashed lines indicate optional relationships. Stacked tables (grouped by thick lines) have common relationships with other tables, except for relationship arrows connecting along the right edge. Examples: 1) Each segment is related to one segment_definer; 2) Each segment_definer is (generally) related to many segments; 3) A frameset is related to one frameset_chanlist entry and to one frameset_writer; 3) A summ_value (or summ_statistics, etc.) entry may or may not be related to a segment and/or a frameset; 4)A single-interferometer event (gds_trigger, sngl_inspiral, etc.) entry may be related to up to one sngl_datasource and/or any number of sngl_transdata entries.

Figure 2. LIGO relational database design for storing metadata and events/triggers generated by pipeline analyses. Dotted box: tables used to summarize and index into time-series frame data. Dashed box: tables used to capture events or triggers generated by search analysis pipelines. Small dashed box (gds trigger): table used to capture instrumental artifact triggers for later vetoing of data. Double thin-lined box (process): table used to capture information about software that was used to generate metadata.

metrics that are useful for a first-look survey of the available frame data. Examples include statistics (RMS, mean, peak-to-peak) for key channels. In addition, representative spectra or other graphic elements may be stored as binary large objects (BLOBS). Another intent of the design of the database is to facilitate later use of LIGO data for data mining purposes. A remote client software package has been developed (GUILD[5]) to enable researchers to query and retrieve information stored in the database.

LIGO Data Flow Model

LIGO Laboratory constitutes four geographically isolated sites: two observatories in Washington[6] and Louisiana[7], and two university laboratories at Caltech and MIT[8]. Data acquisition in large volume takes place at the observatories and the data then migrate first to Caltech and MIT, and then a number of LIGO Scientific Collaboration institutions. Figure 3 presents a schematic of this data flow. The large on-line disk arrays systems are used to store the data as they are produced. These arrays are common to the acquisition system and the data analysis systems. Data will become available on-site as soon as they are produced. The on-line global diagnostics system is part of the real-time control and data systems (CDS[9]). Real-time software monitors are used to track instrumental performance and to generate trigger data that are sent

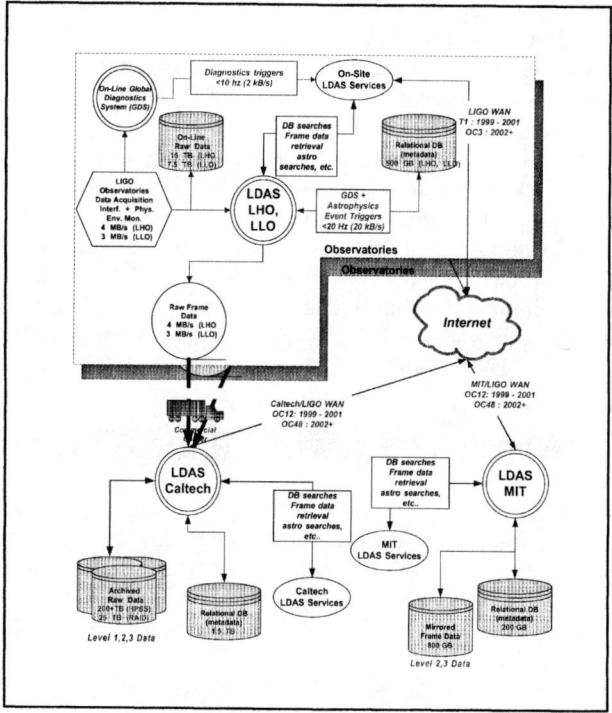

Figure 3. LIGO data flow from observatories to the repositories at Caltech and MIT.

for storage to the data analysis system (LDAS[10]) database server. In addition, pipeline search algorithms running in synchrony with the rate at which data are acquired sift through the data. These employ a number of digital filtering techniques to look for events of astrophysical interest. The complexity and computational efficiency of the search algorithms depends on the frequency-time volume that needs to be searched for different classes of events[11].

Raw frame data will be written to data tape and they will be sent to Caltech, where they will be ingested and made available through a deep archive (HPSS[12]). The rate of growth and volume of the metadata databases will be sufficiently small so that they can be migrated to the archive over the LIGO wide area network (WAN). The MIT databases are small mirrors of the deep archive at Caltech. All data products will be available across the LIGO Scientific Collaboration institutions that are performing data analysis studies.

Initially, the main emphasis of the collaboration-wide astrophysics searches will be development and maintenance of the pipeline searches. Many of the search algorithms require dedicated computational resources running nearly all the time. The Laboratory is providing these resources for coordinated searches that are formed within the LIGO Scientific Collaboration.

Exploratory prototyping by individuals will be mostly performed at the home institutions of members of the LIGO Scientific Collaboration. Data distribution will be provided to enable individuals to have access to datasets for their research. The LIGO data use model is still evolving as this new collaboration learns to look at the data from the instruments that are just now coming on-line.

Access to LIGO science data is available through membership in the LIGO Scientific Collaboration. The collaboration believes that instrumental idiosyncrasies that will always be present in data from the interferometers will require intimate working knowledge of the instruments before the data can be analyzed or interpreted for astrophysical content. This resembles more the model of data access in the high-energy physics community than the astronomical community. The LIGO Scientific Collaboration is open to interested scientific researchers who establish a memorandum of understanding (MOU[13]) with LIGO Laboratory for access to the data archive.

GRID PHYSICS NETWORK (GRIPHYN) AND LIGO

The Grid Physics Network (GriPhyN[14]) is a collaboration among high-energy physics experiments, astronomy, gravitational physics, and computer science whose mission is to mobilize large-scale information technology resources for scientific research. The high-energy physics experiments that are involved include the U. S. members of the CMS[15] and ATLAS[16] collaborations who are developing experiments at CERN for the Large Hadron Collider (LHC[17]). The astronomy members are from the SLOAN Digital Sky Survey (SDSS[18]) and are developing technologies for the National Virtual Observatory (NVO[19]). LIGO Laboratory and its Scientific Collaboration represent gravitational wave astrophysics. The four physics programs each face challenging computational and database needs that place emphasis on massive data movement over high speed networks. They also need large-scale distributed computing resources in order to analyze the data products, which are generated by the experiments. The collaboration includes a number of key computer science institutions that have been pioneering the concepts of a computational grid (in a very rough analogy to the electrical power grid).

Within the GriPhyN data grid concept, there exists a hierarchy of levels, or tiers, of data repositories and computational resources. Referring to Figure 4, at the highest level there is a Tier 0 center (for high-energy physics only). Within the U, S., there will be one Tier 1 center per physics project. This center represents the archive and dedicated computing resources that are required to perform essential data analysis functions common to large groups of researchers. At the next lower level are a number of the Tier 2 centers – regional centers that serve as a data mirrors for the Tier 1 center and that have computational resources that will be available to the respective collaborations. The next level, Tier 3, corresponds to the resources available to individual university research groups. Finally, at the lowest level, Tier 4, there are the individual workstations available to individuals to perform small-scale analysis and prototyping of algorithms.

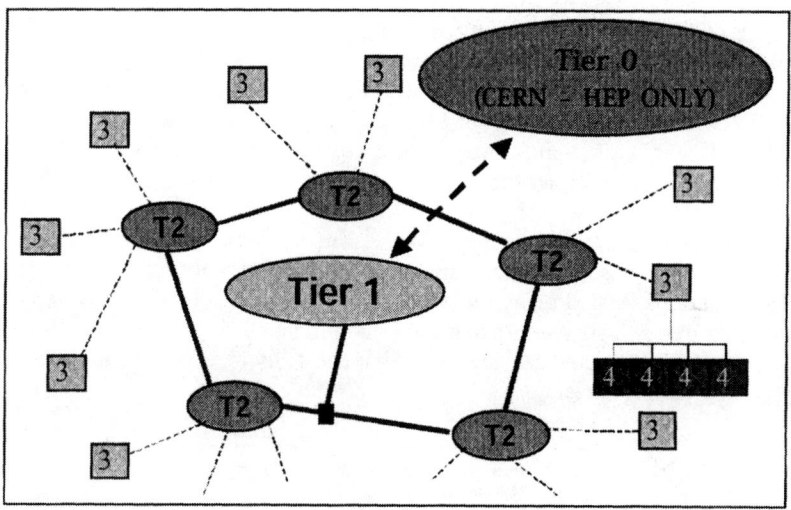

Figure 4. Schematic of the GriPhyN hierarchy of computer and database resources within the U. S.

Within the LIGO Scientific Collaboration, there will be between 3 and 5 Tier 2 centers. These will provide a number of services that will help to off-load the Tier 1 center within the Laboratory. Examples of services that will be provided include the following:

Access to Distributed Computing Power. This provides access to the collaboration of additional computational resources that are not dedicated to specific tasks.

Researchers needing a computational environment suitable for algorithm development or code validation will have a facility at their disposal. In addition, within GriPhyN, application interfaces will be developed to enable individuals to perform less general but commonly needed menu and parameter-driven processing requests that are useful to condition data. These functions are being built into the LIGO Laboratory's data analysis environment and they can be ported to GriPhyN. Last, massively parallel, compute-intensive background jobs can be run at Tier 2 centers if they are not time-critical. One example of this is an all-sky pulsar search (*"Pulsar@GriPhyN"* project) using compute cycles in an opportunistic fashion to process the LIGO data in order to perform a blind search (location unknown, spin frequency, and other dynamical parameters unknown). One challenge of this class of activities is the research required for rendering large libraries of analysis software portable within GriPhyN.

Providing and Tracking LIGO Virtual Data. The richness of the LIGO raw data implies there are many transformations that are possible to generate derived data. Examples include Fourier transformations, regressions, bandpass limited filtering and decimation. In general, the full spectrum of signal processing tools may be applied to LIGO datasets. Typically, these types of transformations are computationally costly while they are also generally useful. Every endeavor should be made to store and re-use "popular" transformed datasets. The ability to do this requires a grid infrastructure built upon such elements as data, catalog, reduced data archives, and data mirrors. While subsets of this functionality are called out in the design of the LIGO Laboratory's data analysis system, a general extension of these concepts to a national computing grid is a challenging computer science research topic. This research activity is being pursued by a group of LIGO Scientific Collaboration researchers who have teamed with USC's Information Sciences Institute to identify the needed enhancements to the existing functionality in order to support LIGO applications within GriPhyN.

In summary, the challenges facing LIGO researchers who want to access and use the data archive are in some aspects unique. The detector produces a datacube of raw data that need to be filtered many times in order to produce events (refer to Figure 5). The expected true signals are rare and false alarm rates likely to remain high during the early searches. The same data are processed repeatedly in different ways according to the source type of interest.

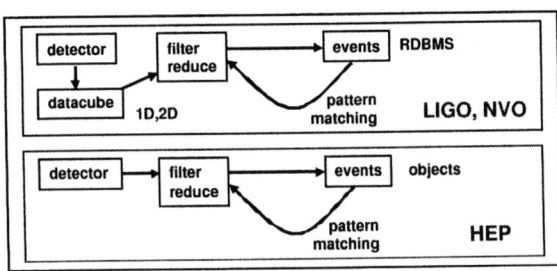

Figure 5. LIGO data analysis challenges involve a multidimensional *datacube* on which a large number of filtering operations are needed in order to generate potential events of interest.

ACKNOWLEDGMENTS

The work reported herein represents the collective efforts of the LIGO team who have been working hard in order to make LIGO become a reality. Without their tremendous spirit and dedication, I would have had little to report. I am indebted to all of them for the materials provided for this contribution. I wish to acknowledge my GriPhyN collaborators and especially Roy Williams of Caltech's Center for Advanced Computing Research (CACR) and Ewa Deelman from USC's Information Sciences Institute (ISI) for the conceptual matter relating to LIGO applications on GriPhyN resources.

The LIGO Laboratory is supported by the National Science Foundation under cooperative agreement PHY-9210038.

REFERENCES

[1] http://www.ligo.caltech.edu
[2] http://www.ligo.caltech.edu/LIGO_web/lsc/lsc.html
[3] http://www.ligo.caltech.edu/docs/T/T970130-D.pdf
[4] http://www.ligo.caltech.edu/docs/T/T990101-02.pdf
[5] http://www.ldas-sw.ligo.caltech.edu/doc_index/user_apis.html
[6] http://www.ligo-wa.caltech.edu/
[7] http://www.ligo-la.caltech.edu/
[8] http://space.mit.edu/LIGO/Welcome.html
[9] http://blue.ligo-wa.caltech.edu/
[10] http://www.ldas-sw.ligo.caltech.edu/
[11] See P. Brady's contribution to these proceedings for information on the scientific searches that will be conducted with LIGO data by the LIGO Scientific Collaboration. See also Brady's talk: http://www.physics.drexel.edu/events/astro_conference/
[12] http://www.cacr.caltech.edu/resources/HPSS/
[13] http://www.ligo.caltech.edu/LIGO_web/mou/mou.html
[14] http://www.griphyn.org/
[15] http://uscms.fnal.gov/
[16] http://www.usatlas.bnl.gov/
[17] http://lhc.web.cern.ch/lhc/
[18] http://www.sdss.org/
[19] http://www.astro.caltech.edu/nvoconf/white_paper.pdf

Gravitational wave data analysis in the LIGO Scientific Collaboration

Patrick R Brady

Department of Physics, University of Wisconsin–Milwaukee, PO Box 413, WI 53201

Abstract. The data analysis effort within the LIGO Scientific Collaboration (LSC) is organized around the four main classes of signals: binary inspiral waves, continuous waves, unmodeled bursts and stochastic waves. The detection of each signal type presents its own challenges. In this article, I discuss the practicalities of data analysis activities within the LSC. This is followed by a description of the sources, algorithms and sensitivity of LIGO with an emphasis on data analysis software development within the LSC.

INTRODUCTION

The construction of kilometer-scale interferometers as gravitational wave detectors brings us into a new era of gravitational physics. It also brings new challenges. To avail of the opportunity afforded by the new instruments, we need to unite several communities to pursue a single goal: gravitational-wave astronomy.

At present, theorists provide guidance about the best sources of gravitational waves as they investigate the physics of strong gravitational fields. Theoretical computations predicting gravitational waves from astrophysical sources are summarized elsewhere in this volume. Some information can also be gleaned from electromagnetic observations – for example, observations of the Hulse-Taylor binary pulsar establishes the existence of binary neutron star systems which will coalesce in less than a Hubble time [1]. Nevertheless, the available information is limited due to our scarce first-hand knowledge of strong gravitational field physics. Nobody has seen a neutron star or black hole up close.

The long-term goal of LIGO data analysis is to develop and exploit gravitational-wave detection as an astrophysical probe [2]. The interferometers being commissioned in the LIGO facilities will have amplitude sensitivity ~ 100 times better than previous instruments, thus increasing the volume of the universe accessible to gravitational wave observations by $\sim 10^6$. Known sources are probably still too weak or too rare to be detected without a little luck. Therefore the first scientific analyses will probably constrain event rates and signal strengths from known sources using LIGO data while simultaneously searching for signals from unexpected sources. Four types of astrophysical signals have been identified in the LSC white paper on data analysis [2]: (i) binary inspirals, (ii) unmodeled bursts, (iii) continuous waves and (iv) stochastic waves. Much of the data analysis effort within the LSC is organized around these four source classes.

In this article, I review the current status of the data analysis effort within the LSC with a view to the scientific searches which form the cornerstones of the effort.

CP575, *Astrophysical Sources for Ground-Based Gravitational Wave Detectors*, edited by J. M. Centrella

ORGANIZATION OF LSC DATA ANALYSIS ACTIVITIES

As with other activities within the LSC, the data analysis effort has been organized through two active working groups: (i) *Astrophysical Source Identification and Signatures* and (ii) *Detector Characterization*. The groups were formed in 1998 and organized under the guidance of a chair-person from the LSC and a LIGO Laboratory liaison. The groups self-organized and developed the following model for their activities.

The groups meet by phone on a (bi-)monthly basis to report on progress. Each group has a web-page and archived mailing list to facilitate communication and interaction. These pages usually include an archive of the minutes of past meetings, important documents including web-based presentations given during the teleconferences, and an archive of e-mail circulated within the group. During the past three years, these two groups have identified the important issues for data analysis using the first generation of interferometers in LIGO. Since there is no existing body of knowledge or experience in gravitational wave data analysis, the approach taken by the groups has been to prioritize the tasks which can be identified during discussions and to enlist volunteers for each task. Usually an institutional member group of the LSC was identified as "responsible" to see that the work on each task is taken to completion. Each task team provides a progress report at the meetings.

To further coordinate the data analysis efforts within the LSC, the chairs of these working groups, together with the LSC spokesperson, have drawn up a white paper on data analysis [2]. This document serves a dual purpose within the LSC. First it identifies essential data analysis development tasks and member institutions who have agreed to work on these tasks. Second it provides a long-term vision of the data analysis activities within the LSC.

LIGO/LSC algorithm library

The white paper also called for a numerical algorithm library – the LIGO/LSC Algorithm Library (LAL) – to provide the algorithms required for data analysis. The LAL specification [3] was drafted during 1999 and software development now follows the standard set out in that document. The standard defines a set of LAL data structures, a method for error handling which allows for damage control should some routine deep inside a search code fail, and I/O methods. Contributed software is organized into packages which must include documentation and a test suite so that code can be automatically verified at build time. The LAL package follows the GNU standard for software development and uses GNU development tools such as autoconf, automake and libtool which simplify support for multiple platforms in a single release of the software. The library itself is built both in static and shared formats and depends on a small number of external software components: FFTW [4] is required, while LAM or MPICH can be used to compile optional MPI based routines. The LAL software librarian, Jolien Creighton, maintains the software archive [5] and develops low-level functionality (e.g.

error-handling macros) with a small team.[1] There are currently 25 packages in LAL; these contributions represent the ongoing code development effort within the LSC. Finally, a *Software Change Control Board* composed of several knowledgeable LSC members must approve changes to the LAL specification or code that has been released. The activities of this board have been limited to date since LAL remains alpha software; this will change when LAL 1.0 is released.

Scientific analysis activities

The construction of hardware and the development of software is a means to an end, not the end itself. The white paper [2] provides a mechanism to coordinate scientific (and development) data analysis activities within the LSC. Such activities are undertaken by working groups who are expected to write a short proposal for internal review. The working groups are open to any member of the LSC who wishes to join the proposed activities. The process is intended to prevent duplication of effort in the Collaboration and to protect the weakest members of the Collaboration – the graduate students. The proposals are submitted to the LSC Spokesperson and the Laboratory Director and reviewed by a panel of their choosing. The review process is expeditious with a turnaround time of order weeks.

CLASSIFICATION OF SOURCES AND ALGORITHMS

Possible sources of gravitational waves may be divided (roughly) into three categories: (i) *burst sources* produce gravitational wave signals lasting for times much shorter than the available observation time, (ii) *continuous wave sources* produce gravitational wave signals lasting for times long compared to the available observing time, and (iii) *stochastic sources* produce a gravitational wave background that is not resolvable into individual signals but is characterized statistical properties of the signal. The burst sources may be further divided into accurately modeled bursts (i.e. the waveform can be accurately computed in advance of detection) and unmodeled, or poorly modeled, bursts. The nature of these sources and LSC activities related to their detection is discussed below.

Coalescing compact binaries

Pairs of orbiting neutron stars or black holes emit gravitational waves (primarily at twice the orbital frequency) over periods of tens to hundreds of millions of years,

---

[1] The work of Jolien Creighton, Teviet Creighton, Duncan Brown and Alan Wiseman over the past 12 months has taken the LAL specification and turned it into a functional library. Much of this work goes unnoticed by the scientific community, but it is an essential part of the successful design and implementation of search codes.

gradually losing energy. Gravitational radiation reaction causes the binary elements to move closer together and the orbit to circularize. The first LIGO interferometers will be sensitive to the gravitational waves emitted during the last 20-30 seconds of the inspiral when the elements orbit each other hundreds of times per second at separations of tens of kilometers before plunging together.

As the gravitational wave frequency sweeps through the LIGO pass-band, two distinct phases of binary evolution can be identified: (i) the inspiral phase when the orbit slowly shrinks as gravitational radiation is emitted, and (ii) the merger phase which occurs when the orbit becomes dynamically unstable and the stars plunge together in a violent high-energy collision. The gravitational waves produced during the inspiral phase can be accurately computed using post-Newtonian methods [6]. Since these waveforms are accurately known in advance, the optimal detection strategy in additive Gaussian noise[2] is matched filtering [8, 9]. For a binary system, in which the orbit has circularized, the strain at the detector

$$c(t) = \frac{1\,\text{Mpc}}{D} \left[\sin \alpha \, h_s(t - t_0) + \cos \alpha \, h_c(t - t_0) \right] \tag{1}$$

depends on the masses (M_1, M_2) of the elements, a constant α determined by the orbital phase and orientation of the binary system, t_0 the laboratory time at which the signal first enters the detector pass-band, and possibly the spins of individual elements. The amplitudes $h_{s,c}(t - t_0)$ are the two polarizations of the gravitational waveform produced by an inspiraling binary system which is optimally oriented at 1 Mpc. The effective distance D depends on the distance to the source and on its orientation with respect to the detector, which has a non-uniform response. If the source is not optimally oriented, then D is greater than the source-detector distance. The 2PN formulae for $h_{s,c}$ can be found in [6].

This discussion is restricted to non-spinning binaries for simplicity; the situation is more complicated when the elements are spinning [10]. Further discussion of spinning binaries can be found elsewhere [11]. Because the inspiral waveforms depend upon the source masses $M = (m_1, m_2)$ a bank of template waveforms placed closely enough [12] in parameter space to detect any signal in the interesting mass range is used for detection. For each mass pair M_k in the template bank two real signals are constructed:

$$X_k^{s,c}(t) = N_k^{s,c} \int_{-\infty}^{\infty} \frac{\tilde{h}(f)\tilde{h}_{s,c}^*}{S_h(|f|)} e^{-2\pi i f t} \, df. \tag{2}$$

These are the time-domain outputs of optimal filters matched to the waveform of the kth mass-pair M_k. The denominator $S_h(|f|)$ is the one-sided power spectral density of $h(t)$. The normalization factor $N_k^{s,c}$ is chosen so that, in the absence of any signals, the mean value of $[X_k^{s,c}(t)]^2$ is unity. The signal to noise ratio (SNR) for the kth template waveform

---

[2] If the noise is not Gaussian, one can construct a locally optimal matched filter as described by Kassam [7]. For example, when the time series contains spurious Poisson distributed noise bursts, the locally optimal filter is the matched filter supplemented with a set of vetoes designed to discard bursts of the expected type.

TABLE 1. Summary of current wisdom on binary inspiral event rates. Listed are the event rate estimates provided by the astrophysics community and the number of events that might be detected by LIGO per year.

Source	Event rate	Number of events	
	$(\mathrm{Mpc})^{-3}(\mathrm{yr})^{-1}$	LIGO-I	LIGO-II
NS/NS	$6 \times 10^{-10} < R < 10^{-5}$	$2 \times 10^{-5} < N < 0.3$	$0.1 < N < 2000$
NS/BH	$6 \times 10^{-9} < R < 10^{-6}$	$4 \times 10^{-4} < N < 0.3$	$2.5 < N < 1600$
BH/BH	$0 < R < 8 \times 10^{-8}$	$0 < N < 0.3$	$0 < N < 2900$

Source: S.A. Hughes, ASIS sourcelist web page [15].

is then

$$\rho_k(t) = \mathrm{SNR} = \sqrt{\left[X_k^s(t)\right]^2 + \left[X_k^c(t)\right]^2}, \tag{3}$$

arrived at by maximizing over the phase α of the binary system. Evaluation of the SNR requires five Fourier transforms to be computed per template. This is not difficult, however, Owen and Sathyaprakash [13] estimate the number of templates needed to search for binaries with component masses down to $M_1 = M_2 = 0.2 M_\odot$ to be 5.6×10^5 and 1.9×10^6 for the LIGO-I and LIGO-II respectively. If data analysis proceeds at the same rate as data acquisition, a binary inspiral search requires 99 Gflops and 250 Gflops respectively. While this computational power is achievable using commodity hardware in a Beowulf design, it is planned to use a hierarchical search strategy to reduce the computational cost of these searches [14].

The implementation of data analysis methods to search for binary inspiral waveforms are well in hand within in the LSC. A group led by Bruce Allen has carried out a prototype analysis using data from the 40m interferometer at Caltech. This instrument was only sensitive enought to detect inspirals in out Galaxy, however the analysis demonstrated a pipeline analysis in which an upper limit on event rate can be determined [16]. Code to perform matched filtering searches is already included in LAL-0.6 (the `find-chirp` package); template generation code (the `inspiral` and `templates` packages) will be available in LAL-0.7 which is scheduled for release in March 2001. A group has also submitted a proposal to the LSC to analyze data from one of the engineering runs. The goal of this project will be twofold: (i) determine an upper limit on binary inspiral events directly from gravitational-wave data and (ii) to develop a data analysis pipeline that can be used when the science run begins. The analysis will also test, for the first time, hierarchical [14], multi-interferometer [17] methods, and an alternative *Fast Chirp Transform* (the `fct` package) detection scheme [18].

For black hole binaries, the phases of the evolution are (i) inspiral, (ii) the intermediate binary black hole (IBBH) phase, (iii) merger, and (iv) black hole ringdown. With these additional phases come new challenges. The IBBH regime is poorly understood; several methods to compute waveforms have been suggested [19], but little is known about the accuracy of the methods. It remains a challenge for theorists to accurately model the binary evolution through this phase. The ringdown phase presents interesting challenges for data analysis. The extremely short duration of this phase means that matched filters can be easily excited by spurious non-Gaussian detector noise [20]. Nevertheless, confidence in a detection will ultimately be established by considering each of these phases and checking for consistency between parameters detected in each phase.

Other burst sources

The uncertainty in the event rates for the well understood sources makes LIGO-I an excellent platform for opportunistic data analysis. Moreover, ignorance of the physical phenomena that may occur in very strong gravitational fields make it plausible that sources of gravitational waves exist which have not been considered by theorists.

To generate large amounts of gravitational radiation, a source must contain high density material (or large curvature) which has rapidly changing multipole moments. The strongest waves tend to come from sources whose quadrupole moments show large variations. Stellar core collapse and supernova explosions have long been considered possible sources in this class [8, 21].

The core of a supernova explosion hosts the birth of a neutron star. Even if the supernova explosion and collapse is spherical, the newborn neutron star will be very hot. During the first ~ 1 s of its life, the material in the star may boil vigorously. The resulting convective turnover produces ~ 100 gravitational wave cycles – thus the waves are emitted at 100 Hz which is near the optimal sensitivity for LIGO. The strength of a burst may be characterized using the characteristic strain $h_c \approx h\sqrt{n}$ where h is the instantaneous gravitational wave amplitude and n is the number of wave cycles per logarithmic frequency interval. Burrows [22] and Müller and Janka [23] have estimated the strength of these waves suggesting that they may be accessible to advanced LIGO configurations. See Fig. 1 for a comparison between signal strength and the inrumental burst sensitivity.

Elementary conservation arguments further suggest that neutron stars should be rotating extremely quickly when they are first born. Moreover as core collapse proceeds, it is possible that the core's rotation may be large enough to flatten it and trigger formation of a bar through either a dynamical or secular instability. This bar can lead to copious gravitational wave emission. Little is known about the detailed evolutions. Moreover, the gravitational wave frequency may either sweep upwards or downwards depending when the bar first forms [24]. Estimates suggest that the gravitational waves might be detected at distances of ~ 20 Mpc by initial LIGO out to ~ 100 Mpc with enhanced interferometers.

Flanagan and Hughes [25] have argued that a large amount of gravitational radiation is expected to be emitted by the merger of two black holes. For intermediate mass ($\sim 10M_\odot$–$1000M_\odot$) black hole binaries this radiation will be in the frequency band of highest sensitivity for LIGO. Estimates suggest that the detection rate for coalescing binary black holes could therefore be higher than for any other source if their waveform could be computed in advance. Unfortunately, the gravitational radiation from black hole mergers results from highly non-linear self-interaction of the gravitational field. This makes it extremely difficult to obtain gravitational waveforms. Efforts to do so have met with only limited success thus far. Binary black hole mergers will therefore not be amenable to detection by matched filtering, at least for the first gravitational wave

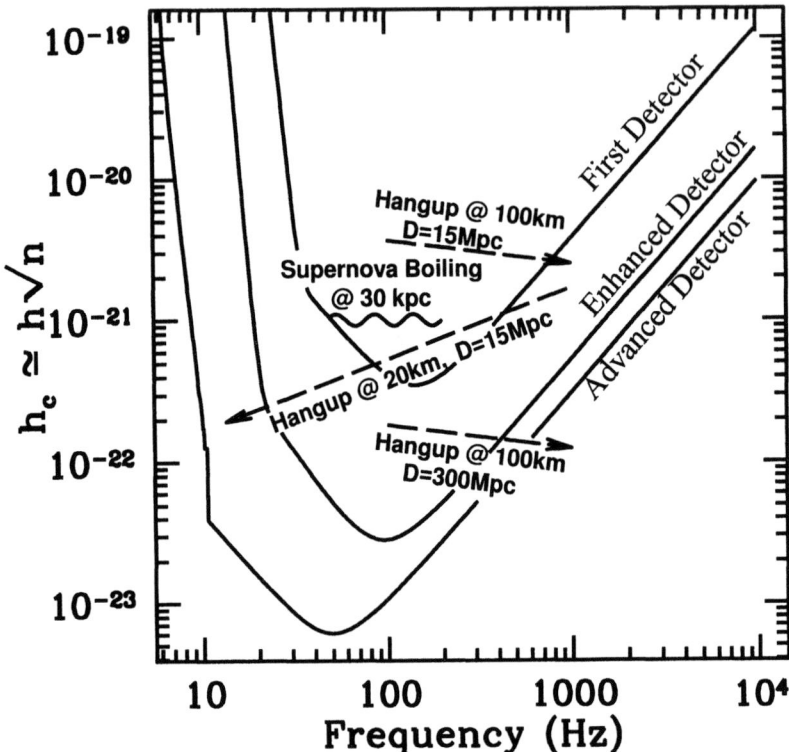

FIGURE 1. LIGO sensitivity to burst sources. Shown are signal strengths for supernova core hang-up and supernova boiling – a possible convective phase in the neutron star evolution just after formation in a supernova collapse.

searches in the 2002-2004 time frame.

Matched filtering is not an appropriate technique for detection of these sources since the waveforms are not accurately known. We must use alternative signal detection methods, often called "blind search" methods.

Time-frequency strategies have become standard in many other areas of signal analysis [26] and there is a growing literature on their application to gravitational wave detection [27, 28]. Indeed, the method explored by Anderson and Balasubramanian [28] has been implemented in the LAL timefreq package. All time-frequency methods rely on identifying patterns of excess-power in the time-frequency plane. When the signal's duration and frequency band can be estimated in advance, as with black hole mergers [25], the optimal method of detection was suggested by Flanagan and Hughes (FH) [25] and explored in detail by Brady, *et al.* [29]. The method is obvious: search for power only in the time-frequency volume where you expect signal power. An implementation of the *excess power* method, as it is called, can be found in the LAL burstsearch package. Schutz [30] also investigated this method in the context of the cross-correlation of outputs from different detectors. An autocorrelation filter for unrestricted frequencies was published by Arnaud *et al.* [31, 32] shortly after and independently of FH. A generalization of the excess power filter has also been discussed in the signal analysis literature [33]. Finally a method closely related to the excess power method has recently been explored by Sylvestre [34]; it is under active development for inclusion into LAL. The copmutational cost associated with running any of these algorithms on a single channel is small compared to the requirements of a binary inspiral search; it should be noted however, that burst type searches will likely be carried out on many channels and not just the gravitational wave channel in real searches.

To understand the viability of time-frequency methods, consider particular examples of th excess power method used for detection. The method depends on the expected duration and bandwidth of the gravitational wave as well as on its intrinsic strength. For instance, the method is not competitive with matched filtering in detecting binary neutron star inspirals, since the time-frequency volume for such signals is very large, $\gtrsim 10^4$. In contrast, for binary black hole mergers, signal durations might be of order tens or hundreds of milliseconds, depending on the black hole masses and spins [25]. Thus, the time-frequency volume of a merger signal is expected to be smaller than ~ 100, and its power would need to be more than one tenth as large as the noise power for detectability with the excess power method. The excess power method does reasonably well in this case.

One problem with all these methods is that any signal, astrophysical or not, will trigger a detection in a single interferometer. Experience shows that the noise in interferometric gravitational wave detectors is not usually stationary or Gaussian – the noise floor can drift over extended periods of time and there is usually an excess (over Gaussian statistics) of burst events in the time series. Some confidence in detection can be gained by examing auxilliary channels (environmental monitors or control signals), but detection eventually requires the use of multiple interferometers to veto false signals. By corellating signals between the two LIGO interferometers, i.e. to require they occur within the 10ms time of flight window and have the same time-frequency structure considerable confidence in detection can be achieved. Thus, the optimal detection of unmodeled bursts in will be achieved by combining output from multiple interferometers at different locations. The two LIGO observatories, located approximately 3000 km apart, provide the first level in a hierarchy which may eventually include VIRGO, GEO and TAMA. This multiple detector network will ultimately provide the a mechanism to do gravitational wave astronomy with unmodeled sources.

Continuous wave sources

The most likely sources of quasi-periodic gravitational waves in the frequency bands of terrestrial interferometric detectors are rapidly rotating neutron stars. A rotating neutron star will radiate gravitational waves if its mass distribution (or mass-current distribution) is not symmetric about its rotation axis. Several mechanisms which may produce non-axisymmetric deformations of a neutron star, and hence lead to gravitational wave generation, have been discussed in the literature [35]. For a typical $1.4 M_\odot$ neutron star of radius 10 km and at a distance of $r = 10$ kpc, the characteristic amplitude is

$$h_c = 7.7 \times 10^{-25} \frac{\varepsilon}{10^{-5}} \frac{I_{zz}}{10^{45} \text{ g cm}^2} \frac{10\text{kpc}}{r} \left(\frac{f_0}{1\text{kHz}} \right)^2 \qquad (4)$$

where

$$\varepsilon = \frac{I_{xx} - I_{yy}}{I_{zz}} \qquad (5)$$

and I_{jk} is the moment of inertia tensor of the source. Notice that the amplitude scales quadratically with the gravitational wave frequency f_0. The magnitude of the gravitational ellipticity ε represents the central uncertainty for these sources. A reasonable upper bound $\varepsilon < 10^{-5}$ is determined by the maximum strain that the outer crust of a neutron star can support, although other mechanisms can produce significantly different values.

The detection of gravitational waves from nearly periodic sources is seemingly the most straightforward data analysis problem facing gravitational-wave astronomers. One simply Fourier transforms the data stream and searches for spikes in the power spectrum. The sensitivity of a search then depends only on the total amount of data available since the signal to noise ratio increases as $\sqrt{T}$, where T is the observation time. Characteristic amplitudes for several signals compared to the interferometer noise are shown in Fig. 2. Unfortunately, the long observation times required to detect these waves, of order $\sim 10^7$ s, mean that Earth-motion induced Doppler effects, and intrinsic frequency drifts, degrade the signal-to-noise by spreading power across many frequency bins; therefore it is necessary to correct for these effects before performing the Fourier transforms. The corrections can be implemented using a parameterized model in which one does a search over a discrete set of points in the parameter space [36]. To understand the computational demands imposed by this search over parameter space, consider looking for sources with spindown age ≥ 1000 years and gravitational wave frequency $f_0 \leq 1000$ Hz. If 10^6 s of data are analyzed, then there are approximately 7.5×10^8 independent points in a parameter space that covers the sky and the first derivative of frequency. A brute force search over all parameter values would require a computer capable of over 100 Tflops if the analysis also takes 10^6 s. Detailed discussion of this and other issues related to detecting these sources can be found in Ref. [36].

The computational demands of these searches has led to development of hierarchical methods which reduce the computational requirements without sacrificing sensitivity. Brady and Creighton [37] have presented a detailed discussion of one approach and analyzed its sensitivity; the method relies on breaking the data stream into shorter segments and recombining the power incoherently over the long observation times. An alternative approach which uses the Hough transform to perform the incoherent part of the search has been proposed by Papa and Schutz [38]; this method has some advantages over the one proposed by Brady and Creighton. At the present time, both methods are under active development in the package pulsar for LAL.

In addition to the detection strategies which seek to optimize the sensitivity of the searches through the gravitational wave channel, it is necessary to consider spurious signals which might mimic continuous gravitational waves. The Doppler modulation of the frequency provides a powerful discriminant itself, but the number of trials competes

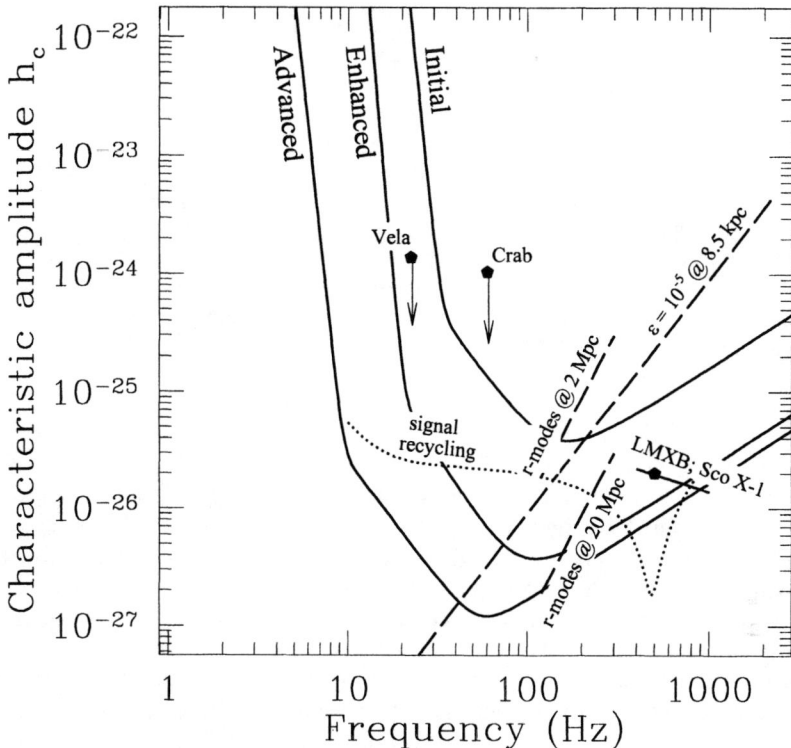

FIGURE 2. LIGO sensitivity to continuous wave sources. Shown are signal strengths crustal deformations, r-modes, low-mass x-ray binaries (LMXB's) and upper-limits on the signal strength from known
isolated pulsars if the gravitational radiation is solely responsible for their spindown.

with this to increase the number of false positives at a given threshold. In addition,
instrumental line noise is ever present in the gravitational wave channel. A group at the
University of Michigan plans to construct a database of known artifacts and to construct
veto techniques based on the time dependent amplitude response of the instrument; their
code is part of the `date` package in LAL.

Two other classes of source may produce continuous waves. First, there are accreting
neutron stars in binary systems. Several such binary systems have been identified via
x-ray observations; the rotation frequencies of the accreting neutron stars are inferred to
be ~ 250–350 Hz (f_{max}=700 Hz). Bildsten [39] has argued that these accreting objects
in low mass x-ray binaries (LMXB's) may emit detectable amounts of gravitational
radiation. Since the positions of these sources are well localized on the sky by their x-ray
emissions, the earth-motion induced Doppler modulations of the gravitational waves can
be precisely determined. The difficulty with these sources is the unknown, or poorly-

TABLE 2. Typical values of $\Omega_0 h_{100}^2$ produced by various early universe phenomena. These numbers should be compared to the LIGO-I sensitivity $\Omega_0 h_{100}^2 = 4.28 \times 10^{-6}$ and the LIGO-II sensitivity $\Omega_0 h_{100}^2 = 8.1 \times 10^{-10}$. The Hubble parameter h_{100} is related to Hubble's constant by $H_0 = 100$ km/(Mpc s) $\times h_{100}$

Source	$\Omega_0 h_{100}^2$
Slow roll inflation	$< 2 \times 10^{-16}$
Stringy inflation	$< 5 \times 10^{-5}$
Standard cosmic strings	$< 1 \times 10^{-8}$
Monopole-based hybrid cosmic strings	$< 2 \times 10^{-8}$
Domain-wall-based hybrid cosmic strings	$< 5 \times 10^{-6}$
First order phase transition (vacuum bubbles)	$< 1 \times 10^{-8}$
Big bang nucleosynthesis	$< 1 \times 10^{-5}$

Source: S.A. Hughes, ASIS source list web page [15].

known, orbits of the neutron stars about their stellar companions, and the stochastic accretion-induced variations in their spin. A preliminary discussion of issues associated with detection of these sources can be found in Ref. [37]; the results suggest that these sources will be accessible to advanced interferometers in LIGO.

Second, newly-formed fast-spinning neutron stars may be copious emitters of gravitational radiation [40]. If the newborn neutron star is rotating sufficiently fast, its r-modes (axial-vector current oscillations whose restoring force is the Coriolis force) are unstable to gravitational radiation reaction. Estimates [41] of the viscous timescales, and the superfluid transition temperature, suggest that the r-modes are stabilized when the star cools below $\sim 10^9$ K and are rotating at ~ 100–200 Hz. Figure 2 shows h_c as a function of frequency at distances $r = 2$ Mpc and 20 Mpc. Sources outside our Galaxy are potentially detectable due to the high gravitational-luminosity of a newborn neutron star with an active r-mode instability if they are long lived signals; recent results suggest that the r-modes may actually produce burst signals and not long-lived quasi-periodic waves.

Livas [42], Jones [43] and Niebauer *et al.* [44] have implemented variants of the coherent search technique without the benefit of the optimization advocated in Ref. [36]. A working group of the LSC has submitted a proposal to develop an analysis pipeline and intends to place upper limits on source strengths using engineering data to be taken in the fall of 2001.

Stochastic background

Stochastic background gravitational waves are produced by many weak incoherent sources. These waves are characterized by their statistical properties – there is no deterministic signal to be searched for. Processes in the early universe should have produced a stochastic background which would have decoupled from matter about 10^{-43} s after the big bang. Observations of these waves would provide a unique probe of the very early universe. (The cosmic microwave background decoupled from matter 10^5 years after the big bang.) The strength of a stochastic background is given by Ω_0, the energy density in the background as a fraction of the amount required to close the universe. Expected values of Ω_0 are given in Table 2 for several early universe phenomena.

The data analysis strategy is quite simple although it relies on having multiple interferometers with noise which is uncorrelated. Since the same stochastic process is responsible for the signal in all the interferometers, one can search for the signal by cross-correlating long segments of data from the different instruments [45]. The optimal analysis strategy does include some weighting functions, known as overlap reduction functions, but these depend only on the location of the detectors and not on some set of unknown parameters. For this reason, the computational burden associated with searches for stochastic background gravitational waves is small compared to most other searches. Using data from all three LIGO interferometers, a search can be run on a single workstation. A `stochastic` package is under active development within LAL.

THE FUTURE

So what does the future hold? Due to lack of space, I have given only a cursory overview of gravitational wave sources and the methods proposed to detect them. The field is new and the prospects are extremely exciting. During the next 12 months, we will see the first scientific analyses executed using engineering data taken simultaneously at the two LIGO observatories. These analyses will undoubtedly reveal surprises about the instruments – it is hoped that we will learn a great deal about the noise and behavior of the interferometers during these analyses. Perhaps, they will also bring surprises about the Universe in which we live. In any case, these are exciting times for the community of physicists involved in gravitational wave detection.

ACKNOWLEDGMENTS

I am grateful to Joan Centrella for organizing the Workshop at Drexel University and her extreme patience waiting for my contribution. This work was supported in part by NSF grant PHY-9970821.

REFERENCES

1. Taylor, J. H., *Rev. Mod. Phys.*, **66**, 711 (1994).
2. B. Allen, *et al.*, "LSC Data Analysis White Paper," unpublished (1999).
3. LSC, "LIGO data analysis system - Numerical Algorithms Library Specification and Style Guide," LIGO Technical Report No. T990030-07-E.
4. FFTW provides is a C subroutine library for computing the Discrete Fourier Transform. It is available from http://www.fftw.org/
5. LAL provides is a C subroutine library for performing gravitational wave data analysis. It is available from http://www.lsc-group.phys.uwm.edu/lal/

6. Blanchet, L., Iyer, B. R., Will, C. M., and Wiseman, A. G., *Class. Quantum Grav.* **13**, 575 (1996); Will, C. M., and Wiseman, A. G., *Phys. Rev. D* **54**, 4813 (1996); Blanchet, L., Damour, T., Iyer, B. R., Will, C. M., and Wiseman, A. G., *Phys. Rev. Lett.* **74**, 3515 (1995).
7. Kassam, S. A., *Signal Detection in Non-Gaussian Noise*, Springer-Verlag, New York, 1988.
8. Thorne, K. S., "Gravitational Radiation" in *Three hundred years of gravitation*, edited by S. W. Hawking and W. Israel, Cambridge University Press, Cambridge, 1987, pp. 330–458.
9. Wainstein, L. A., and Zubakov, V. D., *Extraction of signals from noise*, Prentice-Hall, London, 1962; Finn, L. S., *Phys. Rev. D* **46**, 5236 (1992); Finn, L. S., and Chernoff, D. F., *Phys. Rev. D* **47**, 2198 (1993).
10. Apostolatos, T. A., *Phys. Rev. D* **52**, 605 (1995); Apostolatos, T. A., *Phys. Rev. D* **54**, 2438 (1996).
11. Kalogera, V., *Astrophys. J.* **541**, 319 (2000).
12. Owen, B., *Phys. Rev. D.* **53**, 6749 (1996).
13. Owen, B. J., and Sathyaprakash, B. S., *Phys. Rev. D* **60**, 022002 (1999)
14. Mohanty, S. D., *Phys. Rev. D* **57**, 630 (1998).
15. Hughes, S. A., http://www.tapir.caltech.edu/~hughes/ASIS/sourcelist.html.
16. Allen, B., *et al.*, *Phys. Rev. Lett.* **83**, 1498 (1999).
17. Pai, A., Dhurandhar, S., and Bose, S., "A data-analysis strategy for detecting gravitational-wave signals from inspiraling compact binaries with a network of laser-interferometric detectors," gr-qc/0009078; Finn, L. S., "Aperture synthesis for gravitational-wave data analysis: Deterministic sources," gr-qc/0010033.
18. Jenet, F. A., and Prince, T. A., *Phys. Rev. D* **62**, 122001 (2000).
19. Brady, P. R., Creighton, J. D., and Thorne, K. S., *Phys. Rev. D* **58**, 061501 (1998); Damour, T., Iyer, B. R., and Sathyaprakash, B. S., "Detecting binary black holes with efficient and reliable templates," gr-qc/0012070.
20. Creighton, J. D., *Phys. Rev. D* **60**, 022001 (1999)
21. Thorne, K. S., "Gravitational radiation: A new window onto the universe," gr-qc/9704042.
22. Burrows, A., Hayes, J., and Fryxell, B. A., *Astrophys. J.* **450**, 830 (1995).
23. Müller, E. and Janka, H.-T., *Astron. Astrophys.* **317**, 140 (1997).
24. Lai, D., and Shapiro, S. L., *Astrophys. J.* **442**, 259 (1995). Houser, J. L., J. M. Centrella, J. M., and Smith, S. C., *Phys. Rev. Lett.* **72**, 1314 (1994).
25. Flanagan, E. E., and Hughes, S. A., *Phys. Rev. D* **57**, 4535 (1998).
26. See *Time-Frequency Signal Analysis*, edited by B. Boashash, John Wiley & Sons, Inc., New York, 1992, and references therein.
27. Feo, M., Pierro, V., Pinto, I. M., and Ricciardi, M., in *The Seventh Marcel Grossmann Meeting: Proceedings of the Meeting held at Stanford University, 24-30 July 1994, Part B*, edited by R. T. Jantzen and G. M. Keiser (World Scientific Publishing Co., Singapore, 1994), pp. 1086–1089; Królak, A., and Trzaskoma, P., Class. Quantum Grav. **13**, 813 (1996); Innocent, J.-M., and Torrésani, B., in *Mathematics of Gravitation*, edited by A. Królak (Banach Center Publications, Warsaw, 1997); Gonçalvès, P., Flandrin, P., and Chassande-Mottin, E., in *Second Workshop on Gravitational Wave Data Analysis*, edited by M. Davier and P. Hello (Éditions Frotières, Gif Sur Yvette, France, 1998), pp. 35–46; Chassande-Mottin, E., and Flandrin, P., *ibid.* pp. 47–52; Mohanty, S. D., "A Robust Test for Detecting Non-Stationarity in Data from Gravitational Wave Detectors," gr-qc/9910027.
28. Anderson, W. G., and Balasubramanian, R., *Phys. Rev. D* **60**, 102001 (1999);
29. Anderson, W. G., Brady, P. R., Creighton J. D., and Flanagan, E. E., *Phys. Rev. D* **63**, 042003 (2001)
30. Schutz, B. F., in *The Detection of Gravitational Waves*, edited by D. G. Blair,(Cambridge University Press, Cambridge, England, 1991), pp. 406–452.
31. Arnaud, N., Cavalier, F., Davier, M., and Hello, P., *Phys. Rev. D* **59**, 082002 (1999).
32. Arnaud, N., Cavalier, F., Davier, M., and Hello, P., and Pradier, T., gr-qc/9903025.
33. Fawcett, J., and Maranda, B., *IEEE Trans. Inform. Theory* **37**, 209 (1991).
34. Sylvestre, J., in preparation.
35. Chandrasekhar, S., *Phys. Rev. Lett.* **24**, 611 (1970); Friedman, J. L., and Schutz, B. F., *Ap. J.* **222**, 281 (1978); Bonazzola, S., and Gourgoulhon, E., *Astron. Astr.* **312**, 675 (1996); Zimmermann, M., *Phys. Rev. D.* **21**, 891 (1980); Zimmermann, M., and Szedenits, E., *Phys. Rev. D.* **20**, 351 (1979); Wagoner, R. V., *Astrophys. J.* **278**, 345 (1984).
36. Brady, P. R., and Creighton, T., Cutler, C., and Schutz, B. F., *Phys. Rev. D.* **57**, 2101 (1998).

37. Brady, P. R., and Creighton, T., *Phys. Rev. D* **61**, 082001 (2000)
38. Papa, M. A., Schutz, B. F., Frasca, S., and Astone, P., "Detection of continuous gravitational wave signals: pattern tracking with the Hough transform," in Proceedings of the LISA Symposium (1998); Papa, M. A., Astone, P., Frasca, S., and Schutz, B. F., "Searching for continuous waves by line identification," in the proceedings of Gravitational Wave Data Analysis Workshop (1997); Papa, M. A., private communication.
39. Bildsten, L., *Astrophys. J.* **501**, L89 1998, astro-ph/9804325.
40. Andersson, N., *Astrophys. J.* **502**, 708 (1998), gr-qc/9706075; Friedman, J. L., and Morsink, S. M., *Astrophys. J.* **502**, 714 (1998), gr-qc/9706073; Lindblom, L., Owen, B. J., and Morsink, S. M., *Phys. Rev. Lett.* **80**, 4843 (1998).
41. Owen, B. J., *et al.*, Phys. Rev. **D58**, 084020 (1998), gr-qc/9804044.
42. Livas, J. C., Ph.D. thesis, Massachusetts Institute of Technology, 1987.
43. Jones, G. S., Ph.D. thesis, University of Whales, 1995.
44. Niebauer, T. M., *et al.*, Phys. Rev. D **47**, 3106 (1993).
45. Allen, B., and Romano, J. D., *Phys. Rev. D* **59**, 102001 (1999).

LIGO's "Science Reach"

Lee Samuel Finn

Center for Gravitational Physics and Geometry,[1] The Pennsylvania State University, University Park, Pennsylvania 16803

Abstract. Technical discussions of the Laser Interferometer Gravitational Wave Observatory (LIGO) sensitivity often focuses on its effective sensitivity to gravitational waves in a given band; nevertheless, the goal of the LIGO Project is to "do science." Exploiting this new observational perspective to explore the Universe is a long-term goal, toward which LIGO's initial instrumentation is but a first step. Nevertheless, the first generation LIGO instrumentation is sensitive enough that even non-detection — in the form of an upper limit — is also informative. In this brief article I describe in quantitative terms some of the science we can hope to do with first and future generation LIGO instrumentation: it short, the "science reach" of the detector we are building and the ones we hope to build.

INTRODUCTION

Technical discussions of the Laser Interferometer Gravitational Wave Observatory (LIGO) sensitivity often focus on its effective sensitivity to gravitational waves in a given band; nevertheless, the LIGO Projects goal is to enable a new kind of astronomy — gravitational wave astronomy — and explore the fundamental physics of gravity. The gravitational waves accessible to LIGO are of astronomical origin, arise in regimes of strong, dynamical gravity, and carry with them an imprint of their origin and gravity's character in that regime. From detection we aim to learn about the fundamental physics of gravity and the astronomical character of the sources, and we build LIGO in service of that goal.

Exploiting this new observational perspective to explore the Universe is a long-term goal, toward which LIGO's initial instrumentation is but a first step. To fully develop this new observational perspective will require detectors of increasing sensitivity and detectors that explore different spectral bands. Nevertheless, the first generation LIGO instrumentation is sensitive enough that even non-detection — in the form of an upper limit — is also informative. In this brief article I describe in quantitative terms some of the science we can hope to do with first and future generation LIGO instrumentation: it short, the "science reach" of the detector we are building and the ones we hope to build.

---

[1] Also Department of Physics, Department of Astronomy and Astrophysics

CP575, *Astrophysical Sources for Ground-Based Gravitational Wave Detectors*, edited by J. M. Centrella
© 2001 American Institute of Physics 0-7354-0014-8/01/$18.00

Strong gravitational wave sources have characteristics that make them weak electromagnetic observations, and *vice versa*. An immediate consequence is that we generally know very little about the most intriguing gravitational wave sources. In choosing the source science that I discuss below, I've deliberately taken a conservative approach and focused my attention where speculation can be minimized. I've also chosen to stay "close to the detector": i.e., to focus on the science that comes from direct detection or upper limits, as opposed to interpretation of observations in terms of detailed astrophysical models.

Still, we must remember that "[i]n the field of observation, chance favors the prepared mind" [31]: historically, the opening of a new spectrum to observation has lead to the revision of theoretical models, the development of new ideas, and, frequently, major serendipitous discoveries. LIGO is also a reach into the unknown and it is there the most exciting prospects lay.

A familiarity with the anticipated noise character of the initial and advanced LIGO designs is critical to an appreciation of their sensitivity to different sources of gravitational radiation; for that background I refer the reader to Peter Fritschel's contribution to this volume. In the remainder of this article I discuss first compact binary inspiral, focusing on the effective volume of space that LIGO can survey for these sources, second on the limits that LIGO can place the crustal deformation of young pulsars, thirdly on the energy density in a stochastic gravitational wave background, fourthly on the limits that can be placed on the gravitational radiation that might accompany γ-ray bursts, and lastly on the limits that can be placed on the energy radiated in a stellar core-collapse supernova.

COMPACT BINARY INSPIRAL

Most LIGO stories begin with compact binary inspiral. Binary systems consisting of neutron stars or stellar mass black holes decay owing to the emission of gravitational radiation. During the last few moments in the life of these binary systems the gravitational radiation, emitted at twice the orbital frequency, races through the LIGO band before the binary components collide and coalesce into a final black hole (or, perhaps, another, more massive neutron star supported against collapse by its angular momentum). The detailed character of the radiation emitted during this *inspiral* phase is generally accessible to theoretical calculation and, correspondingly, has been studied in great detail (cf. [3,4] and references therein).

As inspiral proceeds the binary system becomes more compact and, correspondingly, more relativistic and less accessible to a perturbative, post-Newtonian treatment. Ultimately, there is a "last orbit" before the two components plunge toward each other and coalesce. Whether the transition from inspiral to plunge is due to a dynamical instability, a secular instability, or simply dissipative forces (e.g., radiation reaction) is an open question, as is the detailed character of the radiation during plunge and coalescence.

For black holes, coalescence represents the ultimate expression of dynamical, strong field gravity. It is certainly also accompanied by a burst of radiation [14,15]; however, that burst is extremely difficult to model: the spacetime is dynamical, the fields are strong and the structure of the binary components plays an important role. Modeling *coalescence* (which we take to include the plunge from inspiral) has been a long-term goal of numerical relativity, and recent results suggest that the radiation emitted during this phase alone may be as greater as 1% or more of the systems rest mass [20].

Lastly, the final remnant of the coalescence will be strongly perturbed from equilibrium and will shed that perturbation, in part through gravitational radiation. This *ring down* phase is, by definition, perturbative and the waveform well understood (even if it is not particularly well structured).

The inspiral, coalescence and ring down are consecutive parts of the signature of an inspiraling compact binary system. A complete search for compact binary inspiral will look for all three components and not single out any given component, ignoring the rest. Nevertheless, the inspiral waveform is better understood, will appear in the LIGO detectors at an earlier time, and carry more information about the system, then the signals arising from coalescence or ring down. In the LIGO detectors the inspiral of a binary neutron star system deposits most all of its contribution to the signal-to-noise as the orbital frequency increases from approximately $20 \, \mathrm{s}^{-1}$ to about $100 \, \mathrm{s}^{-1}$ over approximately 20 s [9] immediately preceding the coalescence[2]; additionally, the inspiral waveform carries information about the system's component masses, redshift and luminosity distance [36,10,6,9].

For the purpose of comparing the science reach of alternative detectors, then, we will focus on just the radiation associated with the inspiral of two $1.4 \, \mathrm{M}_\odot$ neutron stars or black holes. Assuming that we use matched filtering to detect the inspiral radiation, and that we are contending only with Gaussian detector noise, a false alarm rate of $10^{-4} \, \mathrm{y}^{-1}$ corresponds to a signal-to-noise threshold of approximately 8 for detection. With this criteria we can calculate the observed rate of binary inspiral $\dot{N}$ given the rate density (on a per co-moving time, co-moving volume, basis) $\dot{n}$ of inspiraling binary systems and taking into account the anisotropic radiation pattern of the sources and antenna pattern of the detectors. The ratio $\dot{N}/\dot{n}$ we *define* to be an *effective* volume surveyed by the detector, which we characterize by an effective radius, r_{eff}:

$$\dot{N} = \frac{4\pi r_{\mathrm{eff}}^3}{3} \dot{n}. \tag{1}$$

Table 1 gives the effective radius r_{eff} for the LIGO initial and (as of this writing) advanced detector concepts. Over the volume surveyed by initial LIGO only the most optimistic scenarios give a reasonable event rate for NS/NS binary inspiral; however, advanced LIGO should observe many inspirals, of different types, per

---

[2] The band shifts slightly to lower frequencies and longer periods between the first and later generation LIGO detectors, corresponding to the greater low-frequency sensitivity of these more advanced detectors.

TABLE 1. The effective volume of space surveyed for *just the compact binary inspiral waveform* by the initial and — proposed — advanced LIGO detector system (sometimes known as LIGO II). The volume is represented by an effective radius r, such that $4\pi r^3/3$ is the survey volume. For both cases survey volumes are given for a single interferometer and for an effective interferometer synthesized by the phase coherent sum of the output of the three LIGO interferometers. For advanced LIGO we assume that the 2 Km Hanford detector is upgraded to a 4 Km detector.

	NS/NS	NS/10 $M_\odot$ BH	10 $M_\odot$ BH/10 $M_\odot$BH
Initial LIGO			
Single IFO	14 Mpc	30 Mpc	72 Mpc
Three IFO	21 Mpc	44 Mpc	104 Mpc
Advanded LIGO			
Single IFO	200 Mpc	420 Mpc	620 Mpc
Three IFO	350 Mpc	740 Mpc	1.0 Gpc

year (see Kalogera, this proceedings). An interesting note is that, in terms of the expected binary inspiral detection rate, it is more likely that a NS/NS inspiral will be detected in 3 h of advanced LIGO operations than in 1 y of initial LIGO operations!

PERIODIC SIGNALS

Shortly after their formation, rapidly rotating neutron stars develop a solid crust. As they evolve, they gain angular momentum through accretion, lose it through radiation, and see it redistributed between the fluid interior and the solid crust. Through these processes, as well as crust fracturing and the neutron star equivalent of "continental drift" [35,33,34], neutron stars cease being axisymmetric and will radiate periodic gravitational waves with an amplitude proportional to the crust strain. The amplitude of the gravitational waves depends on the stars period and the degree of non-axisymmetry. If the star is rotating about a principal axis of its moment of inertia the radiation period will be twice the rotational period and its amplitude proportional to the the difference between the moment of inertia of the other two principal axes.

Theoretical prejudice regarding the maximum sustainable crust stress bound this difference to less than or of order 10^{40} g cm^2: *i.e*, something less than one part in 10^5 of the moment of inertia itself [2].[3] Gravitational wave observations over an

---

[3] Observations of the period, spin-down and rate of spin-down of very old pulsars places limits on $\epsilon \equiv |I_2 - I_1|/I_3$ of less than or of order 10^{-8} for these stars; however, owing to their great age and past history the crust of these stars has been heavily annealed. Thus, this limit is not likely representative of young neutron stars.

extended period will, in principle, be able to bound the *actual* strain in nearby pulsars to an order of magnitude greater than this theoretical bound.

To describe the LIGO detector's science reach in the context of periodic signals, we ask what limit can be placed on triaxiality parameter ϵ ($|I_2 - I_1|/I_3$) for a pulsar, rotating about the principal axis with moment of I_3. This limit will depend on the pulsar period, detector noise at the gravitational wave period (half the pulsar period), the distance to the pulsar, and the pulsar's declination. For the purpose of assessing the science reach, focus attention on a source at a distance of 10 Kpc and average over the source declination. Figure 1 shows the science reach of the initial and advanced LIGO detectors. Three curves are shown: the solid curve shows the reach of the initial LIGO detector, which is clearly in a position to place interesting, but not challenging, upper limits on pulsars with frequencies $\sim$ 100 Hz. The dashed curve shows the reach of the advanced LIGO detector, operating in a broadband mode tuned to maximize the detector's reach for binary inspiral (as described in the previous section).

Finally, the dash-dot curve shows the science reach for an advanced detector whose optical configuration has been tuned to maximize the its sensitivity at the indicated frequency. Here we go beyond what is proposed for the next generation LIGO detectors. That detector will be tunable; however, the ability to reach the limit shown depends on choices made in the mirror coatings, which are different if one optimizes for, say, 500 Hz to 1 KHz frequencies as opposed to 100 Hz to 500 Hz, and also the quality of the optical components. The dash-dot curve is the envelope of best possible behavior: i.e., limited at each frequency by the thermal noise of the proposed, advanced detector. The actual proposed advanced detector, optimized for detection at a particular frequency, will have a science reach bounded from below by the dash-dot curve and above by the dashed curve. Particularly for pulsars closer than the galactic center, the advanced LIGO sensitivity gives us the ability to place challenging upper limits on the crustal deformation in younger pulsars.

γ-RAY BURSTS

γ-ray bursts are sudden, intense flashes of γ-rays lasting anywhere from a fraction of a second to hundreds of seconds. The burst sources are confirmed through observations to be at cosmological distances and involve power outputs of 10^{51}–10^{54} erg/s, which is comparable to the total conversion of the Sun's rest-mass to γ-rays over the course of a few seconds or to the emission over the same period of as much energy as our entire galaxy radiates in 100 y. This is a far larger luminosity than that of any other known astronomical source [25].

The progenitors of γ-ray bursts are not yet well identified and there is good reason to believe they may be a heterogeneous population [21,22,28,24]. Two classes of progenitors are currently favored: neutron star binary coalescence or hypernovae or collapsars [37,29,30,17,18,26]. All models converge on the formation of a several

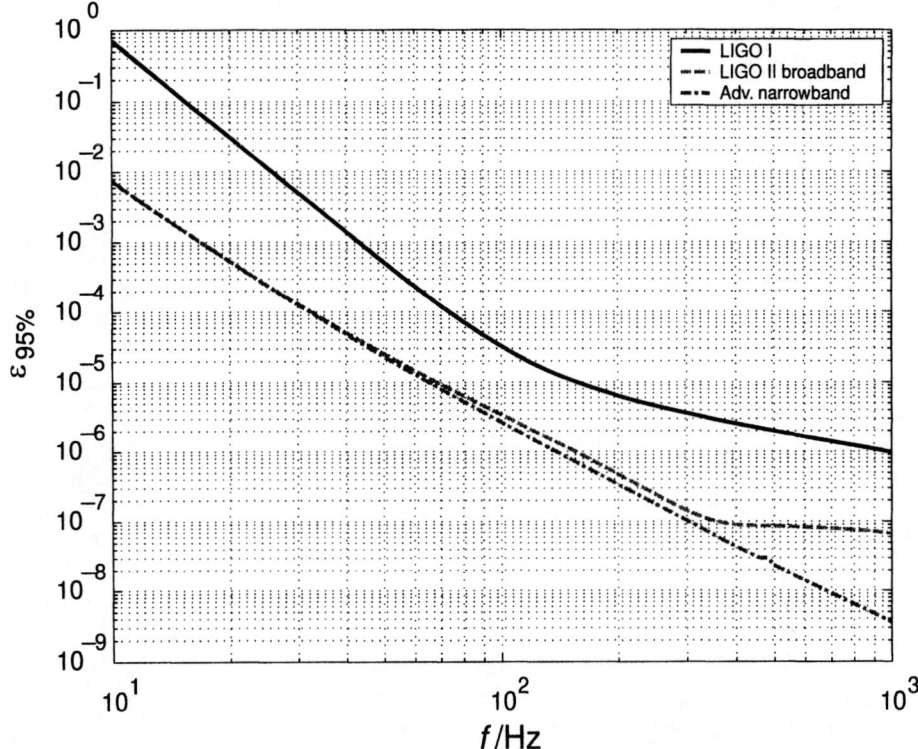

FIGURE 1. The "science reach" of the initial and advanced LIGO detectors, when applied to the determination of a pulsar's triaxiality parameter ϵ. Shown is the upper limit (95% confidence) that can be set on ϵ in a 1 y observation of a pulsar at a distance of 10 Kpc, as a function of the gravitational wave frequency f. The solid curve is the upper limit that can be set by the initial LIGO detectors; the dashed curve the limit that can be set by the proposed advanced LIGO detectors, operating in a broadband mode, and the dot-dashed curve the limit that can be set by the proposed advanced LIGO detector if its sensitivity at any given frequency were tuned so that the photon shot noise were negligible.

solar mass black hole, surrounded by a debris torus whose accretion provides the energy necessary to power the γ-ray burst. The γ-ray burst itself arises as a result of either internal shocks [32] (owing to velocity variations) in the outgoing fireball and/or the formation of forward and reverse shocks when the expanding fireball impacts on, e.g., the interstellar medium.

The violent formation of a black hole will almost certainly involve a gravitational wave burst. The qualitative character of that burst will likely depend on the progenitor: e.g., the spectrum and timescale of the burst arising from a hypernova, or collapsar, or the coalescence of a compact binary will all likely be different. The interval between the gravitational and γ-ray bursts will depend on whether the γ-ray burst is formed via internal shocks (timescales on order 0.1 s) or the formation of a blast wave and reverse shock as the fireball impacts on an external medium (timescales on order 100 s). Correspondingly, observations of correlated gravitational and γ-ray bursts can *i)* test the general model of γ-ray burst sources (i.e., the formation of a several $M_\odot$ black hole), *ii)* potentially distinguish between different progenitors (e.g., hypernovae, collapsar or compact binary coalescence), *iii)* determine whether the γ-rays are produced via internal or external shocks (via the interval between the γ-ray and gravitational wave burst).

Since GRBs occur at cosmological distances the signal-to-noise ratio (SNR) of any individual GWB will likely be insufficient for direct detection in either the initial or advanced LIGO detectors. Nevertheless, it will still be possible to detect a *statistical association* between gravitational wave and γ-ray bursts. If GWBs are associated with GRBs, the correlated output of two GW detectors will be different in the moments immediately preceding a GRB (*on-source*) than at other times not associated with a GRB (*off-source*). (While we focus on γ-ray bursts here, any plausible class of astronomical events can serve as a trigger.) A statistically significant difference between on- and off-source cross-correlations would support a GWB/GRB association and represent a detection of gravitational waves by the detector pair. We can measure this difference using Student's t-test without requiring any foreknowledge of the signal waveform, source or source population (though with such a model the effectiveness of the test can be improved). The measured difference can be used to establish a confidence interval (CI) or upper limit (UL) on the rms amplitude of GWBs associated with GRBs. The CI/UL, in turn, constrains any model for model for GRB/GWB pairs [12]. The 95% confidence upper limit $h_{95\%}$ we can set on the strength of the gravitational waves associated with γ-ray bursts, is given by

$$h_{95\%}^2 = \left\{ \begin{array}{l} (1.35 \times 10^{-22})^2 \\ (2.5 \times 10^{-23})^2 \end{array} \right\} \left(\frac{T}{0.2\,\text{s}} \frac{1000}{N_{\text{on}}} \right)^{1/2} \qquad \begin{array}{l} \text{initial LIGO} \\ \text{advanced LIGO} \end{array} \qquad (2)$$

where T is the delay between the gravitational wave and γ-ray burst (0.1 s for the internal shock model), and N_{on} is the number of γ-ray bursts observed (1000 bursts will involve about three years of observations with, e.g., SWIFT). The advanced LIGO bound described above corresponds to the conversion of $\sim 0.3\,M_\odot$ to gravitational waves at $z \simeq 1/2$.

STOCHASTIC SIGNALS

A stochastic gravitational wave signal, in a single detector, is indistinguishable from detector noise. It is only when we have two or more independent detectors that we can begin to distinguish between detector noise and a stochastic signal: the stochastic signal, arising as a superposition of plane waves from different directions, will appear as correlated noise in pairs of detectors, with the correlation rolling off for wavelengths greater than half the distance between the detectors [5,13]. For the LIGO detector-pair the correlation of the two detectors to a stochastic signal has a null at approximately 64 Hz and a second null at approximately twice that frequency. Beyond that second null there is very little power in the correlated response of the detector pair to the incident stochastic signal.

It is convenient to characterize the strength of a stochastic background by its energy density in a logarithmic frequency band, relative to the closure density of the universe ρ_0:

$$\Omega_{GW} \equiv \frac{f}{\rho_0} \frac{d\rho_{GW}}{df}, \tag{3}$$

where ρ_{GW} is the energy density in gravitational waves.

When considering sources of detectable stochastic gravitational radiation it is conventional to consider signals of primordial origin: e.g., radiation arising during an inflationary epoch or from the decay of a cosmic string network (see [23] for an excellent review). Primordial nucleosynthesis places a bound on the contribution to Ω_{GW} from these primordial sources of 10^{-5}: a larger Ω_{GW} would lead to a significantly larger He_4 abundance than is observed. This bound is weak compared to the theoretical predictions on the size of these backgrounds: $\Omega_{GW} \simeq 10^{-14}$ for inflation and $\Omega_{GW} \lesssim 10^{-11}$ for a cosmic string network.

It is important to note, however, that *the nucleosynthesis bound does not address gravitational waves produced after nucleosynthesis,* and there are a number of mechanisms capable of producing significant stochastic gravitational waves at late times. As important, these mechanisms are not exotic ones: they are as simple as the confusion limit of discrete but unresolved conventional astronomical sources of gravitational waves: e.g., core-collapse supernovae, binaries or binary inspiral, or coalescence, etc.. Stochastic sources like these are certainly important for the LISA detector[4] [16] and they may also be for the LIGO detector: it all depends on the unknown event rate and luminosity of the individual sources.

To assess the science reach of the LIGO detector pair to a stochastic gravitational wave background, we ask what upper limit we can expect to place on Ω_{GW} *today,* based on the observed correlation in the two LIGO detectors. Focusing on an observation involving 1/3 y of data and insisting on a 99% confidence, we find [5,13,1,7]

---

[4] LISA will be overwhelmed at frequencies below $10^{-2.5}$ Hz by the radiation from unresolved close white dwarf binaries and other galactic binary systems [19].

$$\Omega_{99\%} = \begin{cases} 3.3 \times 10^{-6}/h^2 & \text{initial LIGO} \\ 2.0 \times 10^{-9}/h^2 & \text{advanced LIGO} \end{cases} \qquad (4)$$

where h is the Hubble constant in units of 100 Km/s/Mpc. Thus, the initial LIGO detectors can improve on the existing in-band limit on a primordial background by a factor of approximately 3, and set a completely new limit on a background from the confusion limit of more conventional astronomical sources. The proposed advanced LIGO will be able to improve on these bounds by a factor of just over 1000, challenging some of the more optimistic proposals for a stochastic background from conventional sources [23].

CORE-COLLAPSE SUPERNOVAE

As a final example, consider the gravitational waves that arise from the collapse of the stellar core in, e.g., a type II supernovae explosion. The gravitational radiation signature of this source is very uncertain: different plausible models for the progenitor give very different gravitational radiation waveforms [38]. The spectrum of the different bursts are more similar: the burst is roughly white to a KHz, with the energy falling very rapidly at high frequencies. Without assuming a waveform for the gravitational wave burst we can compare the cross-correlation of the LIGO detector outputs at the expected time of the gravitational wave burst with its value at other times, not associated with a supernova. The difference of the actual cross-correlation from its mean value can be interpreted, in this model, as a measure of the mean-square h averaged over the detectors band. From the distance to the supernova (and, again, assuming that the spectrum of h is white to 1 KHz) we can convert this mean-square amplitude into a measure of the fraction of the stellar cores rest mass (assumed to be $\simeq 1.4\,M_\odot$) converted into gravitational waves. That efficiency ϵ we take to be the "science reach" of the detectors for radiation from core-collapse supernovae.

The limit we can place on h^2 in the band, and thus on ϵ, is better the better we know when the core collapse took place. If the core-collapse is galactic then we expect to be able to detect the neutrinos directly, allowing us to fix the time of the core collapse to under a second; on the other hand, if the supernova is extragalactic, then we will have to rely on a backward extrapolation of the light curve, which will allow us to fix the time of the collapse to within an hour. For a supernova at a distance of 55 Kpc (the large Magellanic cloud), the 95% upper limit we can place on ϵ is

$$\epsilon_{95\%} = \begin{cases} 2 \times 10^{-4} & \text{initial LIGO} \\ 2 \times 10^{-6} & \text{advanced LIGO} \end{cases} \qquad (5)$$

Galactic supernova occur at the rate of approximately 1 per 30 y. At the distance of the center of the Virgo cluster — 15 Mpc — the rate is on order 3 y^{-1}. At this distance the advanced LIGO can place an upper bound of $\epsilon_{95\%} \simeq 24\%$ on the efficiency: a physical upper bound (i.e., one less than 100%).

Theoretical prejudice — which, on this source, has revised itself by several orders of magnitude many times over the past three decades — currently estimates the efficiency at $\epsilon < 10^{-7}$–10^{-8} [11,8,27]. These estimates are based on two-dimensional calculations and focus only on the earliest parts of the collapse. As such they exclude the radiation that might be associated with axisymmetries in the collapse or in the rapid convective overturn of the hot proto-neutron star, both of which may be significant sources of gravitational radiation. Thus, while LIGO observations don't approach these theoretical prejudices, they are nevertheless important for the challenge they make to the prejudice.

SUMMARY

I like to say that the initial LIGO detectors bound the possible. In the case of binary inspiral they challenge some of the most optimistic scenarios for the rate density of compact binary neutron star and black hole systems; they set meaningful upper limits on the crustal deformation of nearby, young neutron stars; they improve the in-band limits on the strength of a primordial stochastic gravitational wave background and set a first bound on the background strength owing to the confusion limit of more conventional astronomical sources; and, in the event of a galactic supernova while LIGO is "on the air", will set a physically meaningful upper limit on the efficiency with which core collapse supernovae produce gravitational waves.

The proposed advanced LIGO challenges theory. It will survey a volume of space that all but assures it should see several NS/NS binary inspirals per year and also establish the rate density of NS/BH and BH/BH binary systems. It will measure, or further improve the upper limit on, the deformation of the crust of nearby neutron stars; improve the limits on the stochastic background signal by three orders of magnitude beyond initial LIGO; may well explore the γ-ray burst model; and place physical bounds on the supernova efficiency.

It is a pleasure to thank Joan Centrella for organizing an absolutely splendid workshop, and for her seemingly boundless patience in waiting for my contribution to this proceedings. The work described here was supported by National Science Foundation awards PHY 98-00111 and PHY 99-96213 and its predecessor PHY 95-03084.

REFERENCES

1. Bruce Allen and Joseph D. Romano. Detecting a stochastic background of gravitational radiation: signal processing strategies and sensitivies. *Phys. Rev. D*, 59:102001, 15 May 1999.

2. A. M. Alpar and David Pines. Gravitational radiation from a solid-crust neutron star. *Nature (London)*, 314:334, 1985.

3. Luc Blanchet, Thibault Damour, Bala R. Iyer, Clifford M. Will, and Alan G. Wiseman. Gravitational-radiation damping of compact binary systems to second post-Newtonian order. *Phys. Rev. Lett.*, 74(18):3515–3518, 1 May 1995.

4. Luc Blanchet, Guillaume Faye, and Bénédicte Posnot. Gravitational field and equations of motion of compact binaries to 5/2 post-Newtonian order. *Phys. Rev. D*, 58, 1998.

5. Nelson Christensen. Measuring the stochastic gravitational-radiation background with laser-interferometric antennas. *Phys. Rev. D*, 46(12):5250–5266, 15 December 1992.

6. Curt Cutler and Éanna É. Flanagan. Gravitational waves from merging compact binaries: How accurately can one extract the binary's parameters from the inspiral waveform. *Phys. Rev. D*, 49(6):2658–2697, 1994.

7. Gary J. Feldman and Robert D. Cousins. Unified approach to the classical statistical analysis of small signals. *Phys. Rev. D*, 57(7):3873–3889, 1 April 1998.

8. Lee Samuel Finn. Detectability of gravitational radiation from supernovae. In J. R. Buchler, S. L. Detweiler, and J. Ipser, editors, *Nonlinear Problems in Relativity and Cosmology*, pages 156–172. New York Academy of Sciences, New York, 1991.

9. Lee Samuel Finn. Binary inspiral, gravitational radiation, and cosmology. *Phys. Rev. D*, 53(6):2878–2894, 15 March 1996.

10. Lee Samuel Finn and David F. Chernoff. Observing binary inspiral in gravitational radiation: One interferometer. *Phys. Rev. D*, 47(6):2198–2219, 1993.

11. Lee Samuel Finn and Charles R. Evans. Determining gravitational radiation from Newtonian self-gravitating systems. *Astrophys. J.*, 351:588–600, 1990.

12. Lee Samuel Finn, Soumya D. Mohanty, and Joseph D. Romano. Detecting an association between gamma ray and gravitational wave bursts. *Phys. Rev. D*, 60(12):121101(R), 1999.

13. Éanna É. Flanagan. Sensitivity of the Laser Interferometer Gravitational Wave Observatory to a stochastic background, and its dependence on the detector orientations. *Phys. Rev. D*, 48(6):2389–2407, 15 September 1993.

14. Éanna É. Flanagan and Scott A. Hughes. Measuring gravitational waves from binary black hole coalescences. I. Signal to noise for inspiral, merger, and ringdown. *Phys. Rev. D*, 57:4535, 15 April 1998.

15. Éanna É. Flanagan and Scott A. Hughes. Measuring gravitational waves from binary black hole coalescences. II. The waves' information and its extraction, with and without templates. *Phys. Rev. D*, 57:4566, 15 April 1998.

16. William M. Folkner, editor. Number 456 in AIP Conference Proceedings. American Institute of Physics, Woodbury, New York, July 1998.

17. C.L. Fryer, S.E. Woosley, and D.H. Hartmann. Formation rates of black hole accretion disk gamma-ray bursts. *Astrophys. J.*, 526:152–177, November 1999.

18. J. Goodman. Are gamma-ray bursts optically thick? *Astrophys. J. Lett.*, 308:L47–L50, September 1986.

19. D. Hils, P. L. Bender, and R. F. Webbink. *Astrophys. J.*, 360:75, 1990.

20. Gaurav Khanna, John Baker, Reinaldo J. Gleiser, Pablo Laguna, Carlos O. Nicasio,

Hans-Peter Nollert, Richard Price, and Jorge Pullin. Inspiraling black holes: The close limit. *Phys. Rev. Lett.*, 83(18):3581–3584, 1999.

21. C. Kouveliotou, C.Ã. Meegan, G.J̃. Fishman, N.P̃. Bhat, M.S̃. Briggs, T.M̃. Koshut, W.S̃. Paciesas, and G.Ñ. Pendleton. Identification of two classes of gamma-ray bursts. *Astrophys. J. Lett.*, 413:L101–L104, August 1993.

22. D.Q̃. Lamb, C. Graziani, and I.Ã. Smith. Evidence for two distinct morphological classes of gamma-ray bursts from their short time scale variability. *Astrophys. J. Lett.*, 413:L11–L14, August 1993.

23. Michele Maggiore. Gravitatioal wave experiments and early universe cosmology. *Phys. Rev.*, 331:283–367, 2000.

24. A. Mészáros, Z. Bagoly, I.J̃. Horváth, L.G̃. Balázs, and R. Vavrek. A remarkable angular distribution of the intermediate subclass of gamma-ray bursts. *Astrophys. J.*, 539:98–101, August 2000.

25. P. Mészáros. Gamma-ray bursts: Accumulating afterglow implications, progenitor clues, and prospects. *Science*, 291(5501):79–84, January 2001.

26. P. Mészáros and M. J. Rees. Poynting jets from black holes and cosmological gamma-ray bursts. *Astrophys. J. Lett.*, 482:L29–L32, 10 June 1997.

27. R. Mönchmeyer, G. Schäfer, E. Müller, and R. E. Kates. *Astron. Astrophys.*, 256:417, 1991.

28. Soma Mukherjee, Eric D. Feigelson, Gutti Jogesh Babu, Fionn Murtagh, Chris Fraley, and Adrian Raftery. Three types of gamma-ray bursts. *Astrophys. J.*, 508:314–327, November 1998.

29. B. Paczyński. *Astrophys. J. Lett.*, 308:L43–L46, 1986.

30. B. Paczyński. Are gamma-ray bursts in star-forming regions? *Astrophys. J. Lett.*, 494:L45–+, February 1998.

31. Louis Pasteur. speech, 7 Dec 1854, 1854.

32. M. J. Rees and P. Mészaros. Unsteady outflow models for cosmological gamma-ray bursts. *Astrophys. J. Lett.*, 430:L93–L96, 1 August 1994.

33. M. Ruderman. Neutron star crustal plate tectonics. i - magnetic dipole evolution in millisecond pulsars and low-mass x-ray binaries. *Astrophys. J.*, 366:261–269, January 1991.

34. M. Ruderman. Neutron star crustal plate tectonics. ii - evolution of radio pulsar magnetic fields. iii - cracking, glitches, and gamma-ray bursts. *Astrophys. J.*, 382:576–593, December 1991.

35. M. Ruderman. Neutron star crustal plate tectonics. iii. cracking, glitches, and gamma-ray bursts. *Astrophys. J.*, 382:587–593, December 1991.

36. B. F. Schutz. Determining the Hubble constant from gravitational-wave observations. *Nature (London)*, 323:310–311, 1986.

37. S. E. Woosley. Gamma-ray bursts from stellar mass accretion disks around black holes. *Astrophys. J.*, 405:273, March 1993.

38. Thomas Zwerger and Ewald Müller. Gravitational radiation from rotational core collapse. <http://www.mpa-garching.mpg.de/ ewald/GRAV/grav.html>, 1998.

Coalescing Compact Binaries:
Rates, Scenarios, and Simulations

Event Rates for Binary Inspiral

Vassiliki Kalogera* and Krzysztof Belczynski*,†

*Harvard-Smithsonian Center for Astrophysics, Cambridge, MA 02138
†Nicolaus Copernicus Astronomical Center, 00-716 Warszawa, Poland

Abstract.
  Double compact objects (neutron stars and black holes) found in binaries with small orbital separations are known to spiral in and are expected to coalesce eventually because of the emission of gravitational waves. Such inspiral and merger events are thought to be primary sources for ground based gravitational–wave interferometric detectors (such as LIGO). Here, we present a brief review of estimates of coalescence rates and we examine the origin and relative importance of uncertainties associated with the rate estimates. For the case of double neutron star systems, we compare the most recent rate estimates to upper limits derived in a number of different ways. We also discuss the implications of the formation of close binaries with two non–recycled pulsars.

INTRODUCTION

Compact objects, neutron stars (NS) or black holes (BH), formed from relatively massive stars can spiral in and coalesce when found in tight binaries, the orbital evolution of which is driven by gravitational radiation. As angular momentum losses dominate, the orbit shrinks and the two compact objects can eventually merge as they revolve in orbit around each other. The prototype progenitor system of such inspiral events is the binary pulsar PSR B1913+16 (the "Hulse–Taylor" pulsar [1]). Sensitive pulsar timing measurements have revealed that the orbital period decreases at a rate comparable (to better than 1%) to that predicted by general relativity for the emission of gravitational waves [2], [3]. The ultimate coalescence of the two neutron stars seems inevitable.

Although PSR B1913+16 will not reach coalescence for another 300 Myr, similar inspiraling systems in the Milky Way and nearby galaxies are thought to be primary sources of gravitational radiation for ground–based interferometric gravitational–wave detectors, currently under construction or commissioning (e.g., LIGO, VIRGO, GEO600). In addition to NS–NS close binaries, NS–BH and BH–BH binaries are also expected to form through the evolution of massive binaries and to contribute to the detection of inspiral events.

CP575, *Astrophysical Sources for Ground-Based Gravitational Wave Detectors,* edited by J. M. Centrella

The expected detection rate of inspiral events depends on (i) the strength of the expected gravitational–wave signal, (ii) the gravitational–wave detector sensitivity, and (iii) the coalescence rate of each binary population. The first two considerations define a maximum distance D_{max}, out to which different types of inspiral events and mergers could be detected. The coalescence rate for each population is estimated in two steps: first, the Galactic rate, and then its extrapolation out to the maximum distance of interest. Based on the current understanding of the LIGO sensitivities, the maximum distances out to which inspiral events could be detected by LIGO II (and LIGO I) are (approximately), 350 Mpc (20 Mpc) for NS–NS binaries, 700 Mpc (40 Mpc) for NS–BH binaries, and 1500 Mpc (100 Mpc) for BH–BH binaries (assuming $1.4 \, M_\odot$ NS and $10 \, M_\odot$ BH; Sam Finn, private communication). Given our current best knowledge (based on recent redshift surveys) of galaxy distributions out to those distances [4], it can be estimated that, for a LIGO II detection rate of 1 event per year, the following *Galactic* coalescence rates are required: $\simeq 5 \times 10^{-7} \, \mathrm{yr}^{-1}$ for NS–NS binaries, $\simeq 5 \times 10^{-8} \, \mathrm{yr}^{-1}$ for NS–BH binaries, and $\simeq 5 \times 10^{-9} \, \mathrm{yr}^{-1}$ for BH–BH binaries.

Formation rates of *coalescing* compact binaries (systems with tight enough orbits that merge within a Hubble time $\sim 10^{10} \, \mathrm{yr}$) have been calculated so far using two very different methods: either entirely theoretically, based on binary evolution models, or, for NS–NS binaries, empirically, based on the observed NS–NS sample. In what follows we present an up–to–date review of current rate estimates, addressing in detail the most important uncertainties associated with them. We also discuss independent ways of obtaining upper limits to the coalescence rate of NS–NS binaries and possible implications of the formation of systems without recycled pulsars.

THEORETICAL RATE ESTIMATES

The formation rate of coalescing binary compact objects can be calculated, given a sequence of evolutionary stages leading to binary compact object formation. Over the years, a relatively standard picture has been formed describing the birth of such systems based on considerations of NS–NS binaries [5]. More recently, variations of the standard evolutionary channel have also been discussed and suggested [6], mainly based on worries about the fate of neutron stars in situations of hypercritical accretion (not limited to the photon Eddington rate), and their possible collapse into black holes. In all versions, however, the main picture remains the same: the initial binary progenitor consists of two binary members massive enough to eventually collapse into a NS or a BH. The evolutionary path involves multiple phases of stable or unstable mass transfer, common–envelope phases (where one or possibly two stellar cores spiral in the envelopes of evolved stars and eventually lead to the ejection of these envelopes), and accretion onto compact objects, as well as two core collapse events. The final outcome of interest is the formation of binary compact objects in close binary orbits.

Such theoretical modeling has been undertaken by a number of different groups by means of population syntheses. This provides us with *ab initio* predictions of coalescence rates. Monte Carlo numerical techniques are employed in following the evolution of a large ensemble of primordial binaries with certain assumed initial properties through a multitude of channels until compact object binaries are formed. The changes in the properties of the binaries at the end of each stage are calculated based on our current understanding of the various evolutionary processes involved: wind mass loss from massive hydrogen– and helium–rich stars, mass and angular–momentum losses during mass transfer phases, dynamically unstable mass transfer and common–envelope evolution, effects of highly super–Eddington accretion onto NS, and supernova explosions with kicks imparted to newborn NS or even BH. Given our limited understanding of some of these phases, the results of population synthesis are expected to depend on the assumptions made in the treatment of the various processes. Therefore, exhaustive parameter studies are required by the nature of the problem.

Recent studies of the formation of compact objects and calculations of their Galactic coalescence rates ([7], [8], [9], [10], [11], [12]) have explored the input parameter space and the robustness of the results at different levels of (in)completeness. Almost all of these groups have studied the sensitivity of the predicted coalescence rates to the average magnitude of the kicks imparted to compact objects at birth. The range of predicted NS–NS Galactic rates obtained by varying the kick magnitude alone is found in the range $< 10^{-7} - 5 \times 10^{-4} \, \mathrm{yr}^{-1}$. This large range indicates the importance of supernovae (two in this case) in the evolution of massive binaries. Variations in the assumed mass–ratio distribution for the primordial binaries can *further* change the predicted rate by about a factor of 10, while assumptions of the common–envelope phase add another factor of about $10 - 100$. Variation in other parameters typically affects the results by factors of two or less. Predicted rates for BH–NS and BH–BH binaries lie in the ranges $< 10^{-7} - 10^{-4} \, \mathrm{yr}^{-1}$ and $< 10^{-7} - 10^{-5} \, \mathrm{yr}^{-1}$, respectively when the kick magnitude to both NS and BH is varied. Other uncertain factors such as the critical progenitor mass for NS and BH formation lead to variations of the rates by factors of $10 - 50$.

It is evident that recent theoretical predictions for coalescence rates cover a wide range of values (typically 3–4 orders of magnitude), because the various input parameters and assumptions affect strongly the absolute normalization (birth rate) of the modeled populations. Given these results, it seems fair to say that, at least at present, population synthesis calculations have a rather limited predictive power and provide fairly loose constraints on coalescence rates. One way to improve the reliability of such predictions is to study a number of different binary populations (with or without compact objects) and incorporate a number of independent observational constraints, such as star formation rate, supernova rates of different types, binarity of Wolf–Rayet stars, and others. A number of constraints on the population synthesis models could help restricting the predicted coalescence rates in narrower ranges [13].

EMPIRICAL RATE ESTIMATES

The large range of theoretically predicted Galactic coalescence rates of double compact objects motivates us to examine other ways of obtaining rate estimates. The observed sample of coalescing NS–NS binaries found in the Galactic field (PSR B1913+16 and PSR B1534+12) provides us with alternative estimates of their coalescence rate. "Empirical" estimates can be obtained using the observed pulsar and binary properties along with models of selection effects in radio pulsar surveys [14], [15]. For each observed object, a scale factor can be calculated based on the fraction of the Galactic volume within which pulsars with properties identical to those of the observed pulsar could be detected by any of the radio pulsar surveys, given their detection thresholds. This scale factor is a measure of how many more pulsars like the ones detected in the coalescing NS–NS systems exist in our galaxy. The coalescence rate can then be calculated based on the scale factors and estimates of detection lifetimes summed up for all the observed systems. Based on this method the first two studies concluded that the NS–NS Galactic coalescence rate is $\simeq 10^{-6} \, \mathrm{yr}^{-1}$.

Since then, estimates of the NS–NS coalescence rate have known a significant downward revision primarily because of (i) the increase of the Galactic volume covered by radio pulsar surveys with no additional coalescing NS–NS being discovered [16], (ii) the increase of the distance estimate for PSR B1534+12 based on measurements of post-Newtonian parameters [17] (iii) revisions of the lifetime estimates [18], [19]. Recent estimates place the NS–NS rate for our Galaxy in the range $\simeq 1 - 3 \times 10^{-7} \, \mathrm{yr}^{-1}$. Further, it has been realized that a number of upward correction factors must be included, most importantly to account (i) for the beamed nature of pulsar emission and correct for all the binary pulsars with beams that our line of sight does not intersect, and (ii) for the faint end of the pulsar luminosity function and correct for those systems that are too faint to be detected. These two correction (multiplication) factors have so far typically been assumed to be $\simeq 3$ and $\simeq 10$, respectively.

In a study just recently completed [4], we especially focused on all the uncertainties associated with these empirical estimates. We found that the upward correction factor for the faint end of the pulsar luminosity is the most important source of uncertainty. However, it is highly sensitive to the number of observed objects and its distribution function widens dramatically for small–number samples. For a sample of two objects (as the observed one) the faint–pulsar correction factor can vary from very small (close to unity) to as high as $\simeq 200$ (see following subsection). Beyond the issue of faint pulsars, we considered a number of uncertainties and correction factors. Based on recent observational data for both PSR B1913+16 and PSR B1534+12, we found that the beaming correction factor is higher than previously thought ($\simeq 6$) but with a rather small uncertainty ($\simeq 10\%$). Other factors, such as pulsar ages and lifetimes, and spatial distribution, lead to an uncertainty factor of about 2. We estimate the Galactic NS–NS coalescence rate in the range $\simeq 10^{-6} - 5 \times 10^{-4} \, \mathrm{yr}^{-1}$, which is still narrow compared to the range covered by the

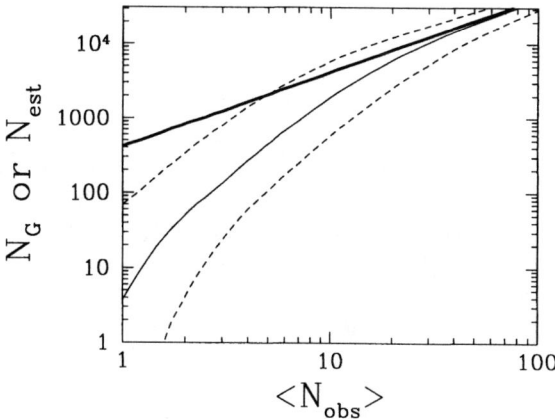

FIGURE 1. Bias of the empirical estimates of the NS–NS coalescence rate because of the small–number observed sample. See text for details.

theoretical estimates.

Small Number Sample and Pulsar Luminosity Function

One important limitation of empirical estimates of the coalescence rates is that they are derived based on *only two* observed NS–NS systems, under the assumption that the observed sample is representative of the true population, particularly in terms of their radio luminosity. Assuming that the recycled pulsars in NS–NS binaries follow the radio luminosity function of young pulsars and that therefore their true Galactic population is dominated in number by low–luminosity pulsars, it can be shown that the current empirical estimates most probably *under*estimate the true coalescence rate. If a small–number sample is drawn from a parent population dominated by low–luminosity (hence hard to detect) objects, it is statistically more probable that the sample will actually be dominated by objects from the high–luminosity end of the population. The result is that the population overall is thought to be brighter than it really is, and therefore, detectable over a larger Galactic volume. Consequently, the empirical estimates based on such a sample will tend to overestimate the detection volume for each observed system, and therefore underestimate the scale factors and the resulting coalescence rate.

This effect can be clearly demonstrated with a Monte Carlo experiment [4] using simple models for the pulsar luminosity function and the survey selection effects. As a first step, the average observed number of pulsars is calculated given a known "true" total number of pulsars in the Galaxy (thick-solid line in Figure 1). As a second step, a large number of sets consisting of "observed" (simulated) pulsars are realized using Monte Carlo methods. These pulsars are drawn from a Poisson distribution of a given mean number ($< N_{obs} >$) and have luminosities assigned

111

according to the assumed luminosity function. Based on each of these sets, one can estimate the total number of pulsars in the Galaxy using empirical scale factors, as is done for the real observed sample. The many (simulated) "observed" samples can then be used to obtain the distribution of the estimated total Galactic numbers (N_{est}) of pulsars. We find that these N_{est} distributions are very strongly skewed and lead to possible correction factors for the faint pulsars in a wide range of values (covering typically a couple of orders of magnitude). The median and 25% and 75% percentiles of this distribution are plotted as a function of the assumed number of systems in the (fake) "observed" samples in Figure 1 (thin–solid and dashed lines, respectively).

It is evident that, in the case of small–number observed samples (less than ~ 10 objects), the estimated total number, and hence the estimated coalescence rate, can be underestimated by a significant factor. For observed samples with an expected number of objects equal to two, for example, the true rate may be much higher by more than a factor as high as $\simeq 200$. This underestimation factor represents an upward correction factor that must be applied to the rate estimated using the observed sample of *coalescing* NS–NS binaries. However, we note that distribution of this correction factor covers a wide range and becomes highly skewed for small number samples (less than about 10 objects), and therefore it is currently quite uncertain. We conclude that correcting for the undetected, faint pulsars in the population cannot be decoupled from the problems of a small–number sample because of the assumption of the observed sample being representative of the population, implicit in the method.

UPPER LIMITS ON THE NS–NS COALESCENCE RATE

Observations of NS–NS systems and isolated pulsars related to NS–NS formation allow us to obtain upper limits on their Galactic coalescence rate in a number of different ways. Depending on how their value compares to the Galactic rate required for a LIGO II detection rate of 1 event per year, such limits can in principle provide us with valuable information about the prospects of gravitational–wave detection.

The absence of any young pulsars detected in NS–NS systems was used to obtain a rough upper limit to the rate of $\sim 10^{-5}\,\mathrm{yr}^{-1}$ [20]. Recently the same basic argument was reexamined in more detail and a more robust upper limit of $\sim 10^{-4}\,\mathrm{yr}^{-1}$ was derived [19].

An upper bound to the NS–NS coalescence rate can also be obtained by combining our theoretical understanding of orbital dynamics (for supernovae with NS kicks in binaries) with empirical estimates of the birth rates of *other* types of pulsars related to NS–NS formation [21]. Progenitors of NS–NS systems experience two supernova explosions. The second supernova explosion (forming the NS that is *not* observed as a pulsar) provides a unique tool for the study of NS–NS formation, since the post–supernova evolution of the system is simple, driven only by gravitational–wave radiation. There are three possible outcomes after the second

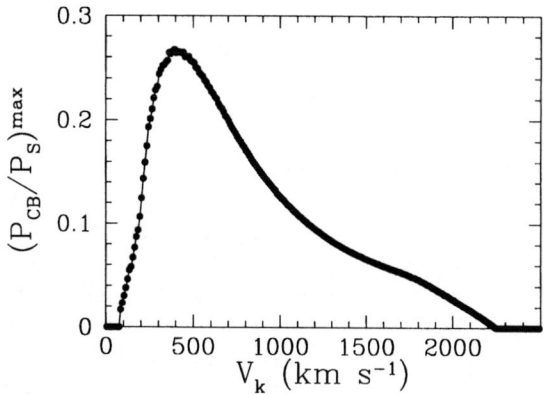

FIGURE 2. Maximum probability ratio for the formation of coalescing NS–NS systems and the disruption of binaries as a function of the kick magnitude at the second supernova.

supernova: (i) a coalescing NS–NS is formed (CB), (ii) a wide NS–NS (with a coalescence time longer than the Hubble time) is formed (WB), or (iii) the binary is disrupted (D) and a single pulsar similar to the ones seen in NS–NS systems is ejected. Based on supernova orbital dynamics we can accurately calculate the probability branching ratios for these three outcomes, P_{CB}, P_{WB}, and P_D. For a given kick magnitude, we can calculate the maximum ratio $(P_{CB}/P_D)^{max}$ for the complete range of pre-supernova parameters defined by the necessary constraint $P_{CB} \neq 0$ (Figure 2). Given that the two types of systems have a common parent progenitor population, the ratio of probabilities is equal to the ratio of the birth rates (BR_{CB}/BR_D).

We can then use (i) the absolute maximum of the probability ratio ($\simeq 0.27$ from Figure 2) and (ii) an empirical estimate of the birth rate of single pulsars similar to those in NS–NS based on the current observed sample to obtain an upper limit to the coalescence rate. The selection of this sample involves some subtleties [21], and the analysis results in $BR_{CB} < 1.5 \times 10^{-5}\,\mathrm{yr}^{-1}$. Note that this number could be increased because of the small–number sample and luminosity bias, which this time affects the empirical estimate of BR_D by a factor of $\simeq 2-6$. Such an upward correction can bring the upper limit in the range $3 - 9 \times 10^{-5}\,\mathrm{yr}^{-1}$.

This is an example of how we can use observed systems other than NS–NS to improve our understanding of their coalescence rate. A similar calculation can be done using the wide NS–NS systems instead of the single pulsars [21].

NON–RECYCLED DOUBLE NEUTRON STARS

We have already pointed out that the empirical methods employed to obtain rate estimates for NS–NS coalescence include the implicit assumption that the properties of the observed sample are representative of the Galactic NS–NS population. This assumption extends to the pulsar properties and their evolutionary history of recycling (spin–up by accretion). Consistent with the pulsars observed in the detected NS–NS systems, it turns out that so far theoretical studies of NS–NS formation have considered systems where one of the neutron stars had the opportunity to be recycled, at least in principle (through stellar winds, Roche–lobe overflow accretion, or even possibly in a common–envelope phase).

Here, we report on a new evolutionary path leading to the formation of close NS–NS binaries, with the unique characteristic that none of the two NS ever had the chance to be recycled by accretion. As we will discuss in more detail, such NS–NS systems have a negligible probability of being detected as binary pulsars, and could represent a "dormant" NS–NS population in galaxies with important implications for gravitational–wave detection of NS–NS inspiral events. The existence of this recently identified [22] evolutionary channel stems from the evolution of helium–rich stars (cores of massive NS progenitors), which has been neglected in most previous studies of double compact object formation. We find that these non-recycled NS–NS binaries are formed from bare carbon–oxygen cores in tight orbits, with formation rates comparable to or maybe even higher than those of recycled NS–NS binaries.

The Method

We study NS–NS binaries formed through a multitude of evolutionary sequences that are not predefined, but instead are realized in Monte Carlo population synthesis calculations.

To describe the evolution of single stars (hydrogen– and helium–rich) from the zero age main sequence (ZAMS) to carbon–oxygen (CO) core formation, we employ analytical formulae from stellar evolution fits [23] However, we have adopted a prescription for the masses of compact objects formed at core–collapse events, based on the relation between CO core masses and final FeNi core masses [24].

Concerning the evolution of interacting binaries, we model the changes of mass and orbital parameters taking into account mass and angular momentum transfer between the stars or loss from the system during Roche–lobe overflow, tidal circularization, rejuvenation of stars due to mass accretion, wind mass loss from massive and/or evolved stars, dynamically unstable mass transfer episodes leading to common–envelope (CE) evolution and spiral–in of the stars. We also account for the *possibility* of hyper-critical accretion onto compact objects during CE phases [6] and effects of asymmetric supernovae (SN) on a binary orbit (mass loss and a kick velocity a newly born compact object receives in SN). More details about the treatment of various evolutionary processes are presented elsewhere [13].

In the synthesis calculations, we typically evolve a few million of primordial binaries to satisfy the requirement that the statistical (Poisson) fractional errors ($\propto 1/\sqrt{N}$) of the final NS–NS population are lower than 10%. The formation rates are calibrated using the latest Type II SN empirical rates and normalized to our Galaxy [25].

In our standard model, primordial binaries follow given distributions: for primary masses $(5 - 100 \, M_\odot)$, $\propto M_1^{-2.7} dM_1$; for mass ratios $(0 < q < 1)$, $\propto dq$; for orbital separations (from a minimum, so both ZAMS stars fit within their Roche lobes, up to $10^5 \, R_\odot$), $\propto dA/A$; for eccentricities, $\propto 2e$. Each of the models is also characterized by a set of assumptions, which, for our standard model, are: (1) *Kick velocities.* We use a weighted sum of two Maxwellian distributions with $\sigma = 175 \, \mathrm{km \, s^{-1}}$ (80%) and $\sigma = 700 \, \mathrm{km \, s^{-1}}$ (20%) [26]; (2) *Maximum NS mass.* We adopt a conservative value of $M_{\mathrm{max}} = 3 \, M_\odot$ [27]; (3) *Common envelope efficiency.* We assume $\alpha_{\mathrm{CE}} \times \lambda = 1.0$, where α is the efficiency with which orbital energy is used to unbind the stellar envelope, and λ is a measure of the central concentration of the giant; (4) *Non–conservative mass transfer.* In cases of dynamically stable mass transfer between non–degenerate stars, we allow for mass and angular momentum loss from the binary [28], assuming that half of the mass lost from the donor is also lost from the system $(1 - f_a = 0.5)$ with specific angular momentum equal to $\beta 2\pi A^2/P$ $(\beta = 1)$; (5) *Star formation history.* We assume that star formation has been continuous in the disk of our Galaxy for the last $10 \, \mathrm{Gyr}$ [29].

Results

We use our population synthesis models to investigate all possible formation channels of NS–NS binaries realized in the simulations. We find that a significant fraction of *coalescing* NS–NS systems are formed through a new, previously not identified evolutionary path. The evolution along this new channel begins with two phases of Roche–lobe overflow. The first, from the primary to the secondary, involves non–conservative but dynamically stable mass transfer and ends when the hydrogen envelope is consumed. The second, from the initial secondary to the helium core of the initial primary, involves dynamically unstable mass transfer, i.e., CE evolution. The post–CE binary consists of two bare helium stars of relatively low masses. As they evolve through core and shell helium burning, the two stars develop 'giant–like" structures, with clear CO cores and convective envelopes. Their radial expansion eventually brings them into contact and the system evolves through a double CE phase (similar to Brown [1995], for hydrogen–rich stars). During this double CE phase, the combined helium envelopes are ejected at the expense of orbital energy. The tight, post–CE system consists of two CO cores, which eventually end their lives as Type Ic supernovae leaving double neutron star system.

The unique qualitative characteristic of this NS–NS formation path is that both NS have avoided recycling. Based on comparison of non–recycled NS–NS *relative* to that of recycled pulsars, for each of our models, we derive a correction factor for empirical estimates of the Galactic NS–NS coalescence rate. Since these estimates account only for NS–NS systems with recycled pulsars, they must be increased

TABLE 1. Galactic NS-NS Coalescence Rates (Myr^{-1})

Model	New NS–NS	Total NS–NS	Rate Increase	Model Description
A	3.8	7.5	2.0	standard model described in text
B1	6.6	7.3	10	zero kicks
B2	7.0	8.4	5.9	single Maxwellian kicks: $\sigma = 50\,km\,s^{-1}$
B3	5.6	9.5	2.5	single Maxwellian kicks: $\sigma = 100\,km\,s^{-1}$
D1	4.3	5.0	6.9	maximum NS mass: $M_{max} = 2\,M_\odot$
D2	2.7	2.7	$\gg 1$	maximum NS mass: $M_{max} = 1.5\,M_\odot$
E1	0.2	0.7	1.4	$\alpha_{CE} \times \lambda = 0.1$
E2	1.6	2.7	2.5	$\alpha_{CE} \times \lambda = 0.25$
E3	3.1	4.8	2.8	$\alpha_{CE} \times \lambda = 0.5$
F4	2.6	7.4	1.5	mass fraction accreted: $f_a = 1.0$

to include any non–recycled systems formed. We have performed an extensive parameter study to assure robustness of our results. In Table 1 we present the formation rates of non–recycled NS–NS binaries and the total NS–NS population with merger times shorter than 10 Gyr, along with the upwards correction factor for the Galactic empirical rate estimates. Results are shown only for models where the derived factor differs from our standard model by more than 25%. We find that these factors are typically $\simeq 1.5 - 3$ but can be higher for some models.

We note that the identification of the formation path for non–recycled NS–NS binaries stems entirely from accounting for the evolution of helium stars and for the possibility of double CE phases, both of which have typically been ignored in previous calculations.

CONCLUSIONS

The current theoretical estimates of NS–NS coalescence rates appear to have a rather limited predictive power. They cover a range of values in excess of 3 orders of magnitude. Most importantly, this range includes the value of $\simeq 5 \times 10^{-7}\,yr^{-1}$ required for a LIGO II detection rate of 1 event per year. This means that at the two edges of the range the conclusion swings from no detection to many per month, and therefore the detection prospects of NS–NS coalescence cannot be assessed firmly. On the other hand, empirical estimates based on the observed sample of coalescing NS–NS systems appear to be more robust. Taking into account recent empirical estimates and the associated uncertainties [4], we find the Galactic NS–NS inspiral rate in the range $10^{-6} - 5 \times 10^{-4}\,yr^{-1}$. If we also include the independently derived upper limit of $10^{-4}\,yr^{-1}$, we expect a detection rate of $2 - 300$ events per year for LIGO II.

It is important to note here that another implicit assumption in derivation of the empirical estimates is that all NS–NS binaries have at some point in their lifetime contained a recycled pulsar with rather long lifetimes ($\sim 10^9\,yr$). However, recent models of NS–NS formation [22] show that there may exist a significant NS–NS population with neutron that never had the chance to be recycled and therefore have very short lifetimes (by 2-3 orders of magnitude, thus preventing their detection). For a variety of population synthesis models, the birth rate of this separate population of coalescing NS–NS binaries is typically comparable or higher than that of the systems with one recycled NS. The total number of coalescing NS–NS systems could be higher by factors of at least 50%, and up to 10 or even higher. Such an increase has important implications for prospects of gravitational wave detection by ground–based interferometers. Using the recent results on the empirical NS–NS coalescence rate [4], we find that the *most optimistic* prediction for the LIGO I detection rate could be raised to at least 1 event per 2–3 years, and the *most pessimistic* LIGO II detection rate could be raised to 3–6 events per year or even higher.

Estimates of the coalescence rate of BH–NS and BH–BH systems rely solely on our theoretical understanding of their formation. As in the case of NS–NS binaries, the model uncertainties are significant and the ranges extend to more than 2 orders of magnitude. However, the requirement on the Galactic rate is less stringent for $10\,M_\odot$ BH–BH binaries, only $\simeq 5 \times 10^{-9}\,\mathrm{yr}^{-1}$. Therefore, even with the pessimistic estimates for BH–BH coalescence rates ($\sim 10^{-7}\,\mathrm{yr}^{-1}$), we would expect at least a few to several detections per year (with LIGO II), which is quite encouraging. We also point out that that a recent examination of formation of close BH-BH through dynamical processes (stellar interactions) in globular clusters leads to detection rates as high as a few per day for LIGO II and 1 event per 2 years for LIGO I [30].

REFERENCES

1. Hulse, R.A., & Taylor, J.H., *ApJ*, **195**, L51 (1975).
2. Taylor J. H., & Weisberg, J. M., *ApJ*, **253**, 908 (1982).
3. Taylor J. H., & Weisberg, J. M., *ApJ*, **345**, 434 (1989).
4. Kalogera, V., Narayan, R., Spergel, D.N., & Taylor, J.H., *ApJ*, submitted (2000) [astro-ph/0012038].
5. Bhattacharya, D. & van den Heuvel, E. P. J., Physical Reports, **203**, 1 (1991).
6. Brown, G.E., *ApJ*, **440**, 270 (1995).
7. Lipunov, V.M., Postnov, K.A., Prokhorov, M.E., *MNRAS*, **288**, 245 (1997).
8. Fryer, C.L., Burrows, A., Benz, W., *ApJ*, **496**, 333 (1998).
9. Portegies-Zwart, S.F. & Yungel'son, L.R., *A&A*, **332**, 173 (1998).
10. Brown, G.E. & Bethe, H., *ApJ*, **506**, 780 (1998).
11. Fryer, C.L., Woosley, S.E., Hartmann, D., *ApJ*, **526**, 152 (1999).
12. Bulik, T., Belczynski, K., & Zbijewski, W., *MNRAS*, **309**, 629 (1999).
13. Belczynski, K., Kalogera, V. & Bulik T., in preparation (2000).
14. Phinney, E.S., *ApJ*, **380**, L17 (1991).
15. Narayan, R., Piran, T., & Shemi, A., *ApJ*, **379**, L17 (1991).
16. Curran, S.J. & Lorimer, D.R., *MNRAS*, **276**, 347 (1995).
17. Stairs, I.H., et al., *ApJ*, **505**, 352 (1998)

18. van den Heuvel, E.P.J. & Lorimer, D.R., *MNRAS*, **283**, 37 (1996).

19. Arzoumanian, Z., Cordes, J.M., & Wasserman, I., *ApJ*, **520**, 696, (1998).

20. Bailes, M., in IAU Symp. 165, Compact Stars in Binaries, ed. J. van Paradijs, E. P. J. van den Heuvel & E. Kuulkers (Kluwer Academic Publishers, Dordrecht), 213, (1996).

21. Kalogera, V. & Lorimer, D.R., *ApJ*, **530**, 890 (2000).

22. Belczynski, K., & Kalogera, V., *ApJ Letters*, submitted (2000) [astro-ph/0012172].

23. Hurley, J.R., Pols, O.R., & Tout, C.A., *MNRAS*, **315**, 543 (2000).

24. Woosley, S.E., in 'Nucleosynthesis and Chemical Evolution", 16th Saas-Fee Course, eds. B. Hauck et al., Geneva Obs., p. 1 (1986).

25. Cappellaro, E., Evans, R., & Turatto, M., *A&A*, **351**, 459 (1999).

26. Cordes, J., & Chernoff, D.F., *ApJ*, **482**, 971 (1997).

27. Kalogera, V., & Baym, G.A., *ApJ*, **470**, L61 (1996).

28. Podsiadlowski, P., Joss, P.C., & Hsu, J.J.L., *ApJ*, **391**, 246 (1992).

29. Gilmore, G., to appear in 'Galaxy Disks and Disk Galaxies", eds. J.G. Funes & E.M. Corsini (San Francisco: ASP) (2001).

30. Portegies-Zwart, S.F. & McMillan, S.L.W., *ApJ*, **528**, L17 (2000).

Black Hole Binaries from Star Clusters

Stephen L. W. McMillan*
Simon F. Portegies Zwart[†1]

* Dept. of Physics, Drexel University, Philadelphia, PA 19104
†Dept. of Physics, Massachusetts Institute of Technology, Cambridge, MA 02139

Abstract. In star clusters, black holes become the most massive objects within a few tens of millions of years; dynamical relaxation then causes them to sink to the cluster core, where they form binaries. These black-hole binaries become more tightly bound by superelastic encounters with other cluster members, and are ultimately ejected from the cluster. The majority of escaping black-hole binaries have orbital periods short enough and eccentricities high enough that the emission of gravitational radiation causes them to coalesce within a few billion years. We predict a black hole merger rate of about 3×10^{-7} per year per cubic megaparsec. Young star clusters in galactic nuclei may also contribute significantly to this total, although their numbers are presently very uncertain. For the first-generation Laser Interferometer Gravitational-Wave Observatory (LIGO-I), this implies one or two detections during the first two years of operation. For LIGO-II, the rate rises to roughly 1 detection per day.

INTRODUCTION

The search for gravitational waves will begin in earnest in January 2002, when LIGO-I becomes fully operational. The appearance of this new and wholly unexplored observational window challenges physicists and astronomers to predict detection rates and source characteristics. Mergers of neutron-star binaries are widely regarded as the most promising sources of gravitational radiation, and estimates of neutron star merger rates (per unit volume) range from $\mathcal{R} \sim 1.9 \times 10^{-7} \, h^3 \, \mathrm{yr}^{-1} \, \mathrm{Mpc}^{-3}$ [1–3], where $h = H_0/100 \, \mathrm{km \, s}^{-1} \, \mathrm{Mpc}^{-1}$, to roughly ten times this value [4,5] (with the lower end of the range apparently preferred, but see also [6]). However, even with optimistic assumptions, we can expect a LIGO-I detection rate of only a few neutron star events per millennium.

Inspiral and merger of black-hole binaries are considerably more energetic events than neutron star mergers, due to the higher masses of the black holes [4,5]. Black-hole binaries can result from the evolution of two stars which are born in a close binary, experience several phases of mass transfer, and subsequently survive two

---

[1)] Hubble Fellow

CP575, *Astrophysical Sources for Ground-Based Gravitational Wave Detectors,* edited by J. M. Centrella
© 2001 American Institute of Physics 0-7354-0014-8/01/$18.00

supernovae [4,7]. However, black holes are much less common than neutron stars, and calculations of event rates from field binaries depend sensitively on many unknown parameters and much poorly understood physics. Binary evolution models generally predict a black hole merger rate $\mathcal{R} \lesssim 2 \times 10^{-9} \, h^3 \, \mathrm{yr}^{-1} \, \mathrm{Mpc}^{-3}$ [4,8,9], substantially lower than the rate for neutron stars (but note that black-hole mergers are expected to be detectable at considerably greater distances, as discussed further below).

We consider here the alternative possibility that black holes become members of close binaries not through internal binary evolution, but rather via dynamical interactions with other stars in a dense stellar system.

MAKING BLACK HOLE BINARIES

Black holes are the products of stars having initial masses exceeding ~ 20–$25 \, \mathrm{M}_\odot$ [10,11]. A Scalo [12] mass distribution with a lower mass limit of $0.1 \, \mathrm{M}_\odot$ and an upper limit of $100 \, \mathrm{M}_\odot$ has 0.071% of its stars more massive than $20 \, \mathrm{M}_\odot$, and 0.045% more massive than $25 \, \mathrm{M}_\odot$. A star cluster containing N stars thus produces $\sim 6 \times 10^{-4} N$ black holes. Known Galactic black holes have masses m_{bh} between $6 \, \mathrm{M}_\odot$ and $18 \, \mathrm{M}_\odot$ [13]. For definiteness, we adopt $m_{\mathrm{bh}} = 10 \, \mathrm{M}_\odot$.

Binary Formation and Dynamical Evolution

A black hole is formed in a supernova explosion. If the progenitor is a single star (i.e. not a member of a binary), the black hole experiences little or no recoil and remains a member of the parent cluster [14]. If the progenitor is a member of a binary, mass loss during the supernova may eject the binary from the cluster potential via the Blaauw mechanism [15], where conservation of momentum causes recoil in a binary which loses mass impulsively from one component.

The Blaauw velocity kick can be as large as the relative orbital velocity of the pre-supernova binary. The escape speed is $\sim 40 \, \mathrm{km \, s}^{-1}$ for a young globular cluster, and somewhat smaller for young "populous" clusters such as R 136 (NGC 2070) in the 30 Doradus region of the Large Magellanic Cloud, or the Arches and Quintuplet systems near the Galactic center [16–19]. Such high recoil velocities are generally achieved only if the binary loses $\gtrsim 50\%$ of its total mass in the supernova, and if its orbital period is initially quite short ($\lesssim 2 \, \mathrm{yr}$) [7]. The binary frequency in globular clusters is between 5 and 40% [20], and less than 30% of binaries have orbital periods smaller than 2 years [21]; we assume that similar distributions apply to binaries in young populous clusters. We estimate that no more than $\sim 10\%$ of black holes are ejected from the cluster immediately following their formation.

After $\sim 40 \, \mathrm{Myr}$ the last supernova has occurred, the mean mass of the cluster stars is $\langle m \rangle \sim 0.56 \, \mathrm{M}_\odot$ [12], and black holes are by far the most massive objects in the system. Mass segregation causes the black holes to sink to the cluster core in a fraction $\sim \langle m \rangle / m_{\text{bh}}$ of the half-mass relaxation time. For a typical globular cluster, the relaxation time is ~ 1 Gyr; for a young populous cluster it is ~ 10 Myr.

By the time of the last supernova, stellar mass loss has also significantly diminished and the cluster core starts to contract, enhancing the formation of binaries by three-body interactions. Single black holes form binaries preferentially with other black holes [23], while black holes born in binaries with a lower-mass stellar companion rapidly exchange the companion for another black hole. The result in all cases is a growing black-hole binary population in the cluster core.

Once formed, the black-hole binaries become more tightly bound through superelastic encounters with other cluster members [22–24]. On average, following each close binary–single black hole encounter, the binding energy of the binary increases by about 20% [25]; roughly one third of this energy goes into binary recoil, assuming equal mass stars. The minimum binding energy of an escaping black-hole binary may then be estimated as

$$E_{b,\text{min}} \sim 36\, W_0 \, \frac{m_{\text{bh}}}{\langle m \rangle} \, kT \,,$$

where $\frac{3}{2}kT$ is the mean stellar kinetic energy and $W_0 = \langle m \rangle |\phi_0|/kT$ is the dimensionless central potential of the cluster [26]. By the time the black holes are ejected, $\langle m \rangle \sim 0.4\,\text{M}_\odot$. Taking $W_0 \sim 5$–10 as a representative range, we find $E_{b,\text{min}} \sim 5000$–$10000\, kT$.

We have tested and refined the above estimates by performing a series of N-body simulations within the "Starlab" software environment [27] (see http::/manybody.org/starlab), using the special-purpose GRAPE-4 computer to speed up the calculations [28]. We find that roughly 40% of black holes are ejected as members of binaries, while 60% are ejected as single stars. These numbers are consistent with the simple picture that each of the last three interactions leading to the binary's escape ejects the other black hole involved. A small fraction (less than 1%) of escaping black holes have a low-mass star as a companion.

The binding energies E_b of the ejected black-hole binaries range from about $1000\, kT$ to $10000\, kT$ in a distribution more or less flat in $\log E_b$, consistent with the assumptions made by [25]. The eccentricities e follow a roughly thermal distribution $[p(e) \sim 2e]$, with high eccentricities slightly overrepresented. About half of the black holes are ejected while the parent cluster still retains more than 90% ($\lesssim 2$ initial relaxation times) of its birth mass; $\gtrsim 90\%$ are ejected before the cluster loses 30% (between 4 and 10 relaxation times) of its initial mass. These findings are in good general agreement with previous estimates that black-hole binaries are ejected within a few gigayears, well before core collapse occurs [23,24].

When more realistic physics—a Scalo mass function, the effects of stellar evolution, and the Galactic tidal field—is included, our model "populous" clusters generally dissolve rather quickly (in ~ 100 Myr). Clusters which dissolve within ~ 40 Myr (before the last supernova) have no time to eject their black holes. However, those that survive beyond this time are generally able to eject at least one close black-hole binary before dissolution.

Based on these considerations, we conservatively estimate the number of ejected black-hole binaries to be about $10^{-4}N$ per star cluster, more or less independent of the cluster lifetime.

Characteristics of the Binary Population

The energy of an ejected binary and its orbital separation are coupled to the dynamical characteristics of the star cluster. For a cluster in virial equilibrium, we have $kT \equiv 2E_{\rm kin}/3N = -E_{\rm pot}/3N = GM^2/6Nr_{\rm vir}$, where M is the total cluster mass and $r_{\rm vir}$ is the virial radius. A black-hole binary with semi-major axis a has $E_b = Gm_{\rm bh}^2/2a$, so

$$\frac{E_b}{kT} = 3N \left(\frac{m_{\rm bh}}{M}\right)^2 \frac{r_{\rm vir}}{a}.$$

Table 1 lists some characteristic parameters for young populous clusters and globular clusters. The first three columns list cluster type, total mass (in solar units) and virial radius (in pc). For globular clusters, mass and virial radius are given as distributions, with a mean and standard deviation around the mean. The corresponding numbers for populous clusters are estimates only—the actual masses and radii are currently very uncertain. The fourth column gives the orbital separation (in solar units) for a $1000\,kT$ binary consisting of two $10\,{\rm M}_\odot$ black holes. The fifth column lists the expected number of black-hole binaries that form within the cluster. The last two columns are discussed in the next section.

TABLE 1. Selected Cluster Parameters

cluster type	M (M$_\odot$) [log]	$r_{\rm vir}$(pc) [log]	$a(1000\,kT)$ [R$_\odot$]	$N_{\rm b}$	$f_{\rm merge}$	MR [Myr^{-1}]
Populous [16]	4.5	-0.4	420	7.9	7.7%	0.0061
Globular [29]	5.5 ± 0.5	0.5 ± 0.3	315	150	51 %	0.0064
Globular*	6.0 ± 0.5	0 ± 0.3	33	500	92 %	0.038

*Estimated parameters for zero-age clusters.

For purposes of our Monte-Carlo estimates of merger rates, the masses and virial radii of globular clusters are assumed to be distributed as independent Gaussians with means and dispersions as presented in the table; this assumption is supported by correlation studies [29]. Table 1 also presents estimates of the parameters of globular clusters at birth (bottom row), based on a recent parameter-space survey of cluster initial conditions [30]; globular clusters which have survived for a Hubble time have lost $\gtrsim 60\%$ of their initial mass and have expanded by about a factor of three. We draw no distinction between core-collapsed globular clusters (about 20% of the current population) and non-collapsed globulars—the present dynamical state of a cluster has little bearing on how black-hole binaries were formed and ejected during the first few gigayears of the cluster's life.

PRODUCTION OF GRAVITATIONAL RADIATION

The merger time of two objects due to the emission of gravitational waves is [31]:

$$t_{\mathrm{mrg}} \approx 150\,\mathrm{Myr} \left(\frac{\mathrm{M}_\odot}{m_{\mathrm{bh}}}\right)^3 \left(\frac{a}{\mathrm{R}_\odot}\right)^4 (1 - e^2)^{7/2}\,.$$

The fraction of black-hole binaries which merge within a Hubble time is given in column six of Table 1, assuming that the binary binding energies are distributed flat in $\log E_b$ between $1000\,kT$ and $10000\,kT$, that the eccentricities are thermal, independent of E_b, and that the universe is 15 Gyr old.

The seventh and final column of Table 1 lists the contribution to the total black hole merger rate (MR) from each cluster category. For populous clusters, the net merger rate is obtained by assuming a steady-state distribution of cluster numbers and a 100 Myr lifetime for each cluster. For globulars, the rate assumes that the mergers are uniformly distributed over a 12 Gyr window (between when the binaries were ejected, on average, and the present time). We return to these assumptions below.

Merger Rate in the Local Universe

Given the black hole merger rate corresponding to each category of star cluster, we now estimate the total merger rate $\mathcal{R}$ per unit volume. Table 2 lists, for various types of galaxies, the space densities ϕ, and S_N, the specific number of globular clusters per $M_v = -15$ magnitude [32]. The total number of globular clusters in the galaxy under consideration is thus

$$N_{GC} = 10^{-0.4(M_v + 15)} S_N\,.$$

The values given for S_N in Table 2 are corrected for internal absorption; the absorbed component is estimated from observations in the far infrared.

TABLE 2. Galaxy Classes and Cluster Content

Galaxy Type	density ϕ $[10^{-3}\,h\,\mathrm{Mpc}^{-3}]$	M_v	$S_N h^2$	GC density $[h^3\,\mathrm{Mpc}^{-3}]$
E–S0	3.49	-20.7	10	6.65
Sab	2.19	-20.0	7	1.53
Sbc	2.80	-19.4	1	0.16
Scd	3.01	-19.2	0.2	0.03
Blue E	1.87	-19.6	14	1.81
Sdm/StarB	0.50	-19.0	0.5	0.01

The estimated number density of globular clusters in the universe is thus

$$\phi_{GC} = 10\ h^3\ \mathrm{Mpc}^{-3},$$

comparable to the result reported by [2]. A conservative estimate of the mean merger rate of black-hole binaries formed in globular clusters is obtained by assuming that globular clusters in other galaxies have characteristics similar to those found in our own. The result is

$$\mathcal{R}_{GC} = 6 \times 10^{-8} \, h^3 \, \text{yr}^{-1} \, \text{Mpc}^{-3}.$$

Irregular galaxies, starburst galaxies, early type spirals and blue elliptical galaxies all contribute to the formation of young populous clusters. In the absence of firm measurements of the numbers of young populous clusters in other galaxies, we simply use the same values of S_N as for globular clusters. The space density of such clusters is then $\phi_{YPC} = 3.5 \, h^3 \, \text{Mpc}^{-3}$ and the black hole merger rate is

$$\mathcal{R}_{YPC} = 2 \times 10^{-8} \, h^3 \, \text{yr}^{-1} \, \text{Mpc}^{-3}.$$

We find that galactic nuclei contribute negligibly to the total black hole merger rate.

Based on the assumptions outlined above, our estimated total merger rate per unit volume of black-hole binaries is

$$\mathcal{R} = 8 \times 10^{-8} \, h^3 \, \text{yr}^{-1} \, \text{Mpc}^{-3}.$$

Normalizing this rate to the assumed total number density of spiral galaxies $\phi_S = 8 \times 10^{-3} h^3 \text{Mpc}^{-3}$, we can write, crudely,

$$\mathcal{R} \sim 10 \, h^2 \, \text{Myr}^{-1} \, / \, \text{galaxy}.$$

Additional Considerations

A number of considerations complicate the above estimate of the black hole merger rate, particularly of the component due to binaries formed in globular clusters, which dominates the total.

Four factors tend to increase the rate above the estimate presented above. First, as already mentioned, our assumed number ($\sim 10^{-4} N$) of ejected black-hole binaries is quite conservative. Second, the correlation between orbital eccentricity and binding energy and the excess of high-eccentricity binaries mentioned earlier both favor more rapid inspiral, causing a larger fraction of the black-hole binaries to merge. Third, the observed population of globular clusters naturally represents only those clusters that have survived until the present day. The study by [30] indicates that $\sim 50\%$ of globular clusters dissolve in the tidal field of the parent galaxy within a few billion years of formation. We have therefore underestimated the total number of globular clusters, and hence the black hole merger rate, by about a factor of two. Finally, a substantial underestimate stems from the assumption that the masses and radii of present-day globular clusters are representative of the initial population. When estimated initial parameters (Table 1, bottom row) are used, the total merger rate increases by a further factor of three.

On the other hand, particularly when more massive clusters are considered, the assumption of a roughly uniform merger rate over the age of the universe becomes problematical. Black hole binaries ejected from the most massive globular clusters (masses $\gtrsim 5 \times 10^6 \, M_\odot$) tend to be very tightly bound, and merge within a few million years of ejection—too long ago to be observed by LIGO, if we imagine that all globular clusters formed in the distant past and the binaries appeared and were ejected shortly after formation. The basic problem is illustrated in Figure 1, which shows the distribution of merger times for fixed black-hole mass and cluster mass and radius, but choosing all other variables randomly from the distributions quoted above. The reader's attention is drawn in particular to the high powers of all parameters appearing in the time scale, which means in practice that almost any desired merger rate can be obtained by an appropriate (and not necessarily extreme) choice of cluster properties.

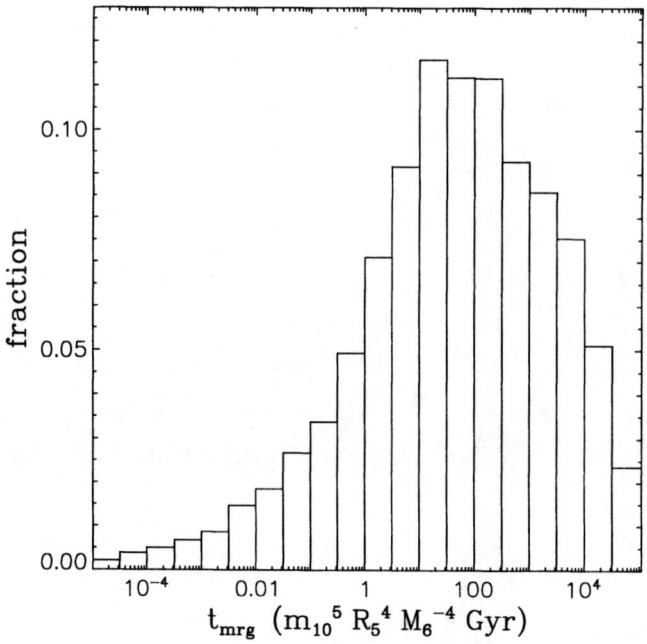

FIGURE 1. Distribution of binary merger times for a given choice of black-hole and cluster parameters. The black hole mass is $10\,m_{10}$ solar masses, the cluster mass is $10^6 \, M_6$ solar masses, and the cluster virial radius is $5\,R_5$ pc.

A more complete calculation, including all these competing affects, is in progress. However, in the interim, we make the following points, which argue that at least some of the high-mass cluster binaries should indeed contribute to the total.

- In general, the longer relaxation times of more massive systems mean that all processes take longer to occur, extending the binary formation time scale.

- Specifically, the Spitzer instability [33] that ultimately leads to runaway black hole binary formation is not necessarily prompt. A cluster must evolve to the point at which the instability sets in (King parameter $W_0 \sim 5$). For a cluster starting with $W_0 \sim 3$, this may require several initial relaxation times, or many gigayears for the most massive clusters.

- The binary ejection process is not necessarily rapid. Following each interaction, binaries are ejected farther and farther into the cluster halo, taking more and more time to relax back into the core before the next encounter can occur. These repeated excursions may extend the process of binary ejection over many cluster relaxation times.

- The binary formation process may be self-regulating, as energy generated by one binary may temporarily support the black-hole subsystem against further collapse, suppressing the formation of new binaries until it is ejected. Again, the effect is to extend the binary formation and ejection time scales.

- Even for prompt formation and ejection, the distribution of merger times (including Gaussian spreads in cluster mass and radius) has a rather extended tail containing $\gtrsim 10\%$ of the total.

- Not all globular clusters are old.

Attempting to take all these effects into account, we (crudely) estimate a net black hole merger rate of

$$\mathcal{R} \sim 3 \times 10^{-7} h^3 \text{ yr}^{-1} \text{ Mpc}^{-3},$$
$$\sim 40 \ h^2 \text{ Myr}^{-1} / \text{ galaxy}.$$

We note that this figure is slightly larger than current best estimates of the neutron-star merger rate.

Estimated LIGO Detection Rate

The maximum distance within which LIGO-I can detect an inspiral event is estimated to be [34]

$$R_{\text{eff}} = 18 \text{ Mpc} \ \left(\frac{M_{\text{chirp}}}{M_\odot}\right)^{5/6}.$$

Here, the "chirp" mass for a binary with component masses m_1 and m_2 is $M_{\text{chirp}} = (m_1 m_2)^{3/5}/(m_1 + m_2)^{1/5}$. For neutron star inspiral, $m_1 = m_2 = 1.4\,M_\odot$, so $M_{\text{chirp}} = 1.22\,M_\odot$, $R_{\text{eff}} = 21$ Mpc, and we obtain the detection rate mentioned in the introduction. For black-hole binaries with $m_1 = m_2 = m_{\rm bh} = 10\,M_\odot$, we find $M_{\rm chirp} = 8.71\,M_\odot$, $R_{\rm eff} = 109$ Mpc, and a LIGO-I detection rate of about $1.5\,h^3$ per year. For $h \sim 0.65$, this results in about 0.5 detection events annually. LIGO-II should become operational by 2007, and is expected to have $R_{\rm eff}$ about ten times greater than LIGO-I, resulting in a detection rate 1000 times higher, or ~ 1 event per day.

DISCUSSION

Many of the problems just addressed pertain specifically to globular clusters, and are not relevant to black-hole binaries formed in dense young clusters, such as those near the Galactic center. Only two such "nuclear" clusters—the Arches and Quintuplet systems—have so far been observed, although 58 cluster candidates have recently been reported within ~ 600 pc in projection of the Galactic center [35]. Detailed N-body studies [36] find that the majority of such clusters are likely to be undetectable against the Galactic background for most of their lives, suggesting that hundreds of Arches-like clusters may in fact exist. The studies also indicate that the masses of those clusters may be larger than had previously been thought, increasing the estimates of the net merger rate to within a factor of ~ 5 of the (corrected) globular cluster rate quoted above. In addition, [37] have suggested that the Arches may also contain an excess of massive stars (and hence black holes), although [27] argue that this observation is a result of dynamical evolution and that the mass function is consistent with that in the solar neighborhood.

Merger rates obtained for black-hole binaries formed in nuclear clusters are presumably continuous, and thus do not suffer from the objections raised earlier to the globular cluster rates. The inferred numbers and lifetimes of nuclear clusters are consistent with the conjecture that the entire Galactic bulge has been created from such systems over the lifetime of the Galaxy. [36] If we assume that all bulge stars form in clusters massive enough that most of their black-hole binaries merge in less than a Hubble time, then a 10^9-star bulge implies a black hole binary merger rate for the Milky Way Galaxy of

$$\mathcal{R} \sim 10\ {\rm Myr}^{-1}\ /\ {\rm galaxy},$$

comparable to the corrected net rate quoted above (taking $h = 0.65$). Nuclear clusters could conceivably dominate the LIGO signal!

By the time the black hole binary is ejected it has experienced ~ 40–50 close encounters with other black holes, as well as a similar number of encounters with other stars or binaries. During each of these latter encounters, there is a small probability that a low-mass star may collide with one of the black holes. Such collisions tend to soften the black hole binary somewhat (see Portegies Zwart et al. 1999), but they are unlikely to delay ejection significantly. A collision between a main-sequence star and a black hole may, however, lead to brief but intense X-ray phase.

Finally, we have assumed that the mass of a stellar black hole is $10\,M_\odot$. Increasing this mass to $18M_\odot$ decreases the expected merger rate by about 50%—higher mass black holes tend to have wider orbits. However, the larger chirp mass increases the signal to noise, and the distance to which such a merger can be observed increases by about 60% and the overall detection rate on Earth increases by about a factor of three. For $6\,M_\odot$ black holes, the detection rate decreases by a similar factor. For black-hole binaries with component masses $\gtrsim 12\,M_\odot$, the first generation of detectors will be more sensitive to the merger itself than to the inspiral phase that precedes it [38]. Since the strongest signal is expected from black-hole binaries with high-mass components, it is critically important to improve our understanding of the merger waveform. Even for lower-mass black holes (with $m_{bh} \gtrsim 10\,M_\odot$), the inspiral signal comes from an epoch when the holes are so close together that the post-Newtonian expansions used to calculate the wave forms are unreliable [39].

Acknowledgments This work was supported by NASA through Hubble Fellowship grant HF-01112.01-98A awarded (to SPZ) by the Space Telescope Science Institute, which is operated by the Association of Universities for Research in Astronomy, Inc., for NASA under contract NAS 5-26555, and by ATP grant NAG5-6964 (to SLWM). SPZ is grateful to Drexel University, Tokyo University and the University of Amsterdam (under Spinoza grant 0-08 to Edward P.J. van den Heuvel) for their hospitality.

REFERENCES

1. Narayan, R., Piran, T., Shemi, A. 1991, ApJ 379, L17
2. Phinney, E. S. 1991, ApJ 380, L17
3. Portegies Zwart, S. F., Spreeuw, F. 1996, A&A 312, L670
4. Tutukov, A. V., Yungelson, L. R. 1993, MNRAS 260, 675
5. Lipunov, V.M., Postnov, K.A., Prokhorov, M.E. 1997, NewA 2, L43
6. Kalogera, V. 2001, these proceedings
7. Portegies Zwart, S. F., Verbunt, F. 1996, A&A 309, 179
8. Portegies Zwart, S. F., Yungelson, L. R. 1998, A&A 332, 173
9. Bethe H. A., Brown G. E. 1999, ApJ 517, 318
10. Maeder, A. 1992, A&A 264, 105 (erratum 1993, A&A 268, 833)
11. Portegies Zwart, S. F. and Verbunt, F., Ergma, E. 1997, A&A 321, 207
12. Scalo, J. M. 1986, Fund. of Cosm. Phys., 11, 1
13. Cowley, A. P. 1992, ARA&A 30, 287
14. White, N. E., van Paradijs, J. A. 1996, ApJ 473, L25
15. Blaauw, A. 1962, BAN 15, 265
16. Massey P., Hunter D. A. 1998, ApJ, 493, 180
17. Nagata, T., Woodward, C. E., Shure, M., and Kobayashi, N. 1995, AJ, 109, 1676
18. Nagata, T., Woodward, C. E., Shure, M., Pipher, J.L., and Okuda, H. 1990, AJ, 109, 1676

19. Okuda, H., Shibai, H., Nakagawa, T., Matsuhara, H., Kobayashi, Y., Kaifu, N., Nagata, T., Gatley, I., and Geballe, T.R. 1990, ApJ, 351, 89
20. Rubenstein, E. P. 1997, PASP 109, 933
21. Rubenstein, E. P., Bailyn, C. 1999, ApJ 513, L33
22. Heggie, D. C. 1975, MNRAS 173, 729
23. Kulkarni, S. R., Hut, P., McMillan, S. L. W. 1993, Nature, 364, 421
24. Sigurdsson, S, Hernquist, L. 1993, Nature, 364, 423
25. Hut, P., McMillan, S., Romani, R. W. 1992, ApJ 389, 527
26. King, I. R. 1966, AJ, 71, 64
27. Portegies Zwart, S. F., McMillan, S. L. W., Makino, J., Hut, P. 2001, submitted to ApJ
28. Makino, J., Taiji, M., Ebisuzaki, T., Sugimoto, D. 1997, ApJ 480, 432
29. Djorgovski, S., Meylan, G. 1994, AJ 108, 1292
30. Takahashi, K., Portegies Zwart, S. F. 2000, ApJ, 535, 759
31. Peters, P. C., Mathews, J. 1963, Phys. Rev. D, 131, 345
32. van den Bergh, S. 1995, AJ 110, 2700
33. Spitzer, L. 1986, "Dynamical Evolution of Globular Clusters" (Princeton)
34. K. Thorne, private communication
35. Dutra, C.M, & Bica, E. 2000, A&A, in press (astro-ph/0006409)
36. Portegies Zwart, S. F., McMillan, S. L. W., Makino, J., Hut, P. 2001, ApJ, 546, L101
37. Figer, D. F., McLean, I. S., and Morris, M. 1999, ApJ, 514, 202
38. Flanagan, É. É., Hughes, S. A. 1998, Phys. Rev. D 57, 4535
39. Brady, P., Creighton, J., Thorne, K. 1998, Phys. Rev. D. 57, 1111

Hydrodynamics of Neutron Star Mergers

Joshua A. Faber and Frederic A. Rasio

Department of Physics, M.I.T., 77 Massachusetts Ave., Cambridge, MA 02139

Abstract. The final burst of gravitational radiation emitted by coalescing binary neutron stars carries direct information about the neutron star fluid, and, in particular, about the equation of state of nuclear matter at extreme densities. The final merger may also be accompanied by a detectable electromagnetic signal, such as a gamma-ray burst. In this paper, we summarize the results of theoretical work done over the past decade that has led to a detailed understanding of this hydrodynamic merger process for two neutron stars, and we discuss the prospects for the detection and physical interpretation of the gravity wave signals by ground-based interferometers such as LIGO. We also present results from our latest post-Newtonian SPH calculations of binary neutron star coalescence, using up to 10^6 SPH particles to compute with higher spatial resolution than ever before the merger of an initially irrotational system. We discuss the detectability of our calculated gravity wave signals based on power spectra.

INTRODUCTION

Coalescing binary neutron stars (NS) are among the most promising sources of gravitational radiation that should be detectable by future generations of gravity wave detectors. LIGO, VIRGO, GEO, and TAMA may ultimately not only serve to test the predictions of the theory of general relativity (GR), but could also yield important information on the interior structure of neutron stars, which cannot be obtained directly in any other way.

Compact binary orbits decay through energy losses to gravitational radiation. So long as the separation between the two NS is large, the binary inspiral is well described by a point-mass treatment, modified to take into account finite-size and relativistic effects, which act only as small corrections. At the end of the inspiral, however, the process is inherently hydrodynamic in nature. Large tidal interactions can drive the system into dynamical instability, at which point a quasi-equilibrium treatment of the binary breaks down, and a numerical treatment is required to accurately model the system.

Essentially all recent calculations agree on the basic picture that emerges for the final coalescence (see [1] and [2] for a complete list of references). As the dynamical stability limit is approached, typically at separations of $r = 3 - 4R_{NS}$, depending on the choice of parameters, the NS undergo a rapid radial plunge and merge in no more than a few rotation periods, much more quickly than would be predicted by a point-mass formula. In many cases, especially for binaries assumed to be initially synchronized, mass shedding sets in immediately after the stars first make contact. Material with a high specific angular momentum located in the outer regions of each NS is shed through the outer Lagrange points of the system, forming spiral arms that encircle the merger remnant left in the center as the NS cores merge. Eventually, the spiral arms also merge into a nearly axisymmetric torus around the dense inner core. For a stiff equation of state (EOS), a core with a significant ellipsoidal (triaxial) deformation can be maintained, and the configuration keeps radiating gravity waves well after the merger is completed. Softer EOS cannot support such a configuration stably, and any remnant produced will relax on a dynamical timescale toward a spheroidal (axisymmetric) configuration, which produces a negligible amount of gravity waves.

The previous statements assume that the merger remnant formed is stable against gravitational collapse to a black hole. Unfortunately, Newtonian calculations are incapable of demonstrating such an effect. Post-Newtonian (PN) simulations can produce configurations that are unstable against collapse, but they are inherently unreliable because in conditions of strong gravity the basic assumptions of the PN expansion break down. Early full GR calculations indicate that merger remnants may very well be stable against collapse, so long as the EOS is assumed to be stiff enough [3]. The mass of the remnant should be nearly twice the mass of a single NS, which is generally taken to be $M_{NS} \approx 1.4 - 1.5 M_{\odot}$, and thus well over the maximum mass for a single, nonrotating NS. However, the remnants formed in binary coalescence are very rapidly and differentially rotating, which can increase the maximum stable mass to a much larger value [4,5].

BINARY NS COALESCENCE CALCULATIONS

Nakamura and collaborators [6,7] were the first group to perform 3-D hydrodynamic calculations of binary NS coalescence, using an Eulerian grid-based code. Rasio and Shapiro [8] used the Lagrangian SPH (Smoothed Particle Hydrodynamics) method to calculate gravitational wave forms from binary coalescence events. Calculations performed since, using both Eulerian methods [10–14] and SPH [15–18] have focused on several aspects of the problem, including the effects of different initial spins, mass ratios, NS EOS, and NS masses. Some groups have incorporated treatments of the nuclear physics involved in the merger [12–14,16–18] in order to study coalescing NS binaries as possible gamma-ray burst sources, and as possible birthplaces for r-process elements.

Much of the early work on coalescing NS binaries assumed Newtonian gravity for simplicity. Later studies added a treatment of the radiation reaction, which is responsible for driving the system towards coalescence, either by adding a frictional drag term to model point-mass inspiral [15–18], or by an exact PN treatment [12–14]. In essence, 2.5PN radiation reaction terms (which scale like $1/c^5$)

are added onto a Newtonian framework, but all lower-order non-dissipative terms are ignored. Unlike adding a frictional drag term which dissipates energy according to the point-mass prediction, the lowest-order treatment of the radiation reaction allows for its effects to be included throughout the entire calculation, including the period after the merger remnant has formed. Unfortunately, however, Newtonian gravity is known to be a poor description of the physical problem at hand. Even NS with stiff EOS generate strong gravitational fields. During the final moments before merger, the velocities found in the system also become relativistic. Thus, the hydrodynamics of the actual coalescence can only be calculated properly by taking into account GR effects.

The Newtonian limit also fails to describe accurately the onset of dynamical instability. PN effects combine nonlinearly with finite-size fluid effects and this can dramatically increase the critical binary separation (and thus lower the frequency) at which dynamical instability sets in. Indeed, the quasi-equilibrium description applies so long as

$$
\left(\frac{dE}{dr}\right)_{equil} > \left(\frac{dE}{dt}\right)_{GW} \left(\frac{dr}{dt}\right)_{infall}^{-1},
\tag{1}
$$

where $E_{equil}(r)$ is the energy of an equilibrium binary configuration at a given separation r. Equilibrium sequences have been calculated for both synchronized and irrotational binaries in Newtonian gravity [19], PN gravity [20], and recently in full GR [21,22] (see also Baumgarte, in this volume). It is generally found that the energy of NS binary configurations reaches a minimum at some critical separation, defining the innermost stable circular orbit (ISCO). Relativistic terms move the ISCO to larger separations, and also reduce the slope of the energy curve just outside the ISCO. Thus, the assumption of quasi-equilibrium, which is used to set up the initial configuration of the binary system, breaks down at a much larger separation than a Newtonian calculation would predict. In addition, NS binaries will already have developed significant infall velocities as they pass through the ISCO calculated for systems in strict equilibrium.

Several groups have been working on full GR calculations of binary NS mergers, but only preliminary results have been reported so far [23]. Proving to be particularly difficult is the extraction of wave forms from the boundaries of large 3-D grids, since extending the grids into the true wave zone would be too expensive computationally. The middle ground between the Newtonian treatments and full GR lies in PN hydrodynamic calculations of binary mergers. The authors [1,24], as well as Ayal et al. [25], have constructed a PN SPH code, described below, for calculating binary mergers. While it too serves only as an approximation to the proper physics which must go into a realistic calculation, the results do provide insight into the relativistic effects that simple Newtonian intuition fails to handle correctly. Additionally, they should serve as valuable checks for future full GR calculations, lending confidence to wave form predictions, and indicating to some degree the difference between real relativistic effects and numerical instabilities.

POST-NEWTONIAN SPH

Our Post-Newtonian calculations use a formalism adapted from that of Blanchet, Damour, and Schaefer [26]. It includes all first-order (1PN) terms of GR, as well as the lowest-order radiation reaction terms (2.5PN). The latter are important because they provide the energy dissipation mechanism which drives the binary system toward dynamical instability and coalescence. Calculating various PN quantities requires the solution of seven additional Poisson equations for 1PN terms and an additional Poisson equation for the radiation reaction, all of which have compact support and thus can be solved by similar methods as the Newtonian gravitational potential. All hydrodynamic quantities in the formalism have been converted from a grid-based Eulerian approach to a particle-based Lagrangian approach, where particles represent distributions of matter in space defined by a spherically symmetric smoothing kernel (rather than point-like objects). Pressure forces and other interactions between particles are handled by summing over neighboring particles. Poisson equations are solved by translating particle-based source term quantities onto a grid, and solving via FFT-based convolution methods [24].

Unfortunately, the consistent use of physically realistic NS parameters is impossible in this PN formalism. The compactness of a NS is given by

$$\frac{GM}{Rc^2} = 0.20 \left(\frac{M}{1.5 M_\odot} \right) \left(\frac{10 \text{ km}}{R} \right), \tag{2}$$

and is typically assumed to fall near $GM/Rc^2 \approx 0.15$ for a reasonable choice of EOS. Since several of the coefficients of the 1PN terms can be quite large, especially for stiff EOS, we find that in many cases the 1PN corrections are larger than the Newtonian terms. Thus, we adapt the formalism, reducing the magnitude of the 1PN corrections by a factor of three (in effect decreasing the mass of the NS to $M_{1PN} = 0.5 M_\odot$), but treating all the radiation reaction terms, of significantly smaller magnitude, at full strength. In essence, we employ a different speed of light for 1PN and 2.5PN terms. By comparison with result including radiation reaction but no 1PN terms whatsoever, we believe we can extrapolate in a qualitative sense toward the proper physically realistic case. By comparison, self-consistent PN calculations performed for NS with artificially small masses [7,25] have the advantage that they can be directly compared to full GR calculations, for which there can be no separation of relativistic terms into separate orders. Unfortunately, these calculations also suffer from a drastic and completely artificial reduction of the radiation reaction effects, which scale like $M^{2.5}$. This produces a significant delay in the onset of dynamical instability past the point where it would be encountered for a physical set of parameters, and can lead to a qualitatively incorrect description of the subsequent merger.

IRROTATIONAL BINARY COALESCENCE

Our most detailed calculation performed to date uses $N = 500,000$ particles per NS, corresponding to the highest spatial resolution ever for a binary coalescence calculation. The spatial resolution (smoothing length) achieved in the central regions of the stars is $h \approx 0.03 R_{NS}$. The calculation was performed using an irrotational initial condition. This is generally thought to be the most realistic case since the viscous tidal locking timescale for two NS is expected to be considerably longer than the inspiral timescale [27]. Corotating (tidally locked) systems are motionless in a frame corotating with the binary, which allows for the use of relaxation techniques that can give a very accurate equilibrium initial condition (thus nearly completely eliminating spurious oscillations around equilibrium during the early phases of the dynamical evolution). Instead, for our irrotational calculation, we model the initial density and velocity profile of the NS as tidally stretched ellipsoids, with parameters drawn from the PN equlibrium calculations of Lombardi, Rasio, and Shapiro [20].

Since all NS in relativistic binary systems are expected to have masses that lie within a very narrow range [28], our calculation uses equal-mass NS. As the NS EOS is still poorly constrained, we choose a simple $\Gamma = 3$ polytropic EOS, i.e., the pressure is given in terms of the rest-mass density by $P = k\rho_*^3$. Stiff EOS, such as this one, are capable of maintaining a long-lived ellipsoidal deformation after the binary merger, with a gravity wave signal that persists on a timescale much longer than the merger timescale [9].

Particle plots showing the evolution of the equal-mass irrotational binary system are shown in Fig. 1. We see that immediately prior to merger a large tidal lag angle develops. The inner edge of each NS leads the axis connecting the centers of mass of the binary components, and the outer regions lag behind. This effect is seen even in Newtonian simulations, but is greatly enhanced by the addition of 1PN correction terms. When first contact is made, a long vortex sheet forms at the interface between the two stars. Unlike the case of synchronized binaries, we do not see significant mass shedding from the outer Lagrange points of the system. The rotational speed of particles on the outer half of each NS is reduced in the irrotational case with respect to the synchronized case, and such particles remain bound and form the outer regions of the eventual merger remnant. At $t = 25$ we do see some hint of mass shedding, but not via the mechanism described above. Particles which have travelled the length of the vortex sheet and retained significant velocities end up being shed from the leading edge of the vortex sheet. It is important to note, though, that the amount of mass shed is extremely small, much less than 1% of the total mass, and that the velocities of the particles ejected are not sufficient to escape the gravitational potential of the remnant. We thus expect them to form an extremely tenuous halo around the central core. At late times, we see the formation of a remnant containing essentially all the mass that was orignally present in the system. As we are using a stiff EOS, the ellipsoidal deformation of the remnant is relatively large, and persists for late times after the

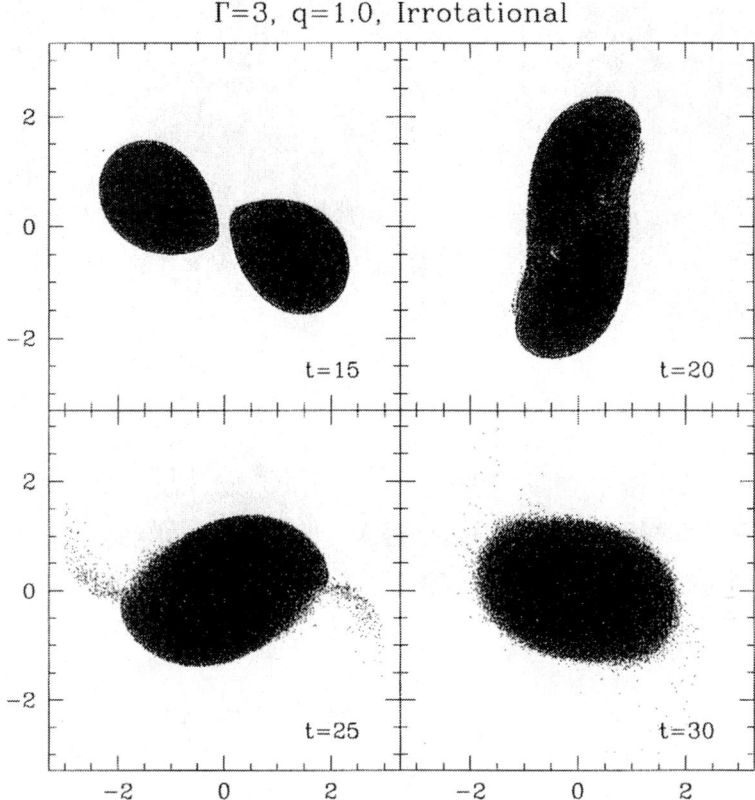

FIGURE 1. Final merger of two identical $\Gamma = 3$ polytropes with an irrotational initial condition. SPH particles are projected onto the equatorial plane of the binary. The orbital rotation is counterclockwise. Spatial coordinates are given in units of the NS radius R. Times are given in units of the dynamical timescale of the system, which here is $t_D = 0.07\text{ms} = 1$. The orbital rotation is in the counterclockwise direction.

merger.

Density contours and velocity profiles in the equatorial plane of the binary are shown in Fig. 2. Velocities are shown in a frame corotating with the material. We see that the initially counterstreaming surfaces of the NS produce a vortex sheet, which is Kelvin-Helmholtz unstable. Large vortices develop along the surface of contact, both in the center of the forming merger remnant and also at a separation which seems to be roughly consistent with the misalignment of the leading edges of each NS, or about $r = 0.5R$. As the merger proceeds, we see that the vortices re-

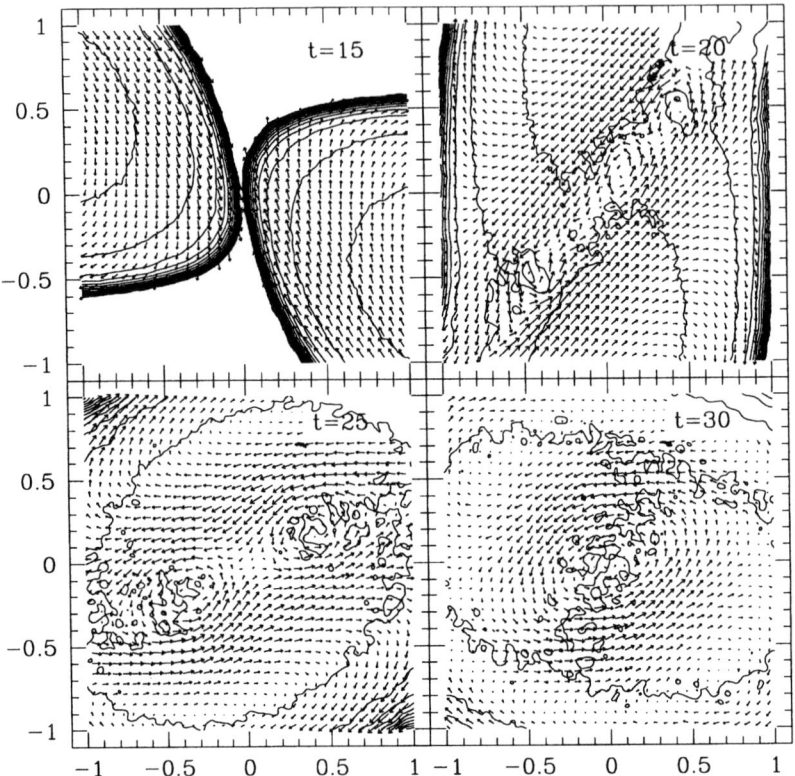

FIGURE 2. Density contours and velocities along the equatorial plane in the corotating frame of the binary, for the same times as in Fig. 1.

main coherent from $t = 20 - 25$, mixing material which was orignally located along the inner parts of each NS. All the while, the cores of the respective NS continue to inspiral toward the center of the merger remnant, until by $t = 30$ they have formed a single core, the vortices having merged together. This produces a characteristic differentially rotating pattern, with the center of the remnant spinning approximately twice as fast as the outer regions.

GRAVITY WAVE SIGNALS AND SPECTRA

We calculate the gravity wave signal for our mergers in the quadrupole approximation. The gravity wave strain h seen by an observer located a distance d from

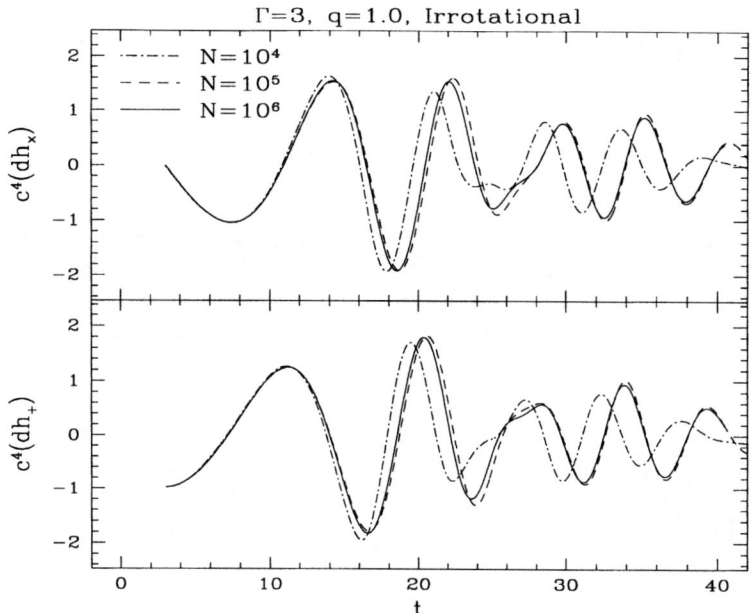

FIGURE 3. Gravity wave signals calculated for coalescences with the same initial parameters but different numerical resolutions. The solid, dashed, and dot-dashed lines correspond to runs with 10^6, 10^5, and 10^4 SPH particles, respectively.

the center of mass of the system along the rotation axis is given for the two polarizations by

$$c^4(dh_+) = \ddot{Q}_{xx} - \ddot{Q}_{yy} \tag{3}$$
$$c^4(dh_\times) = 2\ddot{Q}_{xy} \tag{4}$$

where $\ddot{Q}$, the second time derivative of the quadrupole moment tensor, is given in SPH terms by

$$\ddot{Q}_{ij} = \sum_b m_b(v_i^{(b)}v_j^{(b)} + x_i^{(b)}\dot{v}_j^{(b)} + x_j^{(b)}\dot{v}_i^{(b)}) \tag{5}$$

where the summation is taken over all particles in the calculation.

In Fig. 3, we show the gravity wave signals in both polarizations for the irrotational run described above, as well as for runs with $N = 50,000$ particles and $N = 5,000$ particles per NS. It is immediately apparent that the lowest resolution run shows significant discrepancies from the other two, which agree with each other quite well over the entire time history of the merger. This is a welcome result, given that the vortex sheet appearing at the contact surface is Kelvin-Helmholtz unstable on all wavelengths, including those much smaller than our numerical resolution. Calculations performed at different resolutions do show subtle differences in the exact location and size of the vortices. It is important to note, however, that it is the outer regions of the star, at lower density, that supply material to the vortex sheet. The high density cores of the two NS inspiral during the entire process, and provide the dominant component of the quadrupole moment and thus the gravity wave signal. The path traced out by the NS cores depends sensitively on gravitational forces and properties of the fluid, such as the EOS, but proves to be remarkably insensitive to the details of the flow in the "turbulent" boundary region. The conclusion to be drawn is that numerical convergence for a given set of initial conditions and physical assumptions is possible without requiring excessive computational resources, even for this difficult problem involving small-scale instabilities.

Because the gravity wave signals expected to be seen from binary NS mergers are extremely close to the sensitivity limits of ground-based interferometers, it is important to identify which features in the power spectrum of the gravitational radiation will yield the most information on the physical parameters of NS. Following the method of Zhuge et al. [15], we compute the gravity wave power spectrum per unit frequency interval as

$$\frac{dE_{GW}}{df} = \frac{c^3}{G}\frac{\pi}{2}(4\pi r^2)f^2 \left\langle |\tilde{h}_+(f)|^2 + |\tilde{h}_\times(f)|^2 \right\rangle. \tag{6}$$

Before calculating the power spectra from our simulations, we add a component representing a point-mass inspiral matched to the beginning of our hydrodynamic merger wave form. This produces a spectrum with $dE/df \propto f^{1/3}$ for point-mass inspiral at low frequencies. In essence, what we measure here is the number of orbits spent around a given frequency, weighted by the amplitude of the emission. We identify three frequencies of special interest. The frequency at which dynamical instability sets in is labeled f_{dyn}. The frequency at the peak of the gravity wave luminosity is labeled f_{peak}. Last, the characteristic frequency of gravity wave emission for the merger remnant at late times in the calculation is labeled f_{osc}.

In Fig. 4, we show the gravity wave power spectrum for a Newtonian calculation with a synchronized initial condition, which includes radiation reaction effects but contains no 1PN terms. At low frequencies, we see the power-law behavior from the point-mass inspiral, with no contribution whatsoever from our calculated signal. At the dynamical instability frequency, we see a slight decrease in the gravity wave power, since the rapid plunge causes the binary to sweep up faster through a range of characteristic frequencies. The gravity wave power shows a plateau near the peak emission frequency, when the effects of the rapid infall are balanced by the large increase in gravity wave amplitude. At higher frequencies, there is another slight dip in emission, followed by a sharp peak marking the oscillation frequecncy of the merger remnant. With more careful handling of the late-time behavior of the system, we expect the peak to remain prominent, but our calculation most likely overemphasizes the coherence of the signal.

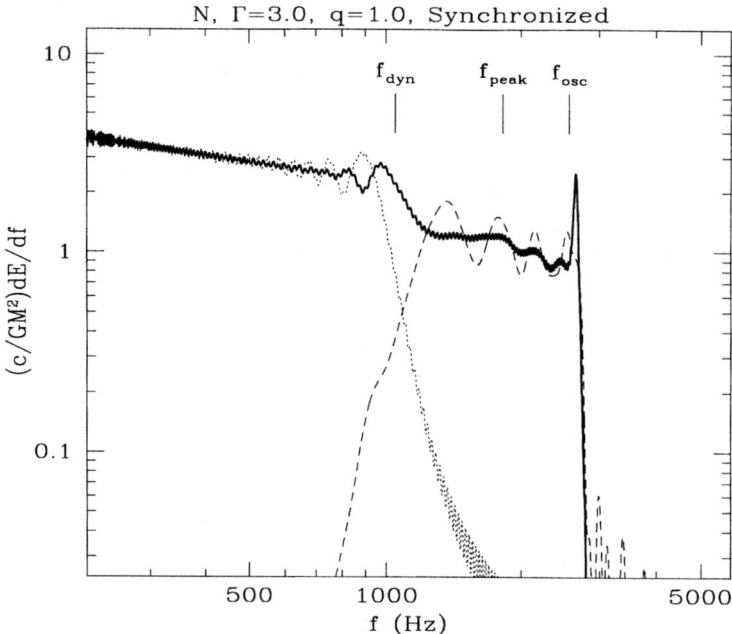

FIGURE 4. Power spectrum calculated from a Newtonian coalescence calculation. The dotted and dashed lines represent the contributions of the point-mass inspiral and our calculated gravity wave signal for the hydrodynamic merger, respectively. The solid line represents the total power.

A strikingly different power spectrum is obtained from our Post-Newtonian, irrotational merger, as shown in Fig. 5. The most significant difference between the two concerns the limit of dynamical instability. Newtonian gravity is strengthened by the addition of relativistic effects, which moves the dynamical stability limit to larger separations (and thus smaller frequencies). Additionally, the final plunge of the two NS is much faster, which significantly reduces the power in the region between f_{dyn} and f_{peak}. At f_{peak}, the power is smaller than in the Newtonian case, but the signal is much more sharply defined. Similarly, the oscillation of the remnant at late times leaves a well defined imprint on the power spectrum, but the amplitude of the peak is approximately an order of magnitude lower than what is found in a Newtonian calculation.

The frequencies of the two peaks seen in the spectrum, representing peak emission and the remnant oscillations, do give a strong clue to the nature of the NS EOS. While the frequency of peak oscillation is essentially the same in all our simulations, the width of the peak is seen to be strongly dependent on the EOS. The softer $\Gamma = 2$ EOS shows a broad peak of emission in the frequency range $f \sim 1500 - 3000$ Hz, whereas the stiffer $\Gamma = 3$ EOS calculations have a peak much more focused around $f = 1800 - 2200$ Hz, regardless of the initial spins. The remnant oscillations break

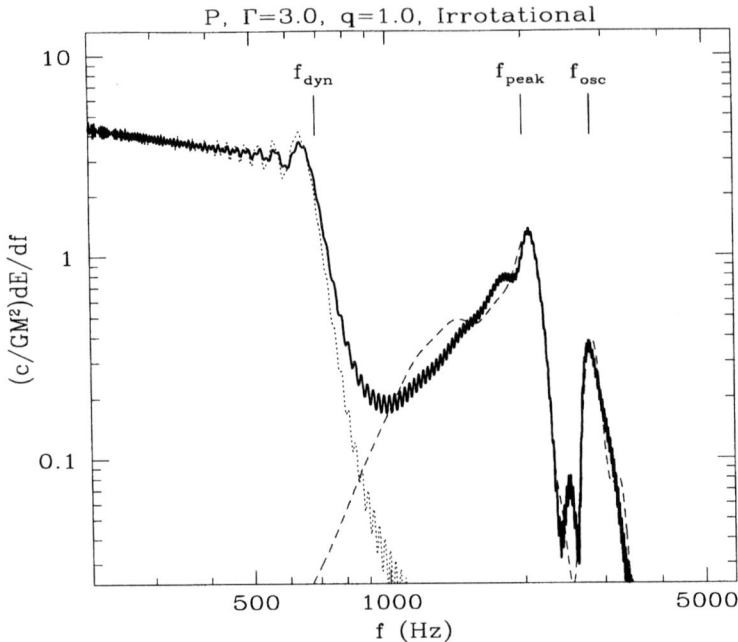

FIGURE 5. Power spectrum from a Post-Newtonian coalescence calculation. Conventions are as in the previous figure.

this degeneracy. The oscillation frequency for the irrotational run is almost 15% less than that of the synchronized run with the same choice of EOS. The stiffer EOS also results in a more rapid oscillation than the softer one, although the frequencies are relatively similar.

In general, several trends can be recognized regarding the strength of gravity wave emission from NS binary coalescence. The large dip at the dynamical instability limit is a general feature of PN calculations, regardless of the choice of initial spins or the EOS. It should be regarded as a consequence of the stronger gravity present in the PN systems. Additionally, PN simulations generally show similar amplitudes to their Newtonian counterparts near the characteristic frequency of peak emission, but significantly less power at higher frequencies, since the effect of strong gravity seems to be a quenching of gravity wave emission after the initial peak. Soft EOS generally show greatly reduced gravity wave emission at high frequencies, since they cannot support a stable, radiating, ellipsoidal configuration, and on relatively short timescales will produce a nearly spheroidal, non-radiating remnant. Finally, for binary systems with unequal-mass components, the magnitude of the gravity wave emission is strongly correlated with the mass ratio q [1,9]. Because the primary in such systems generally remains relatively undisturbed, whereas the secondary is tidally disrupted and accreted onto the primary, a large component of the matter essentially does not contribute to the gravity wave signal. Thus, even if NS masses do not typically lie within a narrow range, there should be a strong bias observationally toward detection of nearly equal-mass systems.

ACKNOWLEDGMENTS

This work was supported in part by NSF Grants AST-9618116 and PHY-0070918 and NASA ATP Grant NAG5-8460. F.A.R. was supported in part by an Alfred P. Sloan Research Fellowship. The computations were supported by the National Computational Science Alliance under grant AST980014N and utilized the NCSA SGI/CRAY Origin2000.

REFERENCES

1. Faber, J.A., Rasio, F.A., and Manor, J.B., *Phys. Rev. D*, accepted, gr-qc/9912097.
2. Rasio, F.A., and Shapiro, S.L., *Class. Quant. Grav.* **16**, R1-R29 (1999).
3. Shibata, M., and Uryu, K., *Phys. Rev. D* **61**, 064001 (2000).
4. Baumgarte, T.W., Shapiro, S.L., and Shibata, M., *Astrophys. J. Lett.* **528** L29-L32 (2000).
5. Rasio, F.A., "The Final Fate of Coalescing Binary Neutron Stars: Collapse to a Black Hole?" to appear in *Black Holes in Binaries and Galactic Nuclei*, edited by L. Kaper, E.P.J. van den Heuvel, and P.A. Woudt, ESO Press.
6. Oohara, K., and Nakamura, T., *Prog. Theor. Phys.* **82**, 535-554 (1989); *ibid.* **83**, 906-940 (1990); Nakamura, T. and Oohara, K., *ibid.* **82**, 1066-1083 (1989); *ibid.* **86**, 73-88 (1991).
7. Shibata, M., Oohara, K., and Nakamura, T., *Prog. Theor. Phys.* **88**, 1079-1095 (1992); *ibid.* **89**, 809-819 (1993).
8. Rasio, F.A., and Shapiro, S.L., *Astrophys. J.* **401**, 226-245 (1992); *ibid.* **438**, 887-903 (1995).
9. Rasio, F.A., and Shapiro, S.L., *Astrophys. J.* **432**, 242-261 (1994).
10. New, K.C.B., and Tohline, J.E., *Astrophys. J.* **490**, 311-327 (1997).
11. Swesty, F.D., Wang, E.Y.M., and Calder, A.C., *Astrophys. J.* **541**, 937-958 (2000).
12. Ruffert, M., Janka, H.-Th., and Schäfer, G., *Astron. Astrophys.* **311**, 532-566 (1996).
13. Ruffert, M., Janka, H.-Th., Takahashi, K., and Schäfer, G., *Astron. Astrophys.* **319**, 122-153 (1997).
14. Ruffert, M., Rampp, M., and Janka, H.-Th., *Astron. Astrophys.* **321**, 991-1006 (1997).
15. Zhuge, X., Centrella, J., and McMillan, S., *Phys. Rev. D* **50**, 6247-6261 (1994); *ibid.* **54**, 7261-7277 (1996).
16. Davies, M.B. et al., *Astrophys. J.* **431**, 742-753 (1994).
17. Rosswog, S. et al., *Astron. Astrophys.* **341**, 499-526 (1999).
18. Rosswog, S. et al., *Astron. Astrophys.* **360**, 171-184 (2000).

19. Lai, D., Rasio, F.A., and Shapiro, S.L., *Astrophys. J. Lett.* **406**, L63-L66 (1993); *Astrophys. J. Suppl.* **88**, 205-252 (1993); *Astrophys. J.* **420**, 811-829 (1994); *ibid.* **423**, 344-370 (1994); *ibid.* **437**, 742-769 (1994).

20. Lombardi, J.C., Rasio, F.A., and Shapiro, S.L., *Phys. Rev. D* **56**, 3416-3438 (1997).

21. Baumgarte, T.W. et al., *Phys. Rev. Lett.* **79** 1182-1185 (1997).

22. Bonazzola, S., Gourgoulhon, E., and Marck, J.-A., *Phys. Rev. Lett.* **82** 892-895 (1999).

23. Baumgarte, T.W., Hughes, S.A., and Shapiro, S.L., *Phys. Rev. D.* **60**, 087501 (1999); Shibata, M., *Phys. Rev. D* **60**, 104052 (1999).

24. Faber, J.A. and Rasio, F.A., *Phys Rev. D* **62**, 064012 (2000).

25. Ayal, S. et al., *Astrophys. J.*, accepted, astro-ph/9910154.

26. Blanchet, L., Damour, T., and Schäfer, G., *Mon. Not. Roy. Astron. Soc.* **242**, 289-305 (1990).

27. Bildsten, L., and Cutler, C., *Astrophys. J.* **400**, 175-180 (1992).

28. Thorsett, S.E., and Chakrabarty, D., *Astrophys. J.* **512**, 288-299 (1999).

Gamma-Ray Bursts and Gravitational Waves

Maximilian Ruffert* and H.-Thomas Janka[†]

*University of Edinburgh, EH9 3JZ, Scotland, U.K.,
[†]Max-Planck Institut für Astrophysik, D-85741 Garching, Germany

Abstract. Several currently popular models for the central engines of gamma-ray bursts are mentioned and some merits and problems discussed. Special emphasis is placed upon the strength of gravitational waves emitted. While merging compact binaries (involving neutron stars or a black hole) promise strong wave signals, collapsars and He-star mergers are more problematic in this respect. We present simulation results for neutron star binary and neutron star - black hole binary coalescences.

INTRODUCTION

Although bursts of gamma-rays are observed routinely at cosmological distances together with their accompanying afterglows through the whole spectral range down to radio waves, it is unknown whether these bursts are preceded by or simultaneous with a detectable burst of gravitational waves. Observationally, no correlations have been found between the currently operating bar detector experiments and the arrival times of observed gamma-ray bursts (GRBs) [1,5,13,24]. Theoretically, no definite predictions are possible yet, since the models for the central engine differ widely concerning this aspect: on the one hand, merging neutron star binaries or binaries including a black hole and a neutron star will certainly emit copious and ultimately detectable amounts of gravitational waves. On the other hand, mergers of less compact binaries involving white dwarfs (or similarly sized objects like the He-core of a star [26]) or the collapse of a collapsar are expected a priori to yield much less gravitational waves due to their slower velocities, lower densities and larger geometrical sizes.

Different geometric sizes, velocities and emission mechanisms of the various models produce different intrinsic timescales. A rough estimate of these follows [9,23,16]:

- orbital-dynamical of NS+NS and NS+BH: 1 ms
- neutrino cooled accretion torus around BH: 0.1 s
- collapsar (disk fed by star): 10 s
- Blandford-Znajek energy extraction from BH: 1000 s
- NS 'flywheel' (r-mode instability, B-field buoyancy): 1 s
- WD+NS, WD+BH, He-merger (accretion disk): 100 s

CP575, Astrophysical Sources for Ground-Based Gravitational Wave Detectors, edited by J. M. Centrella
© 2001 American Institute of Physics 0-7354-0014-8/01/$18.00

TABLE 1. NS+NS and BH+NS merger simulations

Model	Type	Masses $M_\odot$	Spin	t_{sim} ms	t_{ns} ms	d_{ns} km	M_{ns}^{min} $M_\odot$	ΔM_{ej} $M_\odot/100$	L_{GW}^{max} $10^4 \frac{foe}{s}$[a]	rh^{max} 10^4cm	E_{GW} foe
S64	NS+NS	1.2+1.2	solid	10	2.8	15	...	2.0	0.7	5.5	14
D64	NS+NS	1.2+1.8	solid	13	7.3	15	...	3.8	0.4	5.5	13
C64	NS+NS	1.6+1.6	anti	10	3.7	15	...	0.0085	1.2	6.0	23
A64	NS+NS	1.6+1.6	none	10	1.7	15	...	0.23	2.1	8.6	52
B64	NS+NS	1.6+1.6	solid	10	1.6	15	...	2.4	2.1	8.9	37
C2.5	BH+NS	2.5+1.6	anti	10	2.6	11	0.78	0.01	2.3	9.9	32
A2.5	BH+NS	2.5+1.6	none	10	4.3	18	0.78	0.03	2.0	9.9	50
B2.5	BH+NS	2.5+1.6	solid	10	6.0	23	0.78	0.2	2.1	9.6	61
C5	BH+NS	5.0+1.6	anti	15	9.1	76	0.40	2.5	3.9	13.0	50
A5	BH+NS	5.0+1.6	none	20	16.3	65	0.52	2.5	3.2	14.8	102
B5	BH+NS	5.0+1.6	solid	15	10.8	79	0.50	5.6	3.4	14.5	95
C10	BH+NS	10.0+1.6	anti	10	8.0	96	0.65	2.2	7.1	21.9	123
A10	BH+NS	10.0+1.6	none	10	9.3	95	0.60	3.2	6.9	26.2	168
B10	BH+NS	10.0+1.6	solid	10	5.1	97	0.65	10.0	7.3	26.2	163
A	PBH+NS	2.5+1.6	none	15	3	41	1.6	5.5	5.3	19.5	99
B	PBH+NS	2.5+1.6	solid	22	17	65	0.45	> 9	0.9	9.5	29
AR	PBH+NS	2.5+1.6	none	25	17	45	1.6	10	4.5	20.5	230

[a] 1 foe = 10^{51} erg (**fifty one erg**).

The undisputed bimodality of the GRB duration distribution, short bursts are 0.01s – 3s and long bursts take 3s to minutes, is not naturally explained by any of these central engines. Emission from the 'quick' engines (e.g. NS+BH) will need to be stretched by the fireball to reproduce the long bursts, while vice versa, the 'slow' engines' (e.g. collapsar) emission are too long for short bursts. Also the knowledge about occurrence rates for these central engines varies considerably.

Mergers of binary neutron stars (NS+NS) and neutron stars with companion black holes (NS+BH) are known to occur, because the emission of gravitational waves leads to a shrinking orbital separation and does not allow the systems to live forever. The Hulse-Taylor binary pulsar is the most famous example. Eventually, after 10^8–10^{10} years of evolution, these double stars will approach the final catastrophic plunge and will become powerful sources of gravitational radiation. This makes them one of the most promising targets for the upcoming huge interferometer experiments, being currently under construction in Europe (GEO600, VIRGO), Japan (TAMA) and the United States (LIGO). The expected rate of merging events is of the order of 10^{-5} per year per galaxy, estimated, with significant uncertainties, from population synthesis models and empirical data (see, e.g., [10,6]) and references therein). Templates are urgently needed to be able to extract the expected signals from background noise and to interpret the meaning of possible measurements.

Merging compact objects were suggested as possible sources of gamma-ray bursts [3,14], but so far observations cannot provide convincing arguments for this hypothesis. In fact, the exciting detection of afterglows at other wavelengths of the electromagnetic spectrum by the Dutch-Italian *BeppoSAX* satellite and the discovery of lines in the afterglow spectra support the association of gamma-ray bursts (GRBs) with explosions of massive stars in star-forming regions of galaxies at high redshifts (e.g. [11]). However, up to now afterglows were discovered only for GRBs with durations longer than a few seconds. Similar observations for the class of short, hard bursts ($\Delta t_{GRB} \lesssim 2$ seconds) are still missing, and we have neither a proof of their origin from cosmological distances, nor any hint on the astrophysical object they are produced by. NS+NS and NS+BH mergers are therefore still a viable candidate for at least the subclass of short-duration bursts, in particular because such collisions of compact objects must occur at interesting rates, can release huge amounts of energy, and the energy is set free in a volume small enough to account for intensity fluctuations on a millisecond timescale. The ultimate proof of such an association would certainly be the coincidence of a GRB with a characteristic gravitational wave signal.

NUMERICAL MODELS

Doing hydrodynamic simulations of the merging of compact stars is a challenging problem. Besides three-dimensional hydrodynamics, preferably including general relativity, the microphysics within the neutron stars is very complex. A nuclear equation of state (EoS) has to be used, neutrino physics may be important when the neutron stars heat up, magnetic fields might cause interesting effects, etc. Also high numerical resolution is required to account for the steep density gradient at the neutron star surface, and a large computational volume is needed if the ejection of matter is to be followed.

Simplifying approximations are possible if different aspects are the main focus of interest. For example, the surface layers are unimportant and a simple ideal gas EoS ($P = (\gamma - 1)\varepsilon$) and polytropic structure of the neutron stars can be used for parametric studies, when the gravitational wave emission is to be calculated. This is an absolutely unacceptable simplification, however, if mass ejection and nucleosynthesis is to be investigated. For the latter problem as well as for the GRB topic, on the other hand, it is not clear whether general relativistic physics is essential, although, of course, it may be necessary to obtain quantitatively meaningful results. In particular Newtonian models cannot answer the question whether and when a black hole is going to form when two neutron stars merge.

Remarkable progress has been achieved during the past years. General relativistic simulations are within reach now, at least for simple input physics and for certain phases of the evolution of the binary systems, especially during the inspiraling phase [4] and for the first stage of the final dynamical plunge [22,25] although the simulations become problematic once an apparent horizon begins to form. On the other

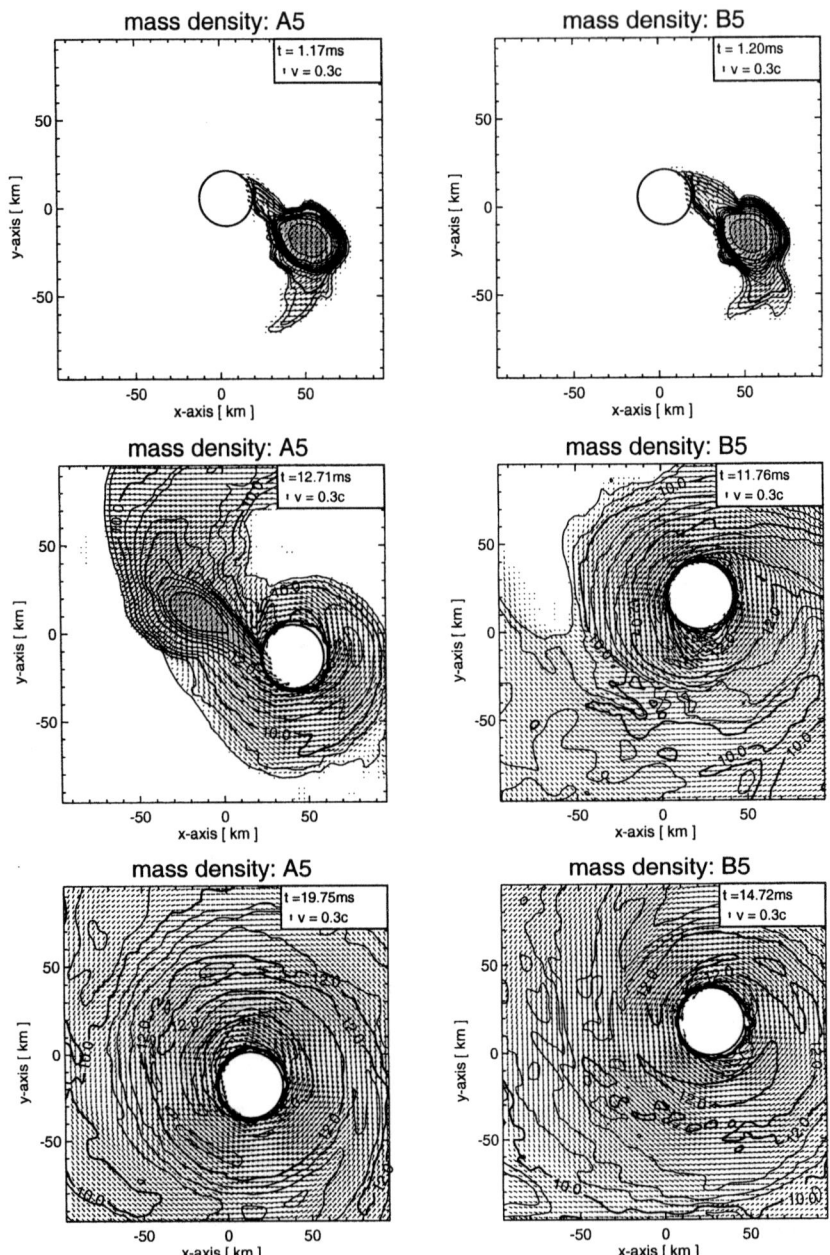

FIGURE 1. Density contours in the orbital plane for the BH+NS merger Models A5 (left column) and B5 (right column). The density (in g/cm^3) is shown logarithmically, the contours being spaced with intervals of 0.5 dex. The arrows indicate the velocity field. The time is given in the upper right corner of the plots.

hand, first attempts have been made to add detailed microphysics, i.e., a physical nuclear EoS and neutrino processes, into Newtonian models [17–21]. Hydrodynamics results were used in post-processing steps to draw conclusions on implications for GRB scenarios [8,9,20] and to evaluate the density-temperature trajectories of ejected matter for nucleosynthesis processes. The results are not finally conclusive, because the simulations suffer from insufficient numerical resolution and associated numerical artifacts.

The usual two main questions to be addressed are
(a) how do the gravitational waveforms and luminosities depend on numerical aspects, and (b) how do the gravitational waveforms and luminosities depend on physical parameters? Numerical aspects cover simple numbers like the number of zones (i.e. resolution), the depth of refinement and initial distance of the binary, but also more tricky approximations like use of a phenomenologic Paczyński–Wiita pseudo-potential [15] instead of the correct general relativistic potential, etc. Genuinely physical free parameters are e.g. the spin of the merging neutron stars and the mass of the black hole, which need to be varied in the simulations in order to ascertain what features of the waveforms are generic. We are well aware that due to the lack of relativistic effects in our code, the actual gravitational waves will surely look different quantitatively (but hopefully not qualitatively) from the ones we obtain and present here. However it is important to first settle the above questions for simplified scenarios before embarking on investigations on as yet unknown complex interactions.

In summary, the numerical model that we implemented contains the following features :

- three-dimensional hydrodynamics, Piecewise Parabolic Method, non-relativistic (except gravitational waves, cf. below), explicit;
- nested Cartesian grids, 4 levels deep, resolution of each grid: 32^3 and 64^3;
- self-gravity calculated by fast Fourier transform; Newtonian potential for gaseous material on grid;
- black hole modelled as vacuum sphere with Newtonian or Paczyński–Wiita pseudo-potential [15].
- realistic equation-of-state by Lattimer & Swesty [12] in tabular form;
- gravitational wave emission and backreaction with Blanchet, Damour and Schäfer formalism [2];
- neutrino emission by leakage scheme, ν_e, $\bar{\nu}_e$, $\nu_x = (\nu_\mu, \bar{\nu}_\mu, \nu_\tau, \bar{\nu}_\tau)$;
- neutrino–antineutrino annihilation is done as post-processing:
- various (baryonic) masses for the NSs: 1.2, 1.6, 1.8 $M_\odot$;
- rigid rotation with various spins of the NSs: 0, $\pm\Omega_{orbit}$;
- computed evolution time: 10–20 ms, initial distance $\approx$ 4 radii.

Table 1 lists initial conditions, properties and some results of a variety of simulations. 'PBH' denotes a BH modeled with a Paczyński–Wiita pseudopotential while the others are all purely Newtonian. 'anti' describes models in which the spin of

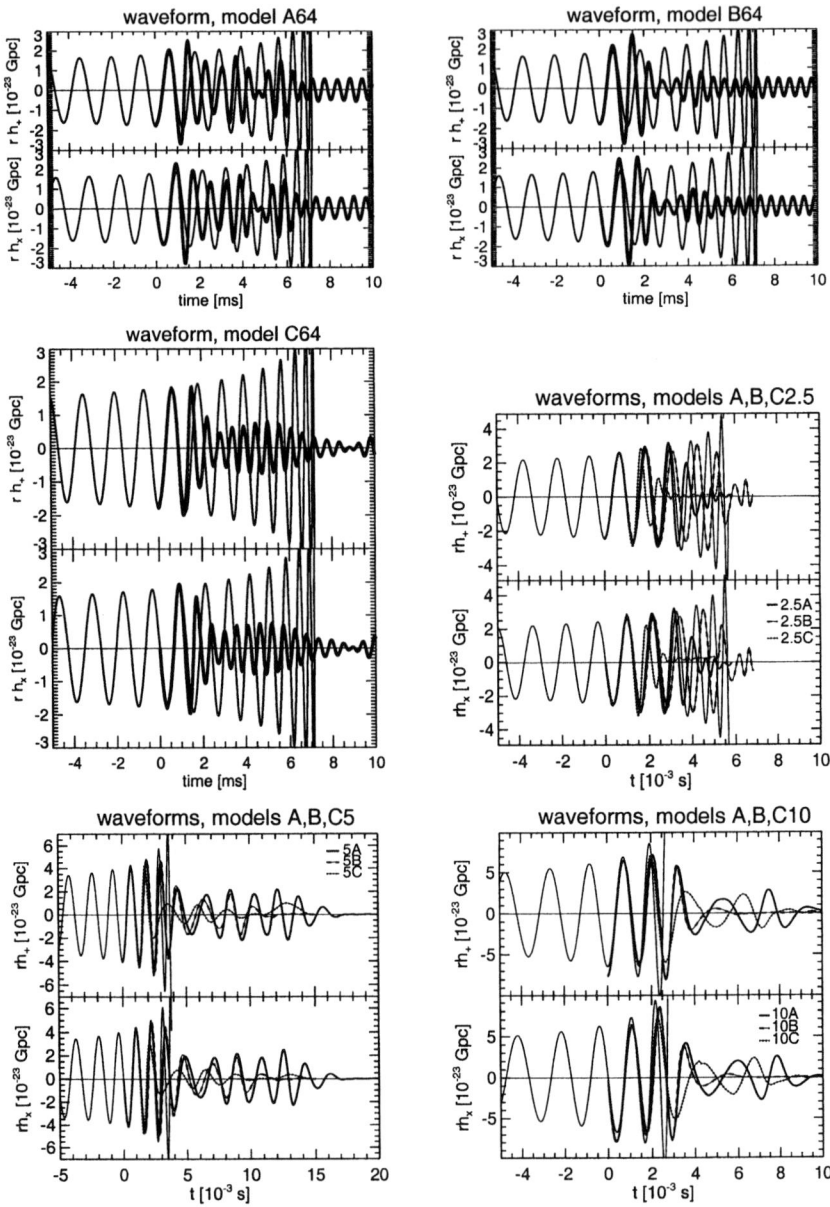

FIGURE 2. Gravitational waveforms for NS+NS merger Models A64, B64 and C64 and all BH+NS merger Models listed in Table 1. Time is measured in ms from the start of the simulation. The thin solid lines correspond to the chirp signal of point masses.

the NS is opposite to the orbit, while 'solid' means synchronously rotating stars. t_{sim} is the duration of the simulation, t_{ns}, d_{ns} and M_{ns}^{min} denote the time just before the NS gets completely tidally shredded by the BH, and the NS's distance and mass at this time, respectively. ΔM_{ej} is the mass ejected out of the system, while the gravitational wave quantities are listed as L_{GW}^{max}, rh^{max} and E_{GW}, which are, respectively, the maximal luminosity, the maximal amplitude and the total energy emitted.

SIMULATIONS

We present a small cross section of the numerical results obtained from our simulations. Fig. 1 shows the evolution of density in the orbital plane for the BH+NS mergerModels A5 (left column) and B5 (right column). Imparting a spin to the neutron star strongly alters the evolution: depending on the direction of the spin with respect to the orbit, coalescence can either be accelerated or delayed with respect to the case in which the NS does not have any intrinsic spin. This is, of course, reflected in the gravitational wave forms emitted (cf. Fig. 2: the ringdown of model A5 takes much longer than that of model B5.

Similar striking differences are visible for the NS+NS models, A64, B64 and C64 ((cf. Fig. 2), with the anti-spin model C64 displaying strong wave beating effects, while the corotating model B64 shows a more constant amplitude sin-wave ringdown. These differences will be crucial to deduce properties of the binary system from gravitational wave measurements. This shows only one example of the rich information present in the gravitational waves.

The right panel of Fig. 3 reinforces what has been said even for simulations including a Paczyński–Wiita pseudopotential for the black hole: While the irrotational case A (NS without spin) settles down quickly, i.e. the NS gets absorbed quickly and efficiently by the BH, the neutron star in model B goes through several mass transfer cycles before being completely disrupted. In this intermediate phase (5ms-18ms) gravitational waves continue to be emitted copiously from the very compact binary. Model AR shown in the left panel of Fig. 3 is prectically identical to model A, except for starting the binary off at a larger initial distance. This panel compares the luminosities of gravitational waves with neutrino emission: while the waves shut off very effectively once all matter is absorbed by the BH, the disk remaining around the black hole slowly heats up and emits neutrinos at an ever higher rate eventually exceeding the gravitational wave luminosity. During this short time, the total neutrino energy involved is still negligible compared to the gravitational waves, but the neutrinos continue to be emitted for a long as the disk survives, possibly as long as 1s.

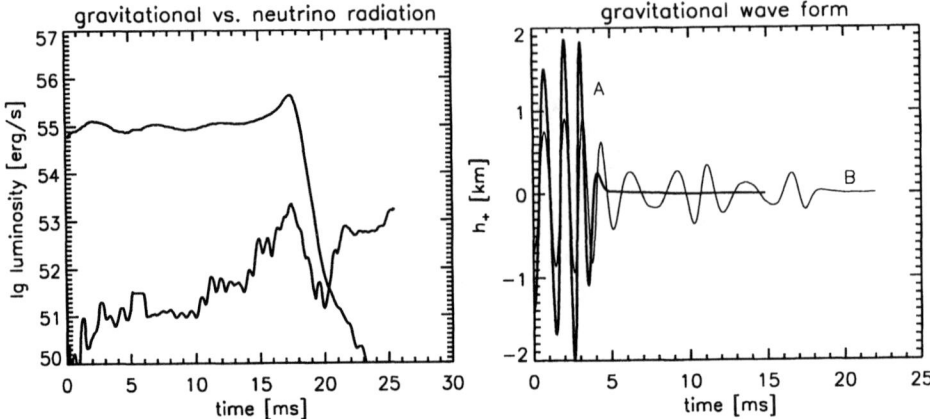

FIGURE 3. Left panel: total luminosities of gravitational and neutrino radiation for BH-+NS model AR. Right panel: gravitational waveforms for BH+NS merger Models A and B. Time is measured in ms from the start of the simulation.

Acknowledgements

MR's travel expenses were refunded by Drexel University. MR is grateful for support by a PPARC Advanced Fellowship. HTJ acknowledges support by the DFG on grant "SFB 375 für Astro-Teilchenphysik".

REFERENCES

1. Astone P., Barbiellini G., Bassan M., et al, *A&ASS* **138**, 603 (1999).
2. Blanchet L., Damour T., Schäfer G., *MNRAS*, **242**, 289 (1990).
3. Blinnikov S.I., Novikov I.D., Perevodchikova T.V., Polnarev A.G., *Sov. Astron. Lett.*, **10**, 177 (1984).
4. Duez M.D., Baumgarte T.W., Shapiro S.L., gr-qc/0009064 (2000).
5. Finn L.S., Mohanty S.D., Romano J.D., gr-qc/9903101 (1999).
6. Fryer C.L., Woosley S.E., Hartmann D.H., *ApJ* **526** 152 (1999)
7. Fryer C.L., Woosley S.E., Herant M., Davies M.B., *ApJ*, **520**, 650 (1999)
8. Janka H.-Th., Ruffert M., *A&A*, **307**, L33 (1996).
9. Janka H.-Th., Eberl Th., Ruffert M., Fryer C., *A&A*, **527**, L39, (1999).
10. Kalogera V., Lorimer D.R., *ApJ*, **530**, 890 (2000).
11. Klose S., astro-ph/0001008 (2000).
12. Lattimer J.M., Swesty F.D., *Nucl. Phys.*, **A535**, 331 (1991).
13. Murphy M.T., Webb J.K., Heng I.S., *MNRAS* **316**, 657 (2000).
14. Paczyński B., *ApJ*, **308**, L43 (1986).
15. Paczyński B., Wiita P.J., *A&A*, **88**, 23 (1980).
16. Popham R., Woosley S.E., Fryer C., 1998, *ApJ*, **518**, 356 (1999).
17. Rosswog S., Davies M.B., Thielemann F.-K., Piran T., *A&A*, **360**, 171 (2000).

18. Ruffert M., Janka H.-Th., Schäfer G., *A&A*, **311**, 532 (1996).
19. Ruffert M., Janka H.-Th., Takahashi K., Schäfer G., *A&A*, **319**, 122 (1997).
20. Ruffert M., Janka H.-Th., *A&A*, **338**, 535 (1998).
21. Ruffert M., Janka H.-Th., *A&A*, **344**, 573 (1999).
22. Shibata M., K. Uryū, astro-ph/9911058 (1999)
23. Spruit H.C., *A&A*, **341**, L1 (1999).
24. Tricarico P., Ortolan A., Solaroli A., et al, gr-qc/0101022 (2001)
25. Uryū K., Shibata M., Eriguchi Y., gr-qc/0007042 (2000)
26. Zhang W., Fryer C.L., astro-ph/0011236 (2000).

Simulations of black hole–neutron star binary coalescence

William H. Lee

Instituto de Astronomía, UNAM
Apdo. Postal 70-264, Cd. Universitaria México D.F. 04510 MEXICO
wlee@astroscu.unam.mx

Abstract.
We show the results of dynamical simulations of the coalescence of black hole–neutron star binaries. We use a Newtonian Smooth Particle Hydrodynamics code, and include the effects of gravitational radiation back reaction with the quadrupole approximation for point masses, and compute the gravitational radiation waveforms. We assume a polytropic equation of state determines the structure of the neutron star in equilibrium, and use an ideal gas law to follow the dynamical evolution. Three main parameters are explored: (i) The distribution of angular momentum in the system in the initial configuration, namely tidally locked systems vs. irrotational binaries; (ii) The stiffness of the equation of state through the value of the adiabatic index Γ (ranging from $\Gamma = 5/3$ to $\Gamma = 3$); (iii) The initial mass ratio $q = M_{NS}/M_{BH}$. We find that it is the value of Γ that determines how the coalescence takes place, with immediate and complete tidal disruption for $\Gamma \leq 2$, while the core of the neutron star survives and stays in orbit around the black hole for $\Gamma = 3$. This result is largely independent of the initial mass ratio and spin configuration, and is reflected directly in the gravitational radiation signal. For a wide range of mass ratios, massive accretion disks are formed ($M_{disk} \approx 0.2 M_\odot$), with baryon–free regions that could possibly give rise to gamma ray bursts.

I INTRODUCTION

The emission of gravitational waves in a binary system will eventually drive the system to coalesce through angular momentum losses, if the decay time is less than the Hubble time. The observed decay for known binary neutron star systems such as PSR 1913+16 matches the prediction of general relativity to high accuracy (Stairs et al. 1998). There are still no observations of black hole–neutron star systems, but they are believed to exist, with corresponding coalescence rates that are comparable to those of double NS systems (see Kalogera & Belczyński, these proceedings).

These systems are candidates for detection by gravitational wave detectors such as LIGO and VIRGO. Although the final coalescence signal will probably be out

CP575, *Astrophysical Sources for Ground-Based Gravitational Wave Detectors*, edited by J. M. Centrella
© 2001 American Institute of Physics 0-7354-0014-8/01/$18.00

of the frequency range of the first observatories, the inspiral phase, during which the stars can be thought of as point masses, will certainly be observable in this respect. For the final coalescence waveform, modeling the hydrodynamics in the system becomes an important issue, as it can affect its evolution in a significant manner. For example, Newtonian tidal effects due to the finite size of the stars, can alone de–stabilize the orbit and make it decay on a dynamical timescale (see Lai, Rasio & Shapiro 1993a).

We have studied the dynamical interactions in close black hole–neutron star binary systems previously for a variety of initial configurations (Lee & Kluźniak 1995, 1999a,b, hereafter papers I & II respectively; Lee 2000, hereafter paper III), and present here an overview of the results these simulations have produced. The results are not only relevant for the production of gravitational waves, but also for the progenitor systems of gamma ray bursts (GRBs, Kluźniak & Lee 1998; Janka et al. 1999), and the production of heavy elements through r–process nucleosynthesis (Lattimer & Schramm 1974, 1976; Symbalisty & Schramm 1982).

II NUMERICAL METHOD AND INITIAL CONDITIONS

A Numerical method

All the computations presented here have been carried out using the Smooth Particle Hydrodynamics (SPH) method. This is a Lagrangian technique, originally developed by Lucy (1977) and Gingold & Monaghan (1977) and is ideally suited for the study of complicated flows in three dimensions, when no assumptions about symmetry in the system are made, and where there are large volumes that are basically devoid of matter. An excellent review has been given by Monaghan (1992). The code is essentially Newtonian, and makes use of a binary tree structure to find hydrodynamical forces and carry out the gravitational force computation, with a multipole expansion to quadrupole order. The viscosity is artificial, its purpose being the modeling of shocks and avoiding the interpenetration of SPH particles. We have settled on the form of Balsara (1995) for our latest work, since it minimizes the effects of shear viscosity on the evolution of the system. This is of particular importance for this work, since massive accretion disks are often formed around the black hole after the initial dynamical encounter.

The black hole is modeled as a Newtonian point mass, producing a potential

$$\Phi = -GM_{BH}/|\vec{r} - \vec{r}_{BH}| \tag{1}$$

at position $\vec{r}$. To model the horizon, an absorbing boundary is placed at the Schwarzschild radius $r_{Sch} = 2GM_{BH}/c^2$. Any particle crossing this boundary is removed from the simulation. Its mass is added to that of the black hole, and the latter's position and momentum are adjusted so as to ensure the conservation of total mass and total linear momentum in the system. The simulations shown here have used between 8,000 and 40,000 SPH particles initially. Since accretion onto the black hole entails a loss of particles, this number decreases during the simulation.

In most of the simulations presented here we have included a back reaction term to mock the effect that the emission of gravitational waves has on the evolution of the system, by draining angular momentum during the orbital evolution. This acceleration is calculated in the quadrupole approximation, assuming the two components are point masses (see e.g. Zhuge, Centrella & McMillan 1996; Davies et al. 1994; Rosswog et al. 1999). This term in the equations of motion is switched off when (and if) complete tidal disruption occurs, or if the mass of the secondary (neutron star) core drops below a certain limit (usually one tenth of the initial neutron star mass), in the cases when the star is not completely shredded by tidal forces (details of the implementation of the back reaction force can be found in Paper III).

The gravitational wave emission is calculated in the quadrupole approximation, by adding the contribution from the fluid as a whole to that of the black hole as a point mass. This gives the waveforms directly from the second derivatives of the inertia tensor. To obtain the luminosity an additional (numerical) derivative is required. For the calculation of power spectra, one can attach a point–mass inspiral signal to the coalescence waveform at earlier times, that matches smoothly at the time the dynamical simulation is started.

B Initial Conditions

To perform dynamical simulations, we first construct a neutron star in hydrostatic equilibrium. The equation of state is taken to be that of a polytrope, with $P = K\rho^\Gamma$ (K and Γ are taken constant throughout the star). We place N SPH particles on a cubic lattice, with masses proportional to the Lane–Emden density at the corresponding radius. This ensures that the spatial resolution is approximately constant, and it helps to model the edge of the star more accurately, since that is where the density gradient is often the largest, for the values of the adiabatic index Γ that we consider. In all the calculations presented here, the neutron star has mass $M_{NS} = 1.4M_\odot$ and radius $R_{NS} = 13.4$ km. For each value of the index $\Gamma = 3; 2.5; 2; 5/3$ we find the value of K that ensures this mass and radius. After placing the particles on the cubic lattice, the star is relaxed in an inertial reference frame, with an artificial damping term in the equations of motion, keeping the specific entropy constant. After relaxation, all our spherical stars satisfy the virial theorem to within one part in 10^3. For the following, distances and masses are measured in units of R_{NS} and M_{NS}, so that time and density are measured in units of $t = 1.146 \times 10^{-4} s (R/R_{NS})^{3/2} (M_{NS}/1.4M_\odot)^{-1/2}$ and $\rho = 1.14 \times 10^{18} kg\ m^{-3} (R/R_{NS})^{-3} (M_{NS}/1.4M_\odot)$

The choice of a polytropic equation of state (instead of a physical equation of state such as that of Lattimer & Swesty 1991) was made in order to explore what effect the compressibility of the fluid has on the global evolution of the system, and on the gravitational wave signal. For polytropes, the mass–radius relationship is $R \propto M^{(\Gamma-2)/(3\Gamma-4)}$. So this means that for $\Gamma > 2$, the neutron star will respond to mass loss by shrinking, while for $\Gamma < 2$ it will expand. A star with $\Gamma = 2$ has a radius that is independent of its mass. As we will show below, this has a crucial effect on the outcome of the coalescence process.

The next step in constructing the initial conditions depends on the spin of the star. Typically, the binary separation is only a few stelar radii when we begin our dynamical calculations, and so the tidal deformation of the neutron star is quite large. We consider two extreme cases of angular distribution in the system.

In the first, the star is tidally locked, so that the same side of the star always faces the black hole companion. This initial condition is easy to set up, since the system is in a state of rigid rotation. Thus we can view the binary in the co–rotating frame and neglect Coriolis forces (since we are interested in an equilibrium configuration), with an artifical damping term in the equations of motion and wait for it to relax to a static configuration. While this relaxation procedure is carried out, the orbital velocity of the co–rotating frame is continously adjusted so that the force on the center of mass of the star is exactly balanced by the centrifugal acceleration.

The second type of initial condition we consider is that of an irrotational binary. In this case, the star has essentially zero spin when viewed from an external, inertial reference frame. This is more complicated to set up, and we have used the method of Lai, Rasio & Shapiro (1993b) for our calculations. They developed a variational method to obtain a solution to this problem by approximating the star as a tri–axial ellipsoid (in this case an irrotational Roche–Riemann ellipsoid). It still experiences tidal deformations, but the shape of the star is fixed in the co–rotating frame, while internal motions with zero circulation take place in its interior.

Realistically, the first kind of initial condition was shown to be nearly impossible by Kochanek (1992) and Bildsten & Cutler (1992), because the viscosity inside neutron stars is not large enough to maintain synchronization during the inspiral phase. Essentially, the stars will coalesce with whatever spin configurations they have when inspiral begins.

In either case, the fact that the stars are *not* point masses has a direct impact on the evolution of the system, and on its configuration immediatly before coalescence. We show in Figure 1 plots of total angular momentum J in the system as a function of orbital separation for several systems. It is clear that there are important deviations from Keplerian point–mass behavior, and even from the result obtained by treating the stars as rigid spheres. The turning points in the curves show the presence of a dynamical instability, which, once reached, can drive orbital decay on a dynamical timescale (for many of the parameters in the runs shown here, this can be as fast as the decay due to the emission of gravitational waves). It is crucial to model the hydrodynamics in these systems if one is to extract information from the gravitational wave signal produced during coalescence.

We show in Table 1 the initial parameters for several of the dynamical runs we

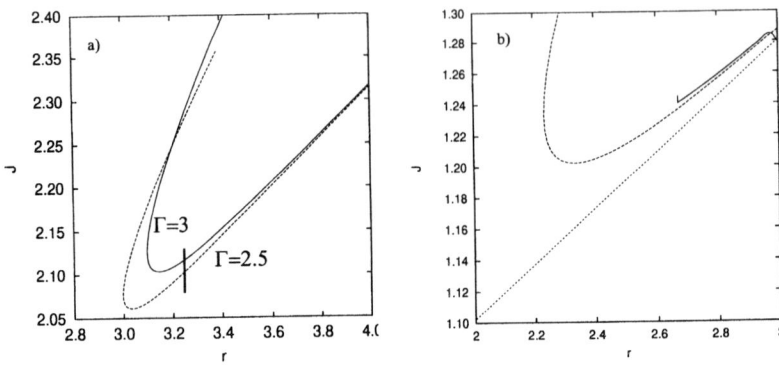

FIGURE 1. Total angular momentum J as a function of binary separation r for various black hole–neutron star binaries. (a) Irrotational binaries with $q = 0.5$ and $\Gamma = 3$ (solid line) and $\Gamma = 2.5$ (dashed line). The thick black vertical line marks the separation used to start dynamical calculations. (b) Tidally locked binary with $\Gamma = 5/3$ and $q = 1$. The solid line is the result of an SPH relaxation sequence, the dashed line results from approximating the neutron star as a compressible tri–axial ellipsoid, and the dotted line from assuming it is a rigid sphere.

have performed. We include the value of the adiabatic index, the spin configuration, the mass ratio, the initial separation and the number of particles initially used to model the neutron star.

III   RESULTS

A   Systems with a stiff equation of state

For systems with stiff equations of state ($\Gamma = 3$ and $\Gamma = 2.5$) the star will respond to mass loss by shrinking slightly, as mentioned above, and tidal effects are more

TABLE 1. Initial set of parameters for dynamical runs.

Run	Γ	Spin[a]	$q = M_{NS}/M_{BH}$	r_i/R_{NS}	N	Reference
AL1.0	3.0	L	1.00	2.78	16,944	Paper I
AL0.31	3.0	L	0.31	3.76	8,121	Paper I
AI0.5	3.0	I	0.50	3.25	38,352	Paper III
AI0.31	3.0	I	0.31	3.76	38,352	Paper III
BI0.31	2.5	I	0.31	3.70	37,752	Paper III
CI0.31	2.0	I	0.31	3.70	37,560	In preparation
DL1.0	5/3	L	1.00	2.70	17,256	Paper II
DI0.31	5/3	I	0.31	3.60	38,736	In preparation

[a] L: tidally locked; I: irrotational

pronounced than for soft equations of state. This is because the star is not as centrally condensed, and the moment of inertia is relatively large (a substantial amount of the star's mass can be found in its outer layers). This effect alone can be large enough to make the orbits dynamically unstable for large enough mass ratios ($q \geq 0.5$, see Lai Rasio & Shapiro 1993a; Paper I).

In any event, angular momenum losses to gravitational waves make the separation decrease, and Roche Lobe overflow occurs. Generally, the orbital evolution of the system is similar for the tidally locked and irrotational cases. A mass transfer stream forms from the neutron star core to the black hole, and there is a rapid episode of accretion, lasting a few milliseconds (peak accretion rates reach a few solar masses per millisecond). What happens next depends on the value of the adiabatic index.

For $\Gamma = 3$, the core of the neutron star responds to mass loss by shrinking enough to cut off the mass transfer stream, and it always survives as a distinct body, remaining in an elliptical orbit around the black hole, with a mass ranging from 0.2 to 0.3 solar masses (see Figure 2). For initial mass ratios $q \geq 0.5$ a massive accretion disk forms, containing a few tenths of a solar mass orbiting at a typical distance of 100 km. For lower mass ratios, there is essentially no disk at our level of resolution (only a handful of SPH particles, amounting to 10^{-3} solar masses are in orbit around the black hole). The orbital eccentricity is $e \simeq 0.2$, and the orbital separation at each subsequent periastron passage is sufficiently small so as to allow secondary episodes of mass transfer to occur, with the gas being either directly accreted by the black hole in the absence of a disk, or feeding it if one was formed during the initial encounter.

For $\Gamma = 2.5$, the neutron star is almost completely disrupted during the initial encounter, with a small core surviving until the second periastron passage a few milliseconds later. Tidal disruption is then complete, and there is always a thick accretion torus around the black hole containing approximately 0.2–0.3 solar masses. This figure is fairly independent of the initial mass ratio (see Figures 2 and 3a).

During the encounters in irrotational binaries, long tidal tails of material are formed (see Figure 3b). This gas is violently ejected through the outer Lagrange point, and some of it (on the order of 10^{-2}–10^{-1} solar masses) has enough mechanical energy to escape the black hole + accretion torus system. This may be relevant for the production of heavy elements through r–process nucleosynthesis (see Rosswog et al. 1999; Freiburghaus, Rosswog & Thielemann 1999). It is interesting to note that these tidal tails are practically nonexistent for the case of the tidally locked binaries (see Figure 5b). This is because the latter events are less violent during the initial encounter and mass transfer episode. This can also be seen by comparing the orbital separation during the coalescence and the secondary episodes of mass transfer. For example, the final separation is on the order of $7\,R_{NS}$ for run AI0.31 (irrotational, see Paper III), but only $4.7\,R_{NS}$ for run AL0.31 (tidally locked, see Paper I). The only difference between these two runs is the inital spin configuration, they both had an initial mass ratio $q = 0.31$ and initial separation $r = 3.76R_{NS}$. The amount of mass transferred from the neutron star core to the

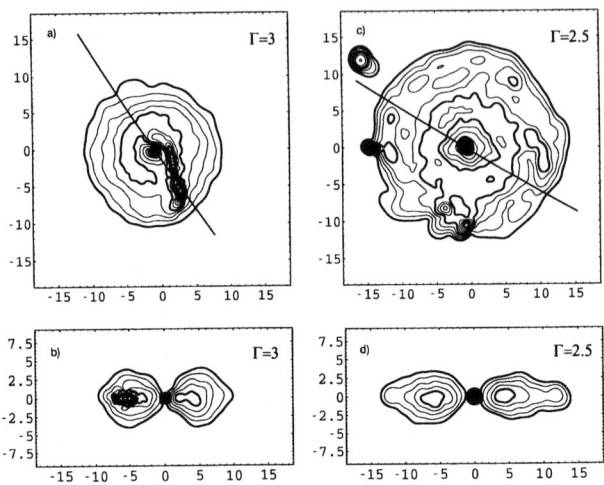

FIGURE 2. Density contours at the end of the dynamical simulations for runs AI0.5 and BI0.31 in (a,c): the orbital plane, and (b,d): in the meridional planes shown in (a,c) by the black lines. All contours are logarithmic and equally spaced every 0.25 dex. Bold contours are plotted at $\log \rho = -5, -4, -3, -2, -1$ (if present), in the units defined in section II B.

black hole during each successive periastron passage is lower in the irrotational case by approximately one order of magnitude. The irrotational encounters are more violent because once the separation becomes small enough, the tidal bulge on the neutron star becomes larger and in a sense, tries to spin up the star. This angular momentum can only come from the orbital component, and thus the decay is faster.

We note that although the star is not immediately disrupted and subsequent mass transfer events occur, this is not a stable or steady state process at all. Gravitational radiation emission tends to decrease the separation, and mass transfer tends to increase it (since the donor is the less massive of the two components and the transfer itself is almost conservative). But these processes appear to balance each other through distinct events and not in a continous fashion (the impossibility of stable mass transfer in such a system was pointed out by Bildsten & Cutler 1992 and Kochanek 1992).

The evolution described above determines what the gravitational radiation signal is like. Since a system with azimuthal symmetry will not radiate gravitational waves, any disk structure will not contribute to such a signal. The one–armed spirals formed through the ejection of gas from the system do not contain enough mass to contribute significantly either, and so the signal at late times is determined by the fate of the neutron star core (see Figure 4). If there is complete tidal disruption, the signal essentially vanishes, whereas if the binary survives, a persistent

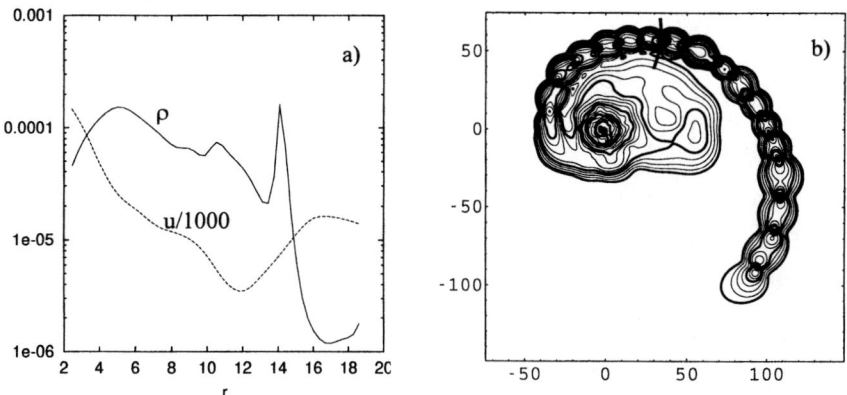

FIGURE 3. (a) Azimuthally averaged profiles for run BI0.31 in the equatorial plane for the density ρ and the specific internal energy u ($u/1000$ is plotted). (b) Density contours in the orbital plane at the end of run AI0.5. All contours are logarithmic and equally spaced every 0.25 dex. Bold contours are plotted at $\log \rho = -8, -7, -6, -5, -4$, in the units defined in section II B. The thick black line across the tidal tail divides gas that is bound to the black hole from that which is on outbound trajectories.

waveform will remain, albeit with an lower amplitude and frequency (since the binary separation has increased and the mass ratio has dropped as well). This is the case for $\Gamma = 3$ for a wide range of mass ratios. The frequency at which the signal amplitude drops marks the onset of intense mass transfer and is a function of the radius of the neutron star. In our Newtonian calculations, this is outside of the LIGO band (between 800 Hz and 1200 Hz). However, general relativistic effects will likely make the orbit unstable at larger separations, and thus the drop would happen at lower frequencies, possibly within the range of LIGO (see Faber & Rasio for binary neutron star calculations using a post Newtonian treatment on this point, these proceedings).

B Systems with a soft equation of state

As before, for systems with a soft equation of state, angular momentum losses to gravitational waves make the binary separation decrease until the star overflows its Roche lobe. This in turn leads to complete tidal disruption, regardless of the initial mass ratio (see Figure 5a). A massive accretion disk is formed, with a few tenths of a solar mass orbiting the black hole at a typical distance of 100 km. A large one–armed spiral forms, with some matter being ejected from the system, both for the irrotational and tidally locked cases (the amount is similar to that observed for the stiff equations of state, between 10^{-2} and 10^{-1} solar masses). It is the soft equation of state, with the ensuing mass–radius relationship, that causes this behavior, as

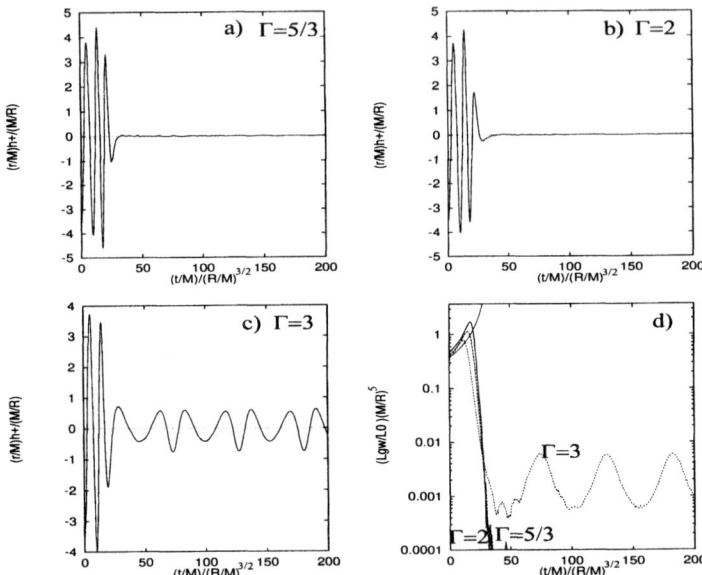

FIGURE 4. Gravitational radiation waveforms (one polarization) for an observer located at a distance r from the system, along the z–axis (perpendicular to the orbital plane) for (a) $\Gamma = 5/3$ (run DI0.31), (b) $\Gamma = 2$ (run CI0.31), (c) $\Gamma = 3$ (run AI0.31). In (d) the luminosity of gravitational radiation is plotted for the same cases. The monotonically increasing curve is the result for two point masses, computed in the quadrupole approximation.

opposed to that observed for larger values of Γ. For $\Gamma = 5/3$, our softest index, mass loss by the neutron star makes it expand, overflowing its Roche lobe even further. The mass transfer process itself is unstable, and it can be strong enough to de–stabilize the orbit by itself, excluding the effects of gravitational radiation back reaction.

By the same arguments as before, the outcome of these coalescence events is reflected in the gravitational radiation signal as an abrupt drop in amplitude (to practically zero) when the star is disrupted (see Figure 4). Again the frequency at which this occurs will give a measure of the radius of the neutron star.

IV DISCUSSION

It is clear that there are serious limitations to the numerical approach we have used to study this type of system, but we nevertheless believe these simulations are useful for many reasons. First, it is clear that hydrodynamical effects can play an important role in the global behavior of the system at small separations, and are crucial to determine the coalescence waveform properly. Second, Newtonian calculations can be used to guide future, more detailed simulations that will incorporate

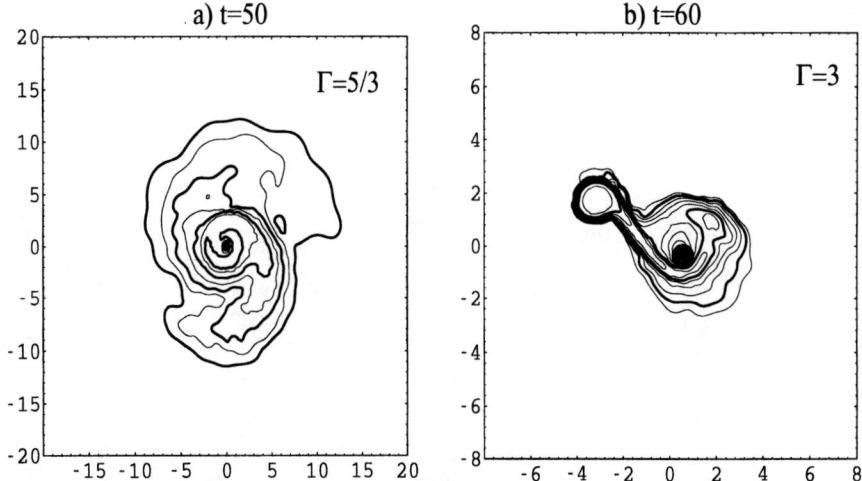

FIGURE 5. (a) Density contour plots in the orbital plane for run DL1.0 ($\Gamma = 5/3$) at $t = 50$. All contours are logarithmic and equally spaced every 0.5 dex. Bold contours are plotted at $\log \rho = -5, -4, -3, -2$, in the units defined in section II B. (b) Same as (a) but for run AL1.0 ($\Gamma = 3$) at $t = 60$, and with bold contours plotted at $\log \rho = -3, -2, -1$.

the effects of general relativity, and can serve as useful guides in this respect.

The main result of all our calculations is that the outcome of the coalescence process is very sensitive to the assumed stiffness of the equation of state. For large values of the adiabatic index $\Gamma \simeq 3$ the star is not disrupted, and there is a remnant left in orbit around the black hole, with an accretion disk as well. The gravitational radiation signal exhibits a drop in amplitude and a return to lower frequencies, but does not vanish completely. The coalescence process is delayed for at least several tens of milliseconds. For lower values of $\Gamma \simeq 2$, when the star does not alter its radius in response to mass loss (or expands, when $\Gamma < 2$), complete tidal disruption is always observed, with massive accretion disks forming around the black hole. The gravitational radiation signal practically vanishes soon after Roche lobe overflow occurs.

In a realistic scenario, the value of Γ will not be uniform throughout the star. However, for the purposes of tidal disruption and the gravitational radiation emitted, it is the value at high densities (roughly above nuclear density) that will determine the evolution of the system. We have performed different tests using a variable Γ (it is specified as a function of density, examples of this approach can be found in Rosswog et al. 2000), taking a stiff equation of state for high densities, and a softer index for the low density regions. The above discussion concerning tidal disruption is valid in this case as well, if one considers the high–density value of the adiabatic index, and since it is the bulk motion of matter that determines the gravitational wave emission, the value of Γ at low densities is unimportant in this respect.

The approach one takes also clearly depends on the problem one wishes to solve. These systems have been suggested as sources for the production of cosmological gamma ray bursts (GRBs) (Paczyński 1986; Goodman 1986; Eichler et al. 1989; Paczyński 1991; Narayan, Paczyński & Piran 1992). Kluźniak & Lee (1998) showed that the conditions during and after coalescence were indeed favorable for this, with the creation of a thick accretion torus and a baryon–free axis in the system, along the rotation axis, that would not hinder the production of a relativistic fireball that could produce a GRB (Mészáros & Rees 1992, 1993). A different equation of state (that of Lattimer & Swesty 1991) has been used by Ruffert & Janka (1996) in the study of double neutron star mergers, and by Janka et al. (1999) in black hole neutron star mergers. The stiffness of this equation of state is not a free parameter, but there is a much greater level of detail in the microphysics, relevant for the implications to GRB models (they additionally included neutrino transport in their calculations). Likewise, regarding the production of heavy elements through r–process nucleosynthesis, a simple ideal gas treatment is inadequate. One must use detailed thermodynamic calcualtions and a realistic equation of state, as Rosswog et al. (1999) and Freiburghaus, Rosswog & Thielemann (1999) have done. Our computations allow us simply to determine how much matter is dynamically ejected from the system during coalescence, a necessary first step if it is to contribute to the observed galactic abundances.

V ACKNOWLEDGEMENTS

It is a pleasure to thank the organizers for a wonderful workshop. This work was supported in part by CONACyT (27987E) and DGAPA–UNAM (PAPIIT-IN-119998).

REFERENCES

1. Balsara D., 1995, J. Comp. Phys.,121, 357
2. Bildsten L., Cutler C., 1992, ApJ, 400, 175
3. Davies M.B., Benz W., Piran T., Thielemann F.K., 1994, ApJ, 431, 742
4. Eichler D., Livio M., Piran T., Schramm D.N., 1989, Nature, 340, 126
5. Freiburghaus C., Rosswog S., Thielemann F.K., 1999, ApJ, 525, L121
6. Gingold R.A., Monaghan J.J., 1977, MNRAS, 181, 375
7. Goodman J., 1986, ApJ, 308, L46
8. Janka H. Th., Eberl T., Ruffert M., Fryer C.L., 1999, ApJ, 527, L39
9. Kluźniak W., Lee W.H., 1998, ApJ, 494, L53
10. Kochanek C., 1992, ApJ, 398, 234
11. Lai D., Rasio F.A., Shapiro S.L., 1993a, ApJ, 406, L63
12. Lai D., Rasio F.A., Shapiro S.L., 1993b, ApJS, 88, 205
13. Lattimer J.M., Schramm D.N., 1974, ApJ, 192, L145

14. Lattimer J.M., Schramm D.N., 1976, ApJ, 210, 549
15. Lattimer J.M., Swesty D., 1991, Nuc. Phys. A, 535, 331
16. Lee W.H., Kluźniak W., 1995, Acta Astron., 45, 705
17. Lee W.H., Kluźniak W., 1999, ApJ, 526, 178 (Paper I)
18. Lee W.H., Kluźniak W., 1999, MNRAS, 308, 780 (Paper II)
19. Lee W.H., 2000, MNRAS, 318, 606 (Paper III)
20. Lucy, 1977, AJ, 82, 1013
21. Mészáros P., Rees M., 1992, MNRAS, 257, 29P
22. Mészáros P., Rees M., 1993, ApJ, 405, 278
23. Monaghan J.J., 1992, ARA& A, 30, 543
24. Narayan R., Paczyński B., Piran T., 1992, ApJ, 395, L83
25. Paczyński B., 1986, ApJ, 308, L43
26. Paczyński B., 1991, Acta Astron., 41, 257
27. Rosswog S., Liebendörfer M., Thielemann F.K., Davies M.B., Benz W., Piran T., 1999, A& A, 341, 499
28. Rosswog S., Davies M.B., Thielemann F.K., Piran T., 2000, A& A, 360, 171
29. Ruffert M., Janka H. Th., 1996, A& A, 311, 532
30. Stairs I.H., Arzoumanian Z., Camilo F., Lyne A.G., Nice D.J., Taylor J.H., Thorsett S.E., Wolszczan A., 1998, ApJ, 505, 352
31. Symbalisty E.M.D., Schramm D.N., 1982, Astrophyscial Letters, 22, 143
32. Zhuge X., Centrella J.M., McMillan S.L.W., 1996, Phys. Rev. D, 54, 7261

Collisions of Black Holes: Cracks in a Hard Problem

Richard A. Matzner

Center for Relativity
The University of Texas
Austin TX 78712

Abstract.
I outline recently published progress in the evolution of binary black hole space-times, using a method that excises the singular regions inside the apparent horizons. This includes a new method for setting data for moving black holes, and the simulation of non-headon (grazing) collisions of binary black holes in which the black hole singularities have been excised from the computational domain. I describe an evolution of up to $t \approx 35m$ involving two equal mass spinning black holes m in which two initially separate apparent horizons are present for $t \approx 3.8m$. At that time a single enveloping apparent horizon forms, indicating that the holes have merged. Apparent horizon area estimates suggest gravitational radiation of about $2\% - 3\%$ of the total mass. The evolutions end after a moderate amount of time because of instabilities.

INTRODUCTION

Gravitational wave detectors [1] will soon begin searching for gravitational radiation from astrophysical binary compact objects. To understand these observations, and to predict parameter regimes in which to search for their radiation, efforts are underway to model the interaction of compact sources.

The direct numerical simulation of interacting spinning *black hole* binaries, in genuinely 3-dimensional (non-headon) trajectories, is a very hard problem. Questions about formulation of the equations, and the stability of the formulations under computational evolution, have only recently been analytically approached, and a complete resolution has not yet been reached. The simulation of black hole interactions, even with excision, has an obvious range of feature sizes that must be resolved, from $O(m)$ near the holes, to the $O(2 \times 20m)$ post merger wavelength of the ringdown radiation of the merged holes. These 1.5 orders of magnitude add to the computational demand in terms of the amount of discretization that must be supported, e.g. the number of points in a finite difference computation. The separation and interaction of longitudinal, gauge, and dynamical variables plays havoc with simple differencing schemes taken over from other fields. (It doesn't

CP575, *Astrophysical Sources for Ground-Based Gravitational Wave Detectors*, edited by J. M. Centrella
© 2001 American Institute of Physics 0-7354-0014-8/01/$18.00

help that the total number of research-hours put into computational relativity is of order one-thousandth of those put into computational hydrodynamics.) This same interaction among aspects of the gravitational field bedevils the development of effective boundary conditions, and the implementation of excision. From the computational viewpoint, the number of variables that must be carried in a computation is over 100 per grid point, and marginally adequate resolution requires discretization sizes of at least 200^3. Such $10Gb$ memory requirements push the evolutions into the "large" (and hence slow) queue in simulations, even on modern parallel computers, and hinder development.

To attack the binary black hole problem in a generic way requires flexible initial data setting with accessible physical content. It requires a way of locating apparent horizons, a stable and effective excision method, a way of moving the holes through the domain (nontrivial in itself), and long enough stability to allow extraction of waveforms.

The results with which I am most familiar are those carried out based on the Binary Black Hole Grand Challenge code, and carried forward by a collaboration of groups at Texas, Penn State, and Pittsburgh using the AGAVE code. I report principally on their results, but I also refer the reader to some complementary work, principally done by the group at the Albert-Einstein-Institut in Golm, Germany [2], [3], [4]. That work is especially interesting because it has achieved rudamentary waveform prediction (not yet achieved by the methods I describe here). I will not pursue other references on analytical work, nor computational work that is not directly related to the binary black hole evolutions.

INITIAL DATA

Any code designed to evolve a general relativistic system will start with a solution to the initial value problem corresponding to the astrophysical situation of interest. In the 3+1 ("ADM", [5]) formulation of general relativity, the problem is defined by the Hamiltonian and momentum constraints [5]. The problem of solving the Hamiltonian and momentum constraints for two black holes has been addressed in the past by several groups (see [6] and references therein). It is an inherently 3-dimensional problem to specify arbitrary boost and spin directions for each hole. The methods in most frequent use [7,8] choose maximal spatial hypersurfaces ($K = 0$, where K is the trace of the extrinsic curvature tensor K_{ij}) and take the spatial 3-metric to be conformally flat ($g_{ij} = \phi^4 \delta_{ij}$, ϕ being the conformal factor). Under these conditions, the Hamiltonian constraint can be decoupled from the momentum constraint. Analytical solutions for K_{ij} for holes with specific linear momenta can be found [9]. This simplification leaves only one elliptic equation for ϕ, which is derived from the Hamiltonian constraint. The inner boundaries (the throats of the black holes) are usually dealt with imposing an isometry condition between two identical asymptotically flat spatial slices, joined by an Einstein-Rosen bridge, though other boundary conditions are sometimes used. Unfortunately, the numerical solution of the equation for ϕ presents a technical challenge at the inner boundaries. Brandt and Brügmann [10] simplified this problem by compactifying the internal asymptotically flat regions to obtain a domain without inner boundaries. However, the main disadvantage of these methods is not numerical, but related to the physical interpretation of black-hole spaces described through a conformally flat 3-metric. There are no space slices for which the spatial 3-metric of a single Kerr (non-zero spin) black hole can be written in a conformally flat way [11], and recent work on sequences of initial data sets for circular orbits casts some doubt on the physical realism of conformally flat approaches to black hole binaries [8]. Matzner, Huq, and Shoemaker [12] proposed a method that bases the initial data on a background metric and extrinsic curvature that is a superposition of Kerr black hole metrics written in ingoing Eddington-Finkelstein coordinates. Thus, for a single Kerr black hole one obtains the exact solution to the problem. Marronetti et al. [13] added to this method a variation on the background fields that eliminates the inner boundary problem, greatly simplifying the numerical treatment of the elliptic equations. The *background* fields generated in this way are also very good approximate solutions to the initial value problem (i.e., the violation of the constraints is small enough to fall below the numerical truncation error for a wide range grid spacings).

I present here a particular solution to the initial value problem for black-hole binaries constructed via Kerr-Schild superposition to obtain physically realistic background 3-metric and extrinsic curvature. (The case studied describes a hyperbolic encounter of two Kerr black holes of mass m, separated by a distance of 11.5 m.) A single Kerr-Schild black hole metric is [14]:

$$g_{\mu\nu} = \eta_{\mu\nu} + 2Hl_\mu l_\nu, \tag{1}$$

where l_μ is an ingoing null vector (i.e: $g^{\mu\nu}l_\mu l_\nu = \eta^{\mu\nu}l_\mu l_\nu = 0$), H is a scalar function of the spacetime coordinates and $\eta_{\mu\nu}$ is the Minkowski spacetime metric.

The initial data method begins specifying a conformal spatial metric which is a straightforward superposition of two Kerr-Schild single hole (spatial) metrics:

$$\tilde{g}_{ij} = \delta_{ij} + 2\ {}_1B\ {}_1H\ {}_1l_i\ {}_1l_j + 2\ {}_2B\ {}_2H\ {}_2l_i\ {}_2l_j \ ,$$
$$\tilde{K} = {}_1B\ {}_1K_i{}^i + {}_2B\ {}_2K_i{}^i \ ,$$
$$\tilde{A}_{ij} = \tilde{g}_{n(i}\ \left({}_1B\ {}_1K_{j)}{}^n + {}_2B\ {}_2K_{j)}{}^n - \frac{1}{3}\delta_{j)}{}^n \tilde{K}\right) \ , \tag{2}$$

where the parenthesis in the subscripts denote symmetrization. The scalar function H, and ingoing null vector congruence l_λ are associated with the solution [12]. For a nonmoving single Kerr black hole, centered at the origin, they are [15]:

$$H = \frac{Mr^3}{r^4 + a^2 z^2} \tag{3}$$

and

$$l_\mu = \left(1, \frac{rx + ay}{r^2 + a^2}, \frac{ry - ax}{r^2 + a^2}, \frac{z}{r}\right), \tag{4}$$

where r is given by

$$\frac{x^2 + y^2}{r^2 + a^2} + \frac{z^2}{r^2} = 1, \tag{5}$$

and a is the Kerr spin parameter. Because of the simple Minkowski structure of the Kerr-Schild form, Lorentz transformations give an immediate way to produce boosted black holes. In the initial data conformal form, Eq.(2), the fields marked with the pre-index 1 (2) correspond to an isolated black hole with specific angular momentum a_1 (a_2) and boosted with velocity v_1 (v_2). The trace-free part of the extrinsic curvature tensor $\tilde{A}_{ij}$ is also constructed from the superposition of the curvatures $_1K_i^j$ and $_2K_i^j$ associated with the black holes 1 and 2 respectively. The attenuation function $_1B$ ($_2B$) is unity everywhere except in the vicinity of hole 2 (hole 1) where it rapidly vanishes so the fields there are effectively those of a single black hole, thus providing an exact solution to the constraints for distances arbitrarily close to the singularities [13].

Following the conformal decomposition presented by York and collaborators [16], relate the physical metric g_{ij} and the trace-free part of the extrinsic curvature A_{ij} to the background fields through a conformal factor:

$$\begin{aligned} g_{ij} &= \phi^4 \tilde{g}_{ij} \\ A^{ij} &= \phi^{-10} \tilde{A}^{ij} \end{aligned} \tag{6}$$

To find a solution to the four constraint eqations, add a longitudinal part to the extrinsic curvature A^{ij}:

$$A^{ij} \equiv \phi^{-10}(\tilde{A}^{ij} + (\tilde{lw})^{ij}) , \tag{7}$$

where w^i is a vector potential to be solved for and [17]

$$(\tilde{lw})^{ij} \equiv \tilde{\nabla}^i w^j + \tilde{\nabla}^j w^i - \frac{1}{3}\tilde{g}^{ij}\tilde{\nabla}_k w^k . \tag{8}$$

Plugging Eqs. (2-8), into the ADM Hamiltonian and momentum constraints, produces four coupled elliptic equations for the fields ϕ and w^i [16]:

$$\tilde{\nabla}^2\phi = (1/8)(\tilde{R}\phi + \frac{2}{3}\tilde{K}^2\phi^5 -$$
$$\phi^{-7}(\tilde{A}^{ij} + (\tilde{lw})^{ij})(\tilde{A}_{ij} + (\tilde{lw})_{ij}))$$
$$\tilde{\nabla}_j(\tilde{lw})^{ij} = \frac{2}{3}\tilde{g}^{ij}\phi^6\tilde{\nabla}_j K - \tilde{\nabla}_j\tilde{A}^{ij} \tag{9}$$

Marronetti and Matzner [18] give a multigrid solution of these equations. Figure 1 is a contour plot of the conformal factor ϕ that results from such a solution,

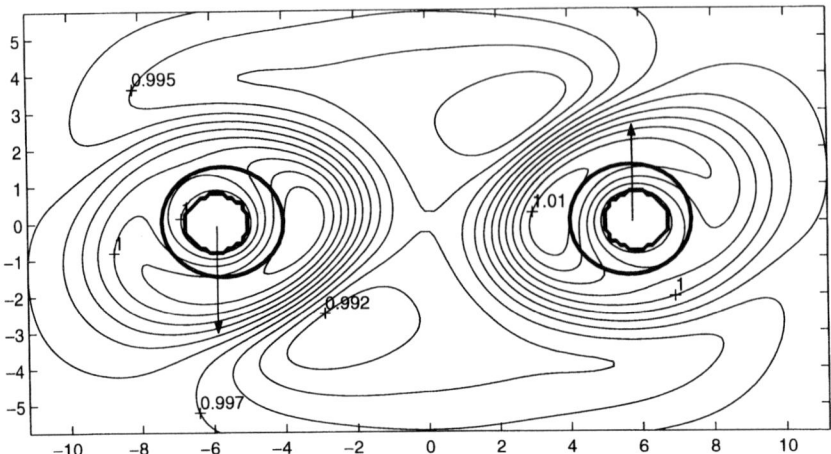

FIGURE 1. Contour plot of the conformal factor ϕ. The inner thick lines limit the "inner" region which will be excised in evolution. The outer thick lines show the apparent horizons; the arrows give the direction of the boosts.

implemented with *Robin* boundary conditions [19]. For this case two black holes with mass m are initialized in the background in a hyperbolic encounter configuration with parameters $\mathbf{r_1} = (5.75m, 0, 0)$, $\mathbf{v_1} = (0, 0.5c, 0)$, and $\mathbf{a_1} = (0, 0, 0.5m)$ corresponding to (black hole)$_1$ coordinate position, velocity, and specific angular momentum. The parameters for (black hole)$_2$ are $\mathbf{r_2} = -\mathbf{r_1}$, $\mathbf{v_2} = -\mathbf{v_1}$, and $\mathbf{a_1} = \mathbf{a_2}$.

SIMULATION

It can be seen that the background solution is already a good approximation to this ($m/8$ discretization) initial data solution, since the conformal factor is close to unity throughout the solution domain. Hence, for the coarse (discretization $\Delta x = m/4$) binary black hole evolution I report here, the background solution, without any solution of the elliptic equations, provides adequate initial accuracy (the analytic estimates of the error are beneath the truncation error). So that is the data setting method taken in the evolutions I report here; only the superposition, not the elliptic solve, is performed. We shall see that we obtain very good constraint data none-the-less.

For the evolution I present a binary black hole simulation of a data set slightly different from that solved above: *spinning* holes, each of mass m, located at $(\pm 5m, \pm m, 0)$, each with Kerr spin parameter a. The holes are boosted in opposite $\hat{\mathbf{x}}$-directions with speed $c/2$, representing a grazing collision with impact parameter of $2m$ [20]. Both holes have $a = 0.5m$ opposite to the orbital angular momentum. In this case the total system has zero angular momentum, and

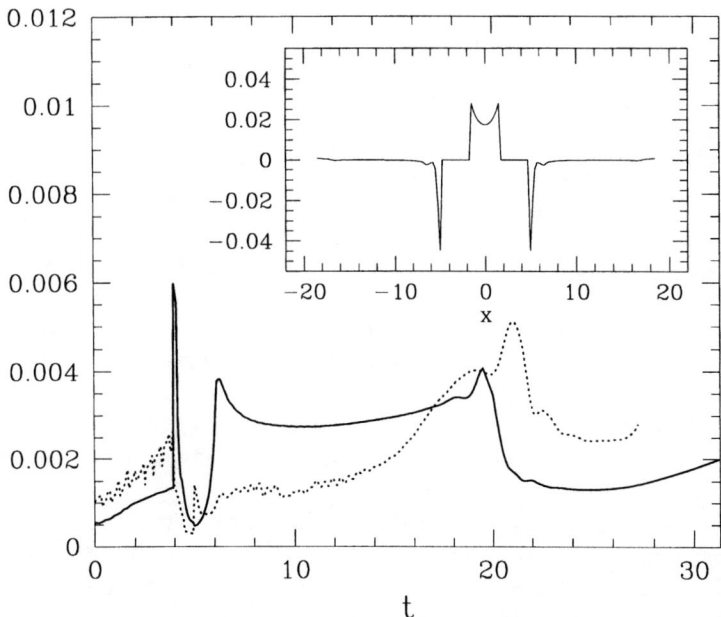

FIGURE 2. For grid ±20m, the Hamiltonian and momentum constraints, on the domain of outer communication (outside the apparent horizon(s) and inside the outer boundary blending zone): the time history of the l_2 (rms) norm of the Hamiltonian (solid line) and the l_2 norm over all three components of the momentum constraint (dotted line). The momentum l_2 is constructed only along coordinate lines (all that is available from this computation); the Hamiltonian l_2 is computed from the whole volume. The sudden change in the errors at $t \approx 4m$ occurs when a single outer apparent horizon envelops the merging holes. Also, the drop at $t \approx 20m$ is due to boundary effects. The inset shows the Hamiltonian constraint along the $x-$axis at time $t = 3m$.

the final black hole will be a Schwarschild black hole. The interior of the equal mass holes and their interior singularities are excised from the computation. (The method is *neither* restricted to equal masses *nor* to parallel spins; see [21] for evolutions of other initial configurations.) The state of the gravitational system (the 3-spatial metric g_{ab}) and its rate of change (the 3-spatial extrinsic curvature K_{ab}) are specified at the initial instant via the initial data construction just described, then stepped to the next instant using an "ADM" [5] form of the Einstein evolution equations [22]. The evolution is unconstrained, and maintenance of the constraint functions with small error is verified throughout the run.

This nonheadon work extends previous work on headon encounters [23–26]. It is comparable to recent results of Brügmann [3], [4]: non-headon black hole evolution through to significant interaction and merger. But the approach described here has the novel feature of singularity excision. Singularity excision allows "natural" motion of the black holes through the computational grid. Since singularity excision provides in principle evolution arbitrarily long into the future, it may be crucial to carrying out long term simulations predicting gravitational waveforms through several wave-cycles.

The total initial ADM mass [5] of the initial configuration is $2.31m$, which agrees very well with the estimate given by the special relativistic limit $m_{ADM} = 2\gamma m$, with $\gamma = (1 - .5^2)^{-1/2} = 1.155$. The total initial ADM angular momentum is zero.

For the discretization used the truncation error is of order 5%. The quality of the data is validated by computing the constraints, normalized to be dimensionless by the factor m^{-2}. Analytically the constraints should be zero everywhere.

As noted, with the parameters of the problem, and with the current discretization and truncation error, the superposed background solution is acceptable with no further elliptic problem solution [13] (i.e. the zeroth order of the elliptic solver). However, it is certain that as the computations progress to larger and better resolved evolutions, it will become mandatory to cycle through the elliptic solve step [18] to obtain satisfactory solution of the constraints. Figure 2 presents the Hamiltonian constraint for the evolution, evaluated at integration timestep $t = 3m$ along the $\hat{x}$−axis, together with a time history of the l_2 norm (over volume outside the horizons, and excluding the outer boundary region) of the Hamiltonian constraint and the normed momentum constraint. The late time rise in the momentum constraint in Figure 2 shows the beginning of the exponential mode that appears at about $t = 36m$ and ends the simulation. There is quite good constraint behavior, of order 0.4%, with peak errors in the Hamiltonian of order 5%, until that time.

EVOLUTION METHODS

The AGAVE code solves the Einstein equations in an ADM 3+1 form via finite difference techniques [22]. A parallel implementation is obtained with the use of MPI, employing the Cactus computational toolkit [27] solely to aid in this task. AGAVE is a major revision of the Binary Black Hole Grand Challenge Alliance Cauchy code [28,29]. The *lapse* function α and *shift* vector β^i express coordinate conditions which are chosen to allow the black holes to move freely. For this simulation, *prior* to the time that a single black hole surrounding the incoming pair is detected, the gauge functions are a superposition of functions from boosted black holes: $\alpha = \alpha_1 + \alpha_2 - 1$, $\beta^i = \beta_1^i + \beta_2^i$, where these functions are centered with the current location of the holes, and with the velocity initially obtained from Newtonian approximation to the trajectories of the holes and subsequently inferred from the history of the locations of the apparent horizons (see below); *after* the detected merger, the lapse and shift are set to those of a single black hole with a mass which is the sum of the original bare masses, and angular momentum which is for this case, zero.

The interior of the black holes is excised (Unruh, quoted in [30]). The apparent

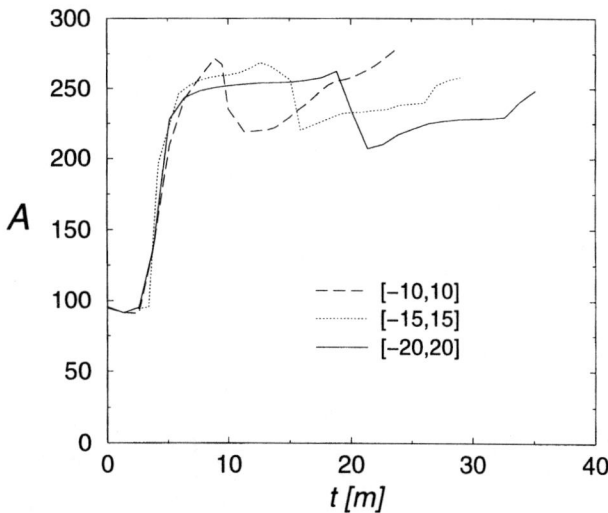

FIGURE 3. The area of the apparent horizon(s) (units of m^2), showing a transition to a single horizon at $t \approx 4m$. For a small domain ($\pm 10m$, dashed line) the simulation runs to $t \approx 26m$ and exhibits strong boundary effects at $t \approx 10m$. In the largest ($\pm 20m$) domain (solid line) boundary effects show at later time, around $20m$. The intermediate ($\pm 15m$) simulation shows boundary effects at $t \approx 15m$. The ($\pm 15m$) simulation is run at higher resolution ($\Delta x = m/6$, instead of $m/4$ for the other simulations). The close agreement of the $\pm 20m$ and the $\pm 15m$ area results encourages confidence in the area results. The $\pm 15m$ run was intentionally stopped at $\approx 30m$. Instabilities cause the measured area to rise abruptly at $t \approx 36m$ and eventually stop the $\pm 20m$ simulation.

horizon surface, locatable at each time-slice, provides a marker for the excision. A combination of two different finite difference methods is used to find the apparent horizon: a direct solver [31], and a curvature flow method [32]. Once the apparent horizon is located, a *mask* function delineates the excluded region (interior to the holes) from the computation. The result is that two holes literally move freely through the computational domain. That domain is a 161^3 lattice, corresponding at our resolution to a cube $(40m)^3$ ($\pm 20m$ in each direction from the centered origin). However, boundary conditions are set by providing (time dependent) Dirichlet boundary conditions for K_{ab} and blending [33,34] outwards from a sphere of radius $19m$ the computational solution of K_{ab} to an analytically given time-dependent solution for K_{ab} at the outer boundary sphere. "Blending" means taking a linear combination of values from the computed and the analytically given solution, over a few (here, four) spatial zones, thus reducing gradients and second derivatives at the boundary. The analytic blending solution is created by superposition of boosted holes given by the initial data construction (with centers and velocities propagated according to the lapse and shift computation), or, after the merger, by the final estimated black hole with post merger lapse and shift.

The discretization of the Einstein equations is consistent to second order accuracy. On the time scale where instabilities do not play a significant role, the convergence rate of this code is ≈ 1.6, reduced from 2 apparently because of extrapolation at the excision boundaries.

RESULTS

The total proper area of the apparent horizon A is shown in Figure 3. The value of A is particularly interesting since it provides a measure of the total mass contained in the apparent horizon. For a given black hole of mass m and spin parameter a its area is $A_{BH} = 4\pi(R_+^2 + a^2)$ (with $R_+ = m + \sqrt{m^2 - a^2}$). Since at early times there is no common apparent horizon the total area is approximately $A \approx A_{BH1} + A_{BH2} = 2A_{BH1}$. As the holes merge the total mass enclosed in the common horizon is (roughly) expected to double, and hence its area would be twice as as big, ie. for a non-spinning final black hole $A \approx 4\pi(2(2m))^2 \lesssim 4A_{BH1}$. Therefore, a plot of A vs. time (like the one in figure 3) shows a considerable 'jump' at the time the holes merge $t \approx 3.8m$. Additionally, effects of the outer boundary can be clearly seen in figure 3. For a ± 10 grid an abrupt 'kink' is seen at $t \approx 10m$ while in the $\pm 15m$ grid the 'kink' appears at $t \approx 15m$, and in the ± 20 grid at $t \approx 20m$. The $\pm 15m$ run was intentionally stopped at $\approx 30m$. At about $t \approx 36m$ ($t \approx 26m$) apparent instabilities in the $\pm 20m$ ($\pm 10m$) grid cause a rapid increase in the computed horizon size and eventually crash the run. Thus at $t \approx 35m$ the solution becomes untrustworthy. During the time that the simulation is free of boundary effects the coincidence of the measured horizon area values for different domains and resolutions (the $(\pm 15m)$ simulation is run at the higher resolution of $\Delta x = m/6$, instead of $m/4$ as for the other simulations) supports confidence in the results. Figures 4A - 4F track the apparent horizons through the merger. A single enveloping black hole appears at $t \approx 3.8m$. The horizon oscillates and grows slightly.

DISCUSSION AND FUTURE DIRECTIONS

The simulations reported here are genuinely, but not excessively, hyperbolic encounters. A Newtonian estimate gives a free fall velocity of $0.4c$ from infinity, as compared with the velocity $0.5c$ specified here. Future work will necessisarily concentrate on generic hyperbolic and elliptic orbits.

Questions remain concerning the late-time stability of the black hole simulations. A number of 1-dimensional simulations are known, which have longer term stability than this 3-dimensional simulation of merged holes. Possible culprits limiting the 3-d longevity include the behavior of the differencing scheme at the inner boundary, and the outer boundary treatment, where new outer boundary algorithms have

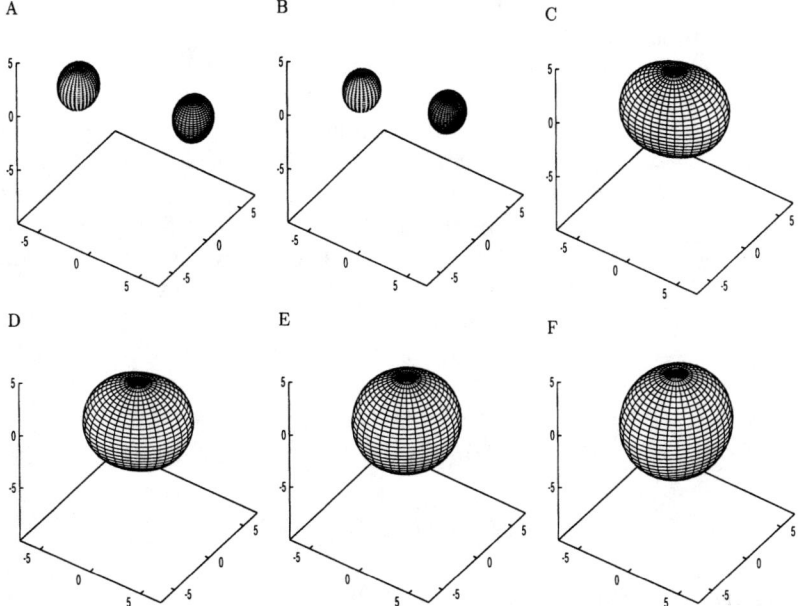

FIGURE 4. Time history of the apparent horizons. The times corresponding to figures 4A-4F are $t = 0$, $2.6m$, $5.1m$, $8.8m$, $13.8m$, $18.8m$. These are coordinate plots; the corresponding areas appear in Figure 3. After the merger the apparent horizon oscillates through a fraction of a cycle.

been shown to be robustly stable in a linearized version of the code [35]. More sophisticated gauges based on elliptic equations for the lapse and the shift include the minimal distortion and minimal shear gauges [36], and other elliptic gauges [12,37], and must be investigated. It should be understood that the excision appears to significantly affect the behavior of simulations employing these conditions, even in cases where appropriate excision boundary conditions are known. This is important since *excised* stable evolution of single black holes simulations seems to be very sensitive to functional (live) gauge choices.

The gauge and boundary conditions for the final merged black hole naively assume that all the initial mass (i.e. $M_{final} = 2m$) and all the angular momentum resides in the final hole: $J_{final} = a_{final} \times M_{final}$ (here zero). These estimates do not take into account the emission of energy and angular momentum during the dynamics, nor the γ factor in the initial mass and angular momentum. However a few experiments indicate that the behavior of the code is robust under changes in the final assumed mass and spin, corresponding to changes in specified (not live) gauges and in boundary conditions.

Of extreme interest is the size of the final apparent horizon. The *total* initial

ADM mass leads to horizon area of $4\pi(2 \times 2.31m)^2 \approx 268m^2$. The post-merger numerically computed apparent horizon area (Figure 3) is about $255m^2$, about 5% smaller. This measure would give a preliminary indication that total energy radiated in this simulation is about 2% − 3%.

The work just described demonstrates the first simulation of binary black hole collision and merger via the excision of singularities. This work: (a) demonstrates the initial motion of black holes through the domain; (b) demonstrates well behaved (convergent) descriptions of the black holes as they evolve; (c) shows that apparent horizon tracking and black hole excision can produce dynamical multi-black hole spacetimes, with reasonably well controlled errors for a considerable length of time (long enough for an accurate modeling of the merger phase); and (d) demonstrates that relatively unsophisticated gauge functions α and β can lead to physically interesting evolution lifetimes. The datasets evolved are useful for validation of the techniques employed, and also provide qualitatively valid datasets in an astrophysical sense for the final "plunge" of the merger.

ACKNOWLEDGMENTS

I thank my colleagues in this work: Steve Brandt, Randall Correll, Roberto Gomez, Mijan Huq, Pablo Laguna, Luis Lehner, Pedro Marronetti, David Neilsen, Jorge Pullin, Erik Schnetter, Deirdre Shoemaker, and Jeffrey Winicour. This work owes much to the Binary Black Hole Grand Challenge, and was supported by NSF PHY/ASC 9318152 (ARPA supplemented), PHY 9310053, PHY9800722, PHY 9800725. I thank the Observatoire de Paris and Los Alamos National Laboratory, where some of this work was carried out. Computations were carried out at the National Center for Supercomputing Applications, at the Albuquerque High Performance Computing Center, and at Los Alamos National Laboratory.

REFERENCES

1. LIGO: A. Abramovici, et al., Science **256**, 325 (1992); GEO: K. Danzmann and the GEO Team, Lecture Notes in Physics **410** 184-209 (1992); VIRGO: C. Bradaschia et al., Nucl. Instrum. Meth., Phys. Res. Sect., **289**, 518 (1990); TAMA: K. Tsubomo, M.-K. Fujimoto and K. Kuroda, TAMA International Workshop on Gravitational Wave Detection (Universal Academic Press, Tokyo, Japan, 1996); LISA: First International LISA Symposium, Class. Quantum Grav. **14**, 1397 (1996).
2. C. Lousto, talk at Numerical Relativity/Black Hole Collisions Workshop at the XX Texas Symposium on Relativistic Astrophysics, Austin TX (December 13, 2000)
3. B. Brügmann, Int. J. Mod. Phys. **D8**, 85 (1999).
4. M. Alcubierre et al., [gr-qc/0012079]
5. R. Arnowitt, S. Deser, and C. Misner, in Gravitation − An Introduction to Current Research, edited by L. Witten (Wiley, New York, 1962).
6. G. B. Cook, To appear in "Living Reviews in Relativity" (2000); [gr-qc/0007085].

7. G. B. Cook,*et al.*, Phys. Rev. **D47**, 1471 (1993); G. B. Cook, Phys. Rev. **D50**, 5025 (1994); T. W. Baumgarte, Phys. Rev. **D62**, 024018 (2000).

8. H. P. Pfeiffer, S. A. Teukolsky and G. B. Cook; [gr-qc/0006084].

9. A. D. Kulkarni, L. C. Shepley and J. W. York Jr. Phys. Lett. **96A**, 228 (1983).

10. S. Brandt and B. Brügmann, Phys. Rev. Lett. **78**, 3606 (1997); [gr-qc/9703066].

11. A. Garat and R. H. Price, Phys. Rev. **D61**, 124011 (2000); [gr-qc/0002013].

12. Richard A. Matzner, M. F. Huq and D. Shoemaker, Phys. Rev. **D59**, 024015 (1999).

13. P. Marronetti,*et al.*, Phys. Rev. **D62**, 024017 (2000); [gr-qc/0001077].

14. R. P. Kerr and A. Schild, in *Applications of Nonlinear Partial Differential Equations in Mathematical Physics*, Proc. of Symposia b Applied Math., Vol. XV11, (1965); R. P. Kerr and A. Schild, in *Atti del Convegno Sulla Relativita Generale: Problemi Dell'Energia E Onde Gravitazionale*, G. Barbera, ed.(1965).

15. R. P. Kerr, *Phys. Rev. Lett.* **11** 237 (1963).

16. J. W. York and T. Piran in *Spacetime and Geometry: The Alfred Schild Lectures*, R. A. Matzner and L. C. Shepley eds. (University of Texas Press, Austin, 1982).

17. Covariant differentiation with respect to g_{ij} ($\tilde{g}_{ij}$) is denoted by ∇_i ($\tilde{\nabla}_i$). Spacetime indices will be denoted by Greek letters and spatial indices by Latin letters.

18. P. Marronetti and Richard A. Matzner, *Physical Review Letters* **85**, 5500 (2000).

19. C. R. Evans, in *Frontiers in Numerical Relativity*, C. R. Evans, L. S. Finn, and D. W. Hobill, eds., (Cambridge University Press, 1989)

20. R. Correll, PhD Dissertation, The University of Texas (1998).

21. S. Brandt,*et al.*, *Physical Review Letters* **85**, 5469 (2000); [gr-qc 0009047].

22. Richard A. Matzner, in *Classical and Quantum Black Holes*, P Fre, V. Gorini., G. Magli, and U. Moschella, editors, Institute of Physics Publishing (Bristol 1999).

23. S. G. Hahn and R. W. Lindquist, *Annals of Physics (NY)* **29** 304 (1964).

24. K. R. Eppley, *Phys. Rev.* **D16** 1609 (1977).

25. L. Smarr, in *Sources of Gravitational Radiation*, edited by L. Smarr (Cambridge University Press, 1978).

26. P. Anninos *et. al.*, *Phys. Rev. Lett.* **71** 2851 (1993); P. Anninos *et. al.* *Phys. Rev.* **D52** 2044 (1995); P. Anninos *et. al.* *Physical Review*, **D52**, 2059-2082 (1995).

27. http://www.mpi-forum.org/; http://www.cactuscode.org/

28. *The Binary Black Hole Grand Challenge Alliance*: G. Cook, et. al. *Phys. Rev. Lett.* **80**, 2512 (1998).

29. M. Huq and Richard A. Matzner, in preparation (2000).

30. J. Thornburg, *Class. Quantum Grav.*, **4**, 1119-1131 (1987).

31. M. F. Huq, M. W. Choptuik, and Richard A. Matzner, *Phys. Rev.* in press (2000).

32. D. Shoemaker, PhD Dissertation, The University of Texas (1999).

33. R. Gomez, at BBH Grand Challenge Workshop, Los Alamos NM (October 1997).

34. *The Binary Black Hole Grand Challenge Alliance*: L. Rezzolla, et. al. *Phys. Rev. Lett.* **80**, 1812 (1998); L. Rezzolla, et. al., *Phys. Rev.* **D59** 064001 (1999).

35. B. Szilagyi *et. al.*. (to appear in *Phys. Rev. D.*) (2000).

36. L. Smarr and J. York, , *Phys. Rev.* **D17** 2529 (1978).

37. P. Brady, J. Creighton, and K. Thorne, *Phys. Rev.* **D58** 061501 (1998).

The Innermost Stable Circular Orbit in Compact Binaries

Thomas W. Baumgarte

Department of Physics, University of Illinois at Urbana-Champaign, Urbana, Il 61801

Abstract. Newtonian point mass binaries can be brought into arbitrarily close circular orbits. Neutron stars and black holes, however, are extended, relativistic objects. Both finite size and relativistic effects make very close orbits unstable, so that there exists an innermost stable circular orbit (ISCO). We illustrate the physics of the ISCO in a simple model problem, and review different techniques which have been employed to locate the ISCO in black hole and neutron star binaries. We discuss different assumptions and approximations, and speculate on how differences in the results may be explained and resolved.

INTRODUCTION

Compact binaries, containing black holes or neutron stars, are among the most promising sources for the new generation of gravitational wave detectors. TAMA has already started taking data, and LIGO, VIRGO and GEO will soon become operational (see, e.g., [1]). With the advent of gravitational wave astronomy arises a need for theoretical templates of gravitational waveforms, which are required for the identification and interpretation of signals in the noisy output of the detectors.

Compact binaries emit gravitational radiation, and therefore loose energy and slowly spiral towards each other. Because of the circularizing effects of gravitational radiation damping, it is reasonable to assume the orbits of close binaries to be quasi-circular. The slow and adiabatic inspiral continues until the binary reaches the innermost stable circular orbit (ISCO), at which the orbits become unstable. At that point, the stars start to plunge towards each other, and coalesce and merge after a dynamical timescale. The ISCO leaves a characteristic signature in the gravitational wave signal of a binary inspiral, and therefore provides an important piece of information for the construction of gravitational wave templates.

In this article, we will explain in a simple point-mass model problem how both tidal and relativistic effects make very close orbits unstable, giving rise to the ISCO. We will then review attempts to determine the ISCO for binary black holes, which, for stellar-mass black holes, is likely to fall into the frequency range of LIGO. In particular, we will compare results from post-Newtonian (PN) and numerical calcu-

CP575, *Astrophysical Sources for Ground-Based Gravitational Wave Detectors*, edited by J. M. Centrella
© 2001 American Institute of Physics 0-7354-0014-8/01/$18.00

lations. We will also discuss some of the qualitative results for binary neutron stars, and will summarize some of the more recent results from relativistic simulations.

THE ISCO IN A SIMPLE MODEL PROBLEM

Most commonly, the ISCO is located with the help of turning-point methods. We motivate this method by applying it to Newtonian point-masses, and introduce tidal and relativistic correction terms to illustrate their effect. Rigorous justifications for turning-point methods have been developed, for example, in [2–4].

Newtonian point-masses

The conditions for a circular orbit can be derived very easily from the Hamiltonian formalism. Consider, for example, a Newtonian point mass binary, for which the energy is the sum of the kinetic and potential energy

$$E = \frac{1}{2}\mu\dot{\mathbf{r}}^2 - \frac{\mu M}{r} = \frac{1}{2}\mu(\dot{r}^2 + r^2\Omega^2) - \frac{\mu M}{r}. \tag{1}$$

Here $M \equiv m_1 + m_2$ is the total mass, $\mu \equiv m_1 m_2/M$ the reduced mass, $\mathbf{r}$ the separation vector, and $\Omega = \dot{\varphi}$ the angular velocity. Rewriting the energy in terms of the conjugate momenta $P = \mu\dot{r}$ and $J = \mu r^2\Omega$ yields the Hamiltonian

$$E = H = \frac{1}{2}\frac{P^2}{\mu} + \frac{1}{2}\frac{J^2}{\mu r^2} - \frac{\mu M}{r}. \tag{2}$$

The four independent variables r, P, φ and J now satisfy the Hamiltonian equations

$$
\begin{aligned}
\dot{r} &= \frac{\partial H}{\partial P} & \dot{P} &= -\frac{\partial H}{\partial r} \\
\Omega = \dot{\varphi} &= \frac{\partial H}{\partial J} & \dot{J} &= -\frac{\partial H}{\partial \varphi} = 0.
\end{aligned}
\tag{3}
$$

The last equation shows that the angular momentum is conserved, since the Hamiltonian is cyclic in φ. To construct a circular orbit, we obviously need $\dot{r} = 0$, which implies $P = 0$, and also $\dot{P} = 0$, so that

$$\left.\frac{\partial H}{\partial r}\right|_J = \left.\frac{\partial E}{\partial r}\right|_J = 0. \tag{4}$$

A circular orbit hence corresponds to an extremum of the energy at constant angular momentum. The orbital frequency of a circular orbit can be determined from

$$\Omega = \frac{\partial H}{\partial J} = \frac{\partial E}{\partial J}. \tag{5}$$

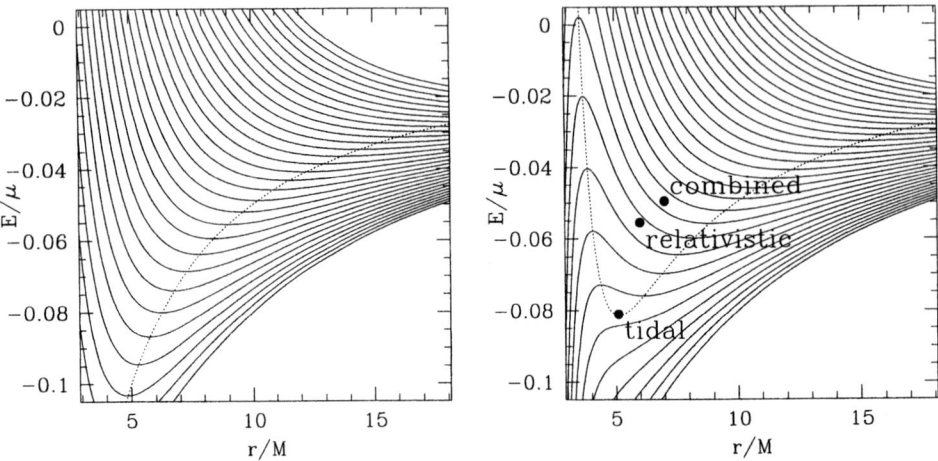

FIGURE 1. Energy as function of separation for a Newtonian point mass binary (left panel), and a model problem including tidal effects (right panel, see text). The solid lines are contours of constant angular momentum J, with the values of J increasing from the bottom to the top. The extrema of these contours correspond to circular orbits (eq. (4)). The dashed line connects the circular orbits and represents the equilibrium energy E_{eq}, and the dots mark the ISCO for tidal, relativistic and combined effects.

The last two equations are crucial for the construction of the ISCO.

Returning to the example of a Newtonian point mass binary, we can construct circular orbits by inserting the energy (2) with $P = 0$ into eqs. (4) and (5). Eq. (4) yields the virial relation

$$\frac{J^2}{\mu r^2} = \frac{\mu M}{r},\tag{6}$$

so that the equilibrium energy of a circular orbit becomes

$$E_{eq} = -\frac{1}{2}\frac{\mu M}{r}.\tag{7}$$

The orbital frequency of these orbits can be found from eq. (5) and satisfies, as expected, Kepler's third law

$$\Omega = \frac{\partial E}{\partial J} = \frac{J}{\mu r^2} = \sqrt{\frac{M}{r^3}},\tag{8}$$

where we have used eq. (6) in the last equality.

Note that Newtonian point masses can be brought into arbitrarily close and arbitrarily tightly bound circular orbits. The orbits are stable for all separations r, indicating that in Newtonian point mass binaries there is no ISCO. As we will see, this is related to the gravitational interaction potential being proportional to $1/r$.

It is also illustrative to construct circular orbits graphically. In the left panel of Fig. 1, the solid lines are contours of the energy (2) at constant angular momentum. The minima of these contours correspond to circular orbits. Connecting these yields the equilibrium energy (7) as a function of separation.

Tidal and relativistic effects

Compact binaries are extended, relativistic objects, of course, so that we have to take into account both tidal and relativistic effects.

For irrotational, identical stars, tidal effects can be modeled by adding a tidal interaction term

$$E_{\text{tidal}} = -2\lambda \frac{\mu M R^5}{r^6} \tag{9}$$

to the energy (2) (compare [5–7]), where R is the radius of the stars at large separations. The coefficient λ depends on how easily the stars can be deformed, and hence on their structure and equation of state (EOS). For polytropic EOSs with polytropic index n we have $\lambda = (3/4) \kappa_n (1 - n/5)$, where the coefficient κ_n is tabulated, for example, in [2]. For an incompressible fluid, $\lambda = 3/4$. Each star induces a quadrupole moment in the companion[1] which scales with $1/r^3$, and the two quadrupole moments' interaction again scales with $1/r^3$, so that $E_{\text{tidal}} \sim r^{-6}$.

From eq. (4), we can now find the equilibrium energy

$$E_{\text{eq}} = -\frac{1}{2} \frac{\mu M}{r} + 4\lambda \frac{\mu M R^5}{r^6}. \tag{10}$$

We immediately see that we can no longer construct arbitrarily tightly bound orbits, and that instead E_{eq} now assumes a minimum for a finite separation r. It is easy to show that any attractive interaction potential proportional to $1/r^n$, $n > 2$, leads to a positive contribution to E_{eq}, and hence the existence of a minimum at finite r.

In the right panel of Fig. 1 we provide a graphical representation for incompressible fluid stars with $m/R = 0.2$, where m is the mass of the individual stars. For large separations, the tidal interaction is very small, and the orbits are similar to those of point masses. For small separations, however, the tidal interaction becomes dominant, and decreases the energy of a configuration with a given separation and angular momentum below that of the point mass binary. For large enough angular momentum, the $J = const$ contours now have a maximum at a small separation in addition to the minimum at a larger separation, while for small angular momenta their contours no longer have any extrema. As a consequence, the equilibrium energy goes through a minimum. Outside of this minimum, the equilibrium energy

---

[1] If the stars are not irrotational, the spin of the individual stars induces an intrinsic quadrupole moment. The tidal interaction energy then scales with a smaller power of r, see, e.g. [5,25]

curve connects minima of the $J = const$ curves, which correspond to *stable* circular orbits, while inside this minimum, the curve connects maxima of $J = const$ curves, which correspond to *unstable* circular orbits. The minimum of the equilibrium curve therefore marks the innermost stable circular orbit, the ISCO[2].

We can similarly mimic relativistic effects by borrowing a relativistic interaction potential

$$E_{\rm rel} = -\frac{M}{\mu}\frac{J^2}{r^3} \tag{11}$$

from the result for test particles in Schwarzschild space times. This interaction again gives rise to an ISCO. We mark the location of this ISCO in the left panel of Fig. 1, as well as the ISCO when computed from the combined tidal and relativistic effects. It is evident from this model calculation that including relativistic effects will move the ISCO to larger separation, and correspondingly smaller values of the orbital frequency Ω.

Obviously, this model problem is very naive, and can at best mimic qualitative effects. However, it does illustrate several important results. Quite in general, the ISCO arises because of contributions to the interaction potential which deviate from a simple $1/r$ scaling. In particular, tidal effects can cause an ISCO even in purely Newtonian systems. In general, both tidal and relativistic effects have to be taken into account for an accurate determination of the ISCO. Lastly, the model problem provides a straight-forward recipe for locating the ISCO: first construct circular orbits and the corresponding equilibrium energy $E_{\rm eq}$ by locating extrema of the energy (eq. (4)), then identify the ISCO with the minimum of $E_{\rm eq}$, and last compute the orbital frequency Ω from eq. (5). We have therefore reduced the problem of accurately determining the ISCO in black hole or neutron star binaries to an accurate determination of their energies, which we will address separately in the following sections.

THE ISCO IN BINARY BLACK HOLES

Determining the ISCO in binary black hole systems has been attempted with two independent approaches: post-Newtonian expansions and numerical calculations in

---

[2] At this point, a word of warning is in order. Relativistic binaries emit gravitational radiation, causing them to slowly spiral towards each other, and they hence do not follow strictly circular orbits. The very concept of an innermost stable *circular* orbit is therefore somewhat ill defined. Also, the minimum in the equilibrium energy identifies the onset of a *secular* instability, while the onset of *dynamical* instability may be more relevant for the binary inspiral (see, e.g., the discussion in [2] and also [3], where it is shown that the two instabilities coincide in irrotational binaries). Moreover, it has been suggested that the passage through the ISCO may proceed quite gradually [8], so that a precise definition of the ISCO may be less meaningful than the above turning method suggests. Ultimately, dynamical evolution calculations will have to simulate the approach to the ISCO and to investigate these issues. For the sake of dealing with a well-defined problem, we will here identify the ISCO with the minimum of the equilibrium energy.

full general relativity. We will outline both approaches, and will then compare the results.

Post-Newtonian Calculations

A large number of researchers has developed post-Newtonian techniques to model compact objects in close binaries (an incomplete list includes [9–13]; see [14] for a particularly pedagogical treatment). Typically, these calculations start by bringing Einstein's equations into the form

$$\Box h^{\alpha\beta} = -16\pi\tau^{\alpha\beta} \equiv -16\pi(-g)T^{\alpha\beta} - \Lambda^{\alpha\beta}. \tag{12}$$

Here, $\Box$ is the flat space wave operator, $h^{\alpha\beta}$ measures deviations from the flat Minkowski metric, g is the determinant of the metric, and the source term $\tau^{\alpha\beta}$ contains both the stress-energy tensor $T^{\alpha\beta}$ as well as non-linear terms in $h^{\alpha\beta}$, which have been absorbed in $\Lambda^{\alpha\beta}$. A formal solution to this equations is the retarded, flat-space Green function

$$h^{\alpha\beta}(t, \mathbf{x}) = 4 \int \frac{\tau^{\alpha\beta}(t', \mathbf{x}')\delta(t' - t + |\mathbf{x} - \mathbf{x}'|)}{|\mathbf{x} - \mathbf{x}'|} d^4x'. \tag{13}$$

Unfortunately, the source term $\tau^{\alpha\beta}$ depends on the solution $h^{\alpha\beta}$, so that this formal solution is of little practical help. However, it can be used to construct a solution iteratively. Starting with a Newtonian point-mass solution, for which $T^{\alpha\beta}_{\text{Newt}}$ is known and for which $h^{\alpha\beta}$ vanishes, we can construct a first iteration from

$$\Box h^{\alpha\beta}_1 = 16\pi T^{\alpha\beta}_{\text{Newt}}. \tag{14}$$

Each solution $h^{\alpha\beta}_n$ can be inserted on the right hand side of eq. (13), which then provides the next iteration $h^{\alpha\beta}_{n+1}$. Each iteration provides a correction over the previous one in the order of $\epsilon \sim v^2 \sim m/R$. The iteration therefore yields a post-Newtonian expansion of the solution $h^{\alpha\beta}$, from which the energy can be constructed in the form of a Taylor expansion

$$E_{\text{eq}} = E_{\text{Newt}} + E_1 \epsilon + E_2 \epsilon^2 + \dots. \tag{15}$$

Here $E_n \epsilon^n$ is the n-th order post-Newtonian correction to the energy.

It turns out, however, that for values close to the ISCO, $\epsilon \sim 1/6$, this expansion converges extremely slowly. The reason is that at a location fairly close to the ISCO, namely at the light radius with $\epsilon \sim 1/3$, some of the functions involved in the expansion diverge and have a pole (see [10]). It is therefore to be expected that a polynomial expansion of the form (15) will converge only very poorly in the neighborhood of that pole. Damour, Iyer and Sathyaprakash [10] provided a solution to this problem by using a resummation technique. Instead of employing the polynomial (15), they use the information contained in the Taylor expansion to construct an expansion in terms of rational functions. To illustrate this with an example, assume that we know the Taylor expansion of an arbitrary function $f(x)$ up to second order. We can then construct an expansion in terms of a rational function

$$f(x) \sim t_0 + t_1\,x + t_2\,x^2 + \ldots \sim \frac{p_0 + p_1\,x + \ldots}{1 + q_1\,x + \ldots} \tag{16}$$

by choosing the coefficients p_0, p_1 and q_1 such that the value and the first two derivatives of the two expansions match at $x = 0$. This expansion is known as a "Padé-approximant", and has been shown to greatly improve the convergence of post-Newtonian expansions at least up to second order.

When $h_n^{\alpha\beta}$ is inserted on the right hand side of eq. (13), the integrals no longer have compact support, so that there is no guarantee that they will converge. Moreover, the use of point-mass sources gives rise to divergent integrals, which have to be re-normalized appropriately. This has been achieved up to second post-Newtonian order, but at third post-Newtonian order some ambiguities remain [11]. Damour, Jaranowski and Schäfer [13] have recently shown that these ambiguities can be expressed by a single dimensionless parameter, ω_{static}, which is currently unknown. They compare three different post-Newtonian approaches (the e-method and j-method based on minimizations of the energy and the angular momentum, and an effective one-body method [12]) and find a quite remarkable result. For one particular choice of ω_{static}, these three independent approaches yield very similar predictions for the angular momentum, energy, and orbital frequency at the ISCO. This self-consistency suggests that the correct values of ω_{static} and the ISCO at 3PN order may have been identified.

Numerical Calculations in General Relativity

A framework for numerically constructing models of binary black holes has been provided by Arnowitt, Deser and Misner's 3+1 (ADM) decomposition of Einstein's equations [15] and York's conformal decomposition [16].

The ADM formalism splits Einstein's equations into constraint and evolution equations. The gravitational fields, described by the spatial metric γ_{ij} and the extrinsic curvature K_{ij}, have to satisfy the constraint equations on each time slice, while the evolution equations describe how they evolve from one time-slice to the next. For the construction of initial data, the two constraint equations, the Hamiltonian constraint and the momentum constraint, have to be solved, which determine only the longitudinal parts of the gravitational fields. The transverse parts of the fields, loosely associated with the gravitational wave degrees of freedom, are unconstrained by the constraint equations, and have to be chosen before the constraints can be solved.

Since the binary inspiral outside the ISCO proceeds very slowly, the gravitational wave content of these spacetimes must be very small. It therefore seems reasonable to attempt to minimize the gravitational wave content by choosing the spatial metric conformally flat, $\gamma_{ij} = \psi^4 f_{ij}$, where ψ the conformal factor and f_{ij} a flat metric. With the further assumption of maximal slicing, $K \equiv \gamma^{ij} K_{ij} = 0$, the Hamiltonian constraint reduces to

$$\hat{\nabla}^2 \psi = -\frac{1}{8} \psi^{-7} \hat{A}_{ij} \hat{A}^{ij}, \tag{17}$$

while the momentum constraint becomes

$$\hat{\nabla}_i \hat{A}^{ij} = 0. \tag{18}$$

Here $\hat{\nabla}$ is the flat space covariant derivative, and $\hat{A}^{ij}$ is the trace-free part of the conformally related extrinsic curvature, $\hat{A}_{ij} = \psi^2 (K_{ij} - \gamma_{ij} K/3)$. Quite remarkably, the momentum constraint (18) is now a linear equation and decouples from the Hamiltonian constraint. Solutions describing a pair of black holes with arbitrary momenta and spins can now be constructed analytically by super-imposing two solutions for single black holes [17]. Given these solutions, the Hamiltonian constraint (17) can then be solved numerically [18,19], which completes the construction of binary black hole initial data.

To construct binary black hole models in circular orbits and to determine their ISCO, turning points of the energy of these solutions have to be located, in complete analogy to the model problem above [20,21][3]. These results have recently been generalized to binaries in which the individual black holes carry spin [22].

Comparison

In Fig. 2, we summarize results from the post-Newtonian and numerical calculations. It is obvious that both the 2PN and 3PN results differ quite significantly from the numerical results, by more than a factor of two in the frequency. This discrepancy is very unsatisfactory, and should be resolved by re-examining both approaches and their assumptions.

The PN approach treats the binary stars as point-masses, and hence neglects tidal effects (even though these could be included, see, e.g., Appendix F of [14]). This approximation is probably fairly poor for binary neutron stars, but may be more adequate for binary black holes. Moreover, including tidal effects would probably move the ISCO to a larger separation and smaller orbital frequency, and would hence increase the discrepancy between the PN and numerical results. However, the convergence properties of the PN expansion and the lacking renormalization of higher-order expansion coefficients remain worrisome.

---

[3] The two approaches of [20] and [21] differ in the topology chosen for the binary black hole: [20] adopts the two-sheeted topology of [18], whereas [21] assumes the three-sheeted topology of [19]. Their results agree to within a few percent, indicating that this choice has little effect on the ISCO.

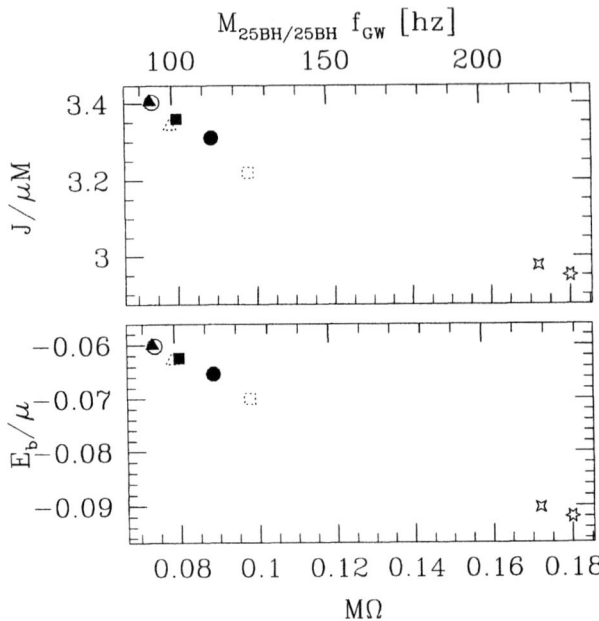

FIGURE 2. Results for the angular momentum, energy and orbital frequency of a black hole binary at the ISCO from post-Newtonian and numerical calculations. The solid triangle, square and circle are the 2PN results of [13] using the effective one-body method, the e-method, and the j-method. The open circle is their 3PN result, with the unknown parameter ω_{static} chosen such that all three methods agree. The four-pointed and six-pointed star are the numerical results of [20] and [21]. The dashed triangle and square are 2PN results using the effective one-body and e-method together with a conformal-flatness assumption. The top label gives the corresponding gravitational wave frequencies for a binary of two 25 $M_\odot$ black holes.

The biggest worry in the numerical calculations is probably the assumption of conformal flatness. The effect of this assumption is evaluated in Appendix B of Ref. [13], where the authors determine the ISCO at 2PN order, using their effective one-body and e-method together with a conformal-flatness assumption. These results are included as the dashed triangle and square in Fig. 2. Obviously, this assumption changes the results somewhat, but not nearly enough to explain the discrepancy between the PN and numerical results. When evaluated for the effective one-body method, it even seems like conformal flatness is quite an adequate approximation.

None of the above more obvious worries seem to be able to completely explain the differences between the PN and numerical results. This suggests that perhaps the two approaches construct sequences which are not physically identical. In our discussion of the model problem above, we have implicitly assumed that the masses

m_1 and m_2 of the binary stars remain constant during the inspiral. In general relativity, however, there is no unique definition of the mass of the individual black holes in a binary, and it is not clear which mass is conserved during the inspiral. It has been conjectured [23] that during the adiabatic inspiral outside of the ISCO the "irreducible mass" [24] of the black hole event horizons is conserved. In the numerical calculations of [18,21], this is approximated by the irreducible mass of the black hole apparent horizons[4]. In the PN approaches the masses m_1 and m_2 of the point sources are kept constant. It is not obvious that the two approximations are equivalent, and it is therefore possible that the two approaches construct physically distinct sequences.

THE ISCO IN BINARY NEUTRON STARS

The ISCO in binary neutron stars has been computed in numerous Newtonian [2,5,25–27], PN [3,7,28] and relativistic calculations [29–31]. We will briefly summarize some of the qualitative findings of these calculations, and will then discuss some of the more recent relativistic results.

Qualitative discussion

For binary neutron stars, the location of the ISCO depends on the physical properties of the individual stars, in particular their EOS and spin. Qualitatively, these effects can be illustrated by re-examining the tidal correction to the energy (9). Computing $E_{\rm eq}$ from (4) and (9), and then locating the minimum of $E_{\rm eq}$, shows that the ISCO occurs at a separation

$$\frac{r_{\rm ISCO}}{R} = (48\lambda)^{1/5} \tag{19}$$

and a frequency

$$m\,\Omega_{\rm ISCO} = \sqrt{\frac{5}{2}}\,(48\lambda)^{-3/10}\left(\frac{m}{R}\right)^{3/2}. \tag{20}$$

Note that $r_{\rm ISCO}$ scales with the stellar radius R, and $m\Omega_{\rm ISCO}$ accordingly with $(m/R)^{3/2}$, where m/R is the stellar compaction (compare [25,7,31]).

The coefficient λ depends on the EOS. For polytropic EOSs, λ increases with decreasing polytropic index n. Eq. (19) therefore implies that $r_{\rm ISCO}/R$ is larger for stiffer EOSs. This explains why an ISCO exists only for stars with sufficiently stiff EOSs; for stars with too soft EOSs the stars merge before they encounter the ISCO. The critical value of n for the existence of an ISCO depends on the rotation

---

[4] Locating an event horizon requires knowledge of a complete spacetime, while an apparent horizon can be located on a single timeslice.

of the stars and on relativistic effects, but is generally believed to be fairly close to $n_{crit} \sim 1$. Eq. (20) also implies that the ISCO frequency is correspondingly smaller for stiffer EOSs.

In the above discussion we have assumed that the stars are irrotational, so that $E_{tidal} \sim r^{-6}$. The effect of individual rotation can be estimated by allowing E_{tidal} to scale with a different power of r (compare [3,5,25]).

Numerical Calculations in General Relativity

Constructing relativistic binary neutron stars requires solving the equations of relativistic hydrodynamics together with the initial value equations of general relativity. This problem can be simplified whenever the fluid flow is stationary, so that the equations of hydrodynamics reduce to a relativistic Bernoulli equation[5].

This approach was first adopted by [29], who constructed corotational models of binary neutron stars in quasi-equilibrium. The initial value equations of general relativity were solved with the assumption that the spatial metric is conformally flat. For moderate (but realistic) compactions ($m/R \lesssim 0.2$), for which the gravitational fields are only moderately strong, the latter assumption has been shown to be quite adequate (compare [32]). Results for various polytropic indices n and stellar compactions m/R can be found in [29]. For stars of 1.4 $M_\odot$, $n = 1$ and $m/R = 0.2$, for example, they find a gravitational wave frequency at the ISCO of $\Omega_{ISCO}^{GW} = 2\,\Omega_{ISCO} \sim 1300\,\text{Hz}$.

The viscosity in neutron star interiors is not believed to be strong enough to maintain corotation during binary inspiral [33]. It is therefore more realistic to assume the binary to be irrotational, in which case the relativistic equations of hydrodynamics can again be reduced to a Bernoulli equation [34]. Constructing irrotational binaries is computationally more complicated than constructing corotational binaries, because one of the boundary conditions has to be imposed on the surface of the stars. The location of the latter changes during the numerical iteration, and is a priori unknown. Nevertheless, several groups have succeeded in constructing relativistic, irrotational models of binary neutron stars [30].

Probably the most careful analysis to date of evolutionary sequences of irrotational binary neutron stars is presented in [31]. As pointed out earlier [30], the authors find that for moderately soft EOSs ($n \gtrsim 2/3$), a cusp forms at the surface of the stars before an ISCO is encountered, at which point the numerical method breaks down. Most likely, the cusp indicates that a Lagrange point forms, and that matter starts to overflow towards the companion. The assumption of a stationary, irrotational fluid flow seems no longer adequate, and will have to be relaxed for the construction of closer binaries. An ISCO appears in these simulations only for fairly stiff EOSs ($n \lesssim 2/3$). For these cases, the authors of [31] quite carefully

---

[5] Note that a solution to the Bernoulli equation is by construction in equilibrium, and yields the equilibrium energy E_{eq}. For binary black holes the latter has to be constructed by finding turning points of the energy E.

analyze the ISCO for various compactions m/R, and show that it is dominated by the hydrodynamical effects of the tidal interaction. Up to moderate compactions ($m/R \lesssim 0.2$), the relativistic effects can be expressed as PN corrections to the Newtonian scaling (20).

SUMMARY

Knowledge of the ISCO may be an important ingredient for the future detection of gravitational wave signals and their interpretation. In this article we illustrate in a simple model problem how turning-point methods can be used to locate the ISCO (see, however, the disclaimer in footnote (2)). We also summarize recent efforts to determine the ISCO in both black hole and neutron star binary systems.

Currently, the biggest challenge for an accurate determination of the ISCO in binary black hole systems seems to be the disagreement between PN and numerical approaches. For binary neutron stars, we seem to be lacking a realistic description of the velocity field in close binaries. Ultimately, the quasi-equilibrium approaches presented in this paper should be backed up with fully dynamical simulations (e.g. [35]).

As a final comment, we point out that in this article we have only discussed binaries containing either black holes or neutron stars. So far black hole-neutron star binaries have only been modeled in Newtonian or PN dynamical simulations [36] and in ellipsoidal model calculations [37]. Such mixed binaries have yet to be constructed in a relativistic framework, and their ISCO (or the onset of tidal disruption) has yet to be determined self-consistently.

It is a pleasure to thank Greg Cook, Stuart Shapiro, and Masaru Shibata for numerous very useful discussions. The author gratefully acknowledges support through a Fortner Fellowship. This work was also supported by NSF Grant PHY 99-02833 at Illinois.

REFERENCES

1. See contribution by B. C. Barish in this volume.
2. D. Lai, F. A. Rasio and S. L. Shapiro, Astrophys. J. Suppl. **88**, 205 (1993).
3. F. C. Lombardi, F. A. Rasio and S. L. Shapiro, Phys. Rev. D **56**, 3416 (1997).
4. T. W. Baumgarte, G. B. Cook, M. A. Scheel, S. L. Shapiro and S. A. Teukolsky, Phys. Rev. D **57**, 6181 (1998).
5. D. Lai, F. A. Rasio and S. L. Shapiro, Astrophys. J. **406**, L63 (1993).
6. D. Lai, Phys. Rev. Lett. **76**, 4878 (1996).
7. D. Lai and A. G. Wiseman, Phys. Rev. D **54**, 3958 (1996).
8. A. Buananno and T. Damour, Phys. Rev. D **62**, 084036 (2000); A. Ori and K. S. Thorne, submitted (2000), also gr-qc/0003032.

9. R. V. Wagoner and C. M. Will, Astrophys. J. **210**, 764 (1976); E. Poisson, Phys. Rev. D **47**, 1497 (1993); L. Blanchet, T. Damour, B.R. Iyer, C.M. Will and A.G. Wiseman, Phys. Rev. Lett **74** 3515 (1995);

10. T. Damour, B. R. Iyer and B. S Sathyaprakash, Phys. Rev. D **57**, 885 (1998).

11. P. Jaranowski and G. Schäfer, Phys. Rev. D **57**, 7274 (1998).

12. A. Buananno and T. Damour, Phys. Rev. D **59**, 084006 (1999).

13. T. Damour, P. Jaranowski and G. Schäfer, Phys. Rev. D **62**, 084011 (2000).

14. C. M. Will and A. G. Wiseman, Phys. Rev. D **54**, 4813 (1996).

15. R. Arnowitt, S. Deser and C. W. Misner, in *Gravitation: An Introduction to Current Research*, edited by L. Witten (Wiley, New York, 1962).

16. J. W. York, Jr., Phys. Rev. Lett. **26**, 1656 (1971).

17. J. Bowen and J. W. York, Jr., Phys. Rev. D **21**, 2047 (1980).

18. G. B. Cook, Phys. Rev. D **44**, 2983 (1991).

19. S. Brandt and B. Brügmann, Phys. Rev. Lett. **78**, 3606 (1997).

20. G. B. Cook, Phys. Rev. D **50**, 5025 (1994).

21. T. W. Baumgarte, Phys. Rev. D **62**, 024018 (2000).

22. H. P. Pfeiffer, S. A. Teukolsky and G. B. Cook, Phys. Rev. D **62**, 104018 (2000).

23. J. D. Bekenstein, in *Black Holes, Gravitational Radiation and the Universe*, edited by B. R. Iyer and B. Bhawal (Kluwer, Dordrecht, 1999).

24. C. Christodoulou, Phys. Rev. Lett. **25**, 1596 (1970).

25. D. Lai, F. A. Rasio and S. L. Shapiro, Astrophys. J. **420**, 811 (1994).

26. F. A. Rasio and S. L. Shapiro, Astrophys. J. **401**, 226 (1992); Astrophys. J. **432**, 242 (1994); Astrophys. J. **438**, 887 (1995); Class. Quantum Grav. **16**, R1 (1999).

27. K. C. B. New and J. E. Tohline, Astrophys. J. **490**, , (3)11 (1997).

28. M. Shibata, K. Taniguchi and T. Nakamura, Prog. Theor. Phys. Suppl. **128**, 295 (1997); M. Shibata, K.-I. Oohara and T. Nakamura, Prog. Theor. Phys. **98**, 1081 (1997).

29. T. W. Baumgarte, G. B. Cook, M. A. Scheel, S. L. Shapiro and S. A. Teukolsky, Phys. Rev. Lett. **79**, 1182 (1997); Phys. Rev. D **57**, 7292 (1998).

30. S. Bonazzola, E. Gourgoulhon and J.A. Marck, Phys. Rev. Lett. **82**, 892 (1999); P. Marronetti, G. J. Mathews and J. R. Wilson, Phys. Rev. D **60**, 087301 (1999); K. Uryu and Y. Eriguchi, Phys. Rev. D **61**, 124023 (2000); E. Gourgoulhon, P. Grandclement, K. Taniguchi, J.-A. Marck and S. Bonazzola, submitted (also gr-qc/0007028).

31. K. Uryu, M. Shibata and Y. Eriguchi, Phys. Rev. D **62**, 104015 (2000).

32. F. Usui, K. Uryu and Y. Eriguchi, Phys. Rev. D **61**, 024039 (2000).

33. L. Bildsten and C. Cutler, Astrophys. J. **400**, 175 (1992); C. S. Kochanek, Astrophys. J. **398**, 234 (1992).

34. S. Bonazzola, E. Gourgoulhon and J.-A. Marck, Phys. Rev. D **56**, 7740 (1997); H. Asada, Phys. Rev. D **57**, 7292 (1998); S. A. Teukolsky, Astrophys. J. **504**, 442 (1998); M. Shibata, Phys. Rev. D **58**, 024012 (1998).

35. M. Shibata and K. Uryu, Phys. Rev. D **61**, 064001 (2000).

36. See contributions by M. Ruffert and W. Lee in this volume.

37. M. Shibata, Prog. Theor. Phys. **96**, 917 (1996); P. Wiggins and D. Lai, Astrophys. J. **532**, 530 (2000); M. Vallisneri, Phys. Rev. Lett. **84**, 3519 (2000).

Cosmology

Stochastic Gravitational Radiation from Phase Transitions

Arthur Kosowsky*, Andrew Mack* and Tinatin Kahniashvili†

*Department of Physics and Astronomy, Rutgers University
Piscataway, New Jersey 08854-8019
†Abastumani Astrophysical Observatory, Tbilisi, Republic of Georgia

Abstract. A stochastic background of gravitational radiation from cosmological processes in the very early Universe is potentially detectable. We review the gravitational radiation which may arise from cosmological phase transitions, covering both bubble collisions and turbulence as sources. Prospects for detecting a direct signal from the electroweak phase transition or other cosmological sources with a space-based laser interferometer are discussed.

INTRODUCTION

Scientific and design goals for gravitational radiation detectors have been driven primarily by point sources of radiation like binary compact objects. While the distribution and properties of such sources will undoubtedly be of cosmological interest, the most interesting gravitational wave signals for cosmology are likely to be stochastic backgrounds produced in the early Universe. Since the Universe is opaque to electromagnetic radiation at redshifts greater than $z = 1300$ due to Thomson scattering off free electrons, gravitational radiation offers the only hope of "imaging" the earliest epochs of the Universe when it was dominated by high-energy fundamental physics processes.

Several possible stochastic sources have been considered. The best motivated are cosmic defects like strings or textures; quantum fluctuations in the fields driving inflation; and phase transitions. More speculative ideas include string-theory inspired scenarios for cosmic evolution at times prior to the apparent initial singularity (so-called "pre-big-bang" cosmology) or explosive preheating at the end of inflation. Here we review the stochastic background of gravitational radiation expected from first-order phase transitions. New results for gravitational radiation from turbulence, which could be produced in a first-order phase transition, are included. The final section considers the detectability of various stochastic sources, with an emphasis on the electroweak phase transition. A thorough review of stochastic gravitational radiation, with extensive material about detector phe-

CP575, *Astrophysical Sources for Ground-Based Gravitational Wave Detectors,* edited by J. M. Centrella
© 2001 American Institute of Physics 0-7354-0014-8/01/$18.00

nomenology, overviews of a variety of sources, and a comprehensive bibliography, has recently appeared [1].

GENERAL CONSIDERATIONS

The time evolution of cosmological gravitational radiation follows directly from the linearized Einstein equations,

$$\ddot{h}_{ij}(\mathbf{k}, \eta) + 2\frac{\dot{a}}{a}\dot{h}_{ij}(\mathbf{k}, \eta) + k^2 h_{ij}(\mathbf{k}, \eta) = 8\pi G a^2 \Pi_{ij}(\mathbf{k}, \eta), \tag{1}$$

where $\mathbf{k}$ is a comoving wave vector, η is conformal time, overdots are derivatives with respect to η, a is the scale factor of the Universe, h_{ij} is the tensor metric perturbation, and Π_{ij} is the tensor piece of the stress-energy tensor. We will also make use of the Hubble parameter, $H = (1/a)da/dt$ and its present value h_{100} in units of 100 km/s/Mpc, and throughout we use natural units with $\hbar = c = k_B = 1$. The right side of Eq. (1) is the source term for generating gravitational radiation, while the left side describes the free propagation of the radiation, including damping due to the expansion of the Universe. Since each comoving $\mathbf{k}$-mode is independent, the physical wavelength of a given gravitational wave increases linearly with the scale factor a, just as with electromagnetic radiation.

Solutions to the homogeneous equation with $\Pi = 0$ are easily obtained. For a radiation-dominated universe, $a \propto \eta$ and

$$h_{ij} \propto j_0(k\eta), \qquad y_0(k\eta) \tag{2a}$$

while for the matter-dominated era, $a \propto \eta^2$ and

$$h_{ij} \propto \frac{j_1(k\eta)}{k\eta}, \qquad \frac{y_1(k\eta)}{k\eta}, \tag{2b}$$

where j_l and y_l are the usual spherical Bessel functions. It is straightforward to show that in both cases, the amplitude of the gravitional waves scales as a^{-1}. This result also follows from the stress-energy tensor for gravitational radiation, whose energy density is given by [2]

$$\rho_{GW} = \frac{1}{32\pi G}\left\langle \frac{\partial h_{ij}}{\partial t}\frac{\partial h^{ij}}{\partial t}\right\rangle \tag{3}$$

where the average is over many wavelengths. So the energy density is proportional to $h_c^2 f^2$, where h_c is the characteristic amplitude of the metric perturbation h_{ij} and f is the (physical) frequency of the wave. The energy density scales like a^{-4} with the expansion of the Universe, while the frequency scales like a^{-1}. It follows that $h_c \propto a^{-1}$ as derived above from the evolution equation. Note this is different from the amplitude scaling of the electric and magnetic fields in an electromagnetic wave, which drop off like a^{-2}.

In practice, only very large stress-energy sources contribute significantly to the stochastic gravitational wave amplitude. If a particular source acts for a period of time shorter than the Hubble time H^{-1}, then the expansion of the Universe can be neglected during the time the source is active and the resulting gravitational wave amplitude can be conveniently calculated in flat space. Then the entire subsequent evolution of the gravitational waves is simply a reduction of amplitude and frequency by the factor

$$\frac{a_*}{a_0} = 8.0 \times 10^{-14} \left(\frac{100}{g_*}\right)^{1/3} \left(\frac{1 \text{ GeV}}{T_*}\right), \tag{4}$$

where a_* and T_* are the scale factor and temperature at the time the source is active, g_* is the total number of relativistic degrees of freedom at the time of the source, and a_0 is the scale factor today. This expression is valid as long as the evolution of the Universe since the source has been adiabatic. If a source exists for a time interval long compared to the Hubble time, then the expansion of the Universe may impact the source evolution and the full evolution equation, Eq. (1), must be used.

A stochastic background of gravitational waves can be characterized by $\Omega_{\text{GW}}(f)$, its energy density per frequency octave in units of the critical density $\rho_c = 3H^2/8\pi G$. From this, we follow Thorne [3] in defining a convenient characteristic amplitude at frequency f as

$$h_c(f) \equiv 1.3 \times 10^{-18} \left[\Omega_{\text{GW}}(f)h_{100}^2\right]^{1/2} \left(\frac{1 \text{ Hz}}{f}\right). \tag{5}$$

We will calculate characteristic energy densities in gravitational wave backgrounds and convert to characteristic amplitudes with this relation.

MODEL PHASE TRANSITIONS

The most interesting potential source of a cosmological gravitational wave background is an early-Universe phase transition. Phase transitions very likely occurred in the early Universe. The standard model of particle physics provides two: the electroweak phase transition at $T \simeq 100$ GeV, at which the electroweak symmetry was spontaneously broken, and the QCD phase transition at $T \simeq 150$ MeV, at which chiral symmetry was broken. If the standard model is actually the result of the breaking of a larger gauge symmetry (i.e. a Grand Unified Theory), then a GUT phase transition at an energy scale of $T \simeq 10^{16}$ GeV is also likely. Other more speculative phase transitions involving the breaking of hypothesized additional symmetries have also been discussed.

The details of a particular phase transition, in particular its order, latent heat, and dynamics, are determined by the effective potential driving the phase transition. In a given particle theory, calculation of an effective potential is in general a difficult, non-perturbative problem of finite-temperature field theory. For purposes of gravitational wave signals, we consider a generic, model first-order phase transition described by several physical parameters; a specific phase transition corresponds to some set of model parameters.

As the Universe cools due to its expansion, phase transitions can occur if a new state with lower energy density becomes physically possible. Assume this occurs at some characteristic temperature scale T_*. The energy density difference between the two phases is roughly the vacuum energy density, ρ_{vac}. If the two phases are separated by a significant potential energy barrier (greater than the characteristic thermal energy density at temperature T_*), then the transition must proceed via the nucleation of new-phase bubbles via quantum or thermal processes. We define $\alpha \equiv \rho_{\text{vac}}/a_R T_*^4$, the ratio of vacuum energy density to thermal energy density (a_R is the radiation constant). Once a bubble of new phase is nucleated, the potential energy difference between the two phases exerts an outward force on the walls of the bubble, causing it to expand. Hydrodynamic forces will tend to resist the bubble expansion, which will quickly attain some equilibrium velocity v. We neglect possible instabilities which might distort the bubble's spherical shape and redistribute energy. As the bubbles expand, some fraction κ of the available vacuum energy is converted to the kinetic energy of the expanding bubble walls, while the rest goes into heat.

Once multiple bubbles have been nucleated in some region of the Universe, they will expand until eventually their walls meet and they merge together, converting the entire region to the lower-energy phase. The energy stored up in the bubble walls is eventually dissipated into heat via hydrodynamical turbulence. If α is large enough, the kinetic energy of the bubble walls will represent a significant fraction of the entire energy density of the Universe, and will have v a significant fraction of c. Note that for $T_* > 10^5$ K, the energy density of the Universe is dominated by radiation, so the sound speed will be $c_s = c/\sqrt{3}$, the relativistic limit. As soon as the spherical symmetry of an individual bubble is broken due to collisions with other bubbles, gravitational radiation will be produced; the high energy densities and velocities involved make bubble collisions a potentially potent source [4–6].

The characteristic frequency of the gravitational radiation is determined by the duration of the phase transition, corresponding to the percolation time of the biggest bubbles with the largest energies. A simple bubble nucleation model takes an exponential bubble nucleation rate per unit volume $\Gamma = \Gamma_0 \exp(\beta t)$ [7]. This form is motivated by the general observation that Γ will typically be the exponential of some nucleation action, and the time dependence of the nucleation action can be approximated as linear at the time of the phase transition. For such a nucleation rate, the duration of the phase transition is roughly $\tau \simeq \beta^{-1}$. The characteristic length scale is just $\beta^{-1}v$, the size of the largest bubbles at the end of the phase transition. In general, $\beta \simeq 4\ln(m_{\text{Pl}}/T_*)H_* \simeq 100 H_*$, where H_* is the Hubble parameter at the time of the phase transition [7]. Thus the characteristic wavelength of gravitational radiation at the time of the phase transition is perhaps a percent of the Hubble radius at that time.

A simple model of a first-order phase transition is comprised of the quantities defined above: the characteristic time scale β^{-1}; the vacuum energy ρ_{vac}; the bubble expansion velocity v; the temperature of the phase transition T_* or equivalently the ratio of vacuum energy to thermal energy α; and the efficiency factor κ.

The gravitational radiation resulting from the violent collision of bubbles in a first-order phase transition has been calculated in detail [8–10]. It was demonstrated that the radiation source is primarily the bubble walls, which for can be treated in the thin-wall approximation if the bubble walls propagate as detonation fronts. The traceless portion of the stress-energy tensor is nonzero only at the position of the spherically expanding bubble walls, and the amplitude of the stress-energy tensor follows simply from the amount of vacuum energy liberated by a particular bubble's expansion. Then the radiation from many randomly nucleated bubbles in a sample volume can be computed numerically. The result is a stochastic background of radiation with energy density

$$\Omega_{\text{GW}}(f)h^2 \simeq 1.1 \times 10^{-6}\kappa^2 \left(\frac{H_*}{\beta}\right)^2 \left(\frac{\alpha}{1+\alpha}\right)^2 \left(\frac{v^3}{0.24 + v^3}\right) \left(\frac{100}{g_*}\right)^{1/3}, \quad (6)$$

peaking at a characteristic frequency

$$f_{\text{max}} \simeq 5.2 \times 10^{-8}\,\text{Hz} \left(\frac{\beta}{H_*}\right) \left(\frac{T_*}{1\,\text{GeV}}\right) \left(\frac{g_*}{100}\right)^{1/6}, \quad (7)$$

with characteristic amplitude

$$h_c(f_{\text{max}}) \simeq 1.8 \times 10^{-14}\kappa \left(\frac{\alpha}{\alpha+1}\right) \left(\frac{H_*}{\beta}\right)^2 \left(\frac{1\,\text{GeV}}{T_*}\right) \left(\frac{v^3}{0.24 + v^3}\right)^{1/2} \left(\frac{100}{g_*}\right)^{1/3}. \quad (8)$$

At frequencies lower than the characteristic peak frequency, the energy density per frequency octave increases like $f^{2.5}$ while above this frequency it drops off more slowly [9]. For very strong phase transitions, the energy spectrum of gravitational radiation has a long tail at high frequencies, due to the presence of numerous small bubbles with significant energy densities. These results hold for a strongly first-order phase transition where the bubbles expand supersonically as detonation fronts. For weaker transitions with deflagration (subsonic) bubble walls, the dynamics are more complex, but the resulting radiation is expected to be much smaller.

Note that the frequency range of the proposed LISA space-based interferometer, 10^{-5} to 10^{-1} Hz, is particularly well-suited to probing the electroweak phase transition at an energy scale of 100 GeV: β/H_* is generically around 100, putting the frequency for the peak of the gravity wave power spectrum somewhere between 10^{-3} and 10^{-4} Hz. The strength of the electroweak phase transition is an open question; this issue is discussed below in the concluding section.

MODEL TURBULENCE

Substantial gravitational radiation can also be produced by a period of turbulence in the early Universe [10]. The following analysis [11] is completely general, without reference to a particular source of turbulence, although the most likely source is a first-order phase transition: once the bubbles of low-temperature phase begin to collide with each other, they stir up the primordial plasma on a characteristic length scale corresponding to the size of the largest bubbles in the transition. The gravitational radiation produced by the turbulence will be in addition to that from the bubbles given above.

We construct a simple model of isotropic cosmological turbulence as follows: at some particular temperature T_* and corresponding enthalpy $w_* = \rho_* + P_*$ the turbulence commences, with energy input into the Universe on a largest length scale L_S. In general, the energy input will be a complicated process; we make the simple idealization that the turbulence will last for a time τ and that the turbulent energy will be distributed as a stationary Kolmogoroff spectrum with

$$E(k) \equiv \frac{1}{w} \frac{d\rho_{\text{turb}}}{dk} \simeq \bar{\varepsilon}^{2/3} k^{-5/3}, \tag{9}$$

where $\bar{\varepsilon}$ is the energy density dissipation rate per unit enthalpy [12]. The turbulent energy injected at the scale L_S will be written as $\epsilon\rho_{\text{vac}}$, with ϵ an efficiency factor. This energy cascades to smaller scales until it is dissipated by viscous damping at a scale determined by the kinematic viscosity ν of the fluid. The damping scale L_D can be obtained from the Reynolds number, which can be approximated as

$$\text{Re} = \left(\frac{L_S}{L_D}\right)^{4/3} \simeq \frac{2}{3}\left(\frac{L_S}{2\pi}\right)^{4/3}\left(\frac{\epsilon\rho_{\text{vac}}}{\nu^3 \tau w}\right)^{1/3}, \tag{10}$$

if the turbulent source lasts for a long time compared to the dynamical time scale of the turbulence on the scale L_S. The critical Reynolds number for the development of a Kolmogoroff spectrum of turbulence is around 2000; generally this number is easily exceeded for early-Universe phase transitions. If the source lasts for a time short compared to the dynamical time at the scale L_S (as is likely with the electroweak phase transition [10]), then fully developed turbulence never appears, but we argue on physical grounds that, as far as gravitational wave production is concerned, this case can be treated like fully-developed turbulence which lasts for one dynamical time on the scale L_S [11].

In analogy with the first-order phase transition calculation, the model turbulence is completely described by the physical quantities defined above: the characteristic time scale τ, which we write as βH_*^{-1}; the energy density which goes into turbulent motions, $\epsilon\rho_{\text{vac}}$; the characteristic length scale L_S, which we write as γH_*^{-1}; the temperature of the phase transition T_*, which determines the enthalpy density w_*; and the kinematic viscosity ν, which fixes the scale L_D at which the turbulence is dissipated into heat.

Given this model, the turbulent stress-energy tensor can be constructed. The starting point is the Fourier-space expression

$$T_{ij}(\mathbf{k}) = \frac{w}{(2\pi)^3} \int d\mathbf{q}\, u_i(\mathbf{q}) u_j(\mathbf{k} - \mathbf{q}) \tag{11}$$

where $\mathbf{u}$ is the relativistic velocity of the fluid at a given point and w is the fluid enthalpy, assumed to be independent of position. The source for gravitational radiation can be obtained via a projection tensor [13],

$$\Pi_{ij}(\mathbf{k}) = \left(P_{il} P_{jm} - \frac{1}{2} P_{ij} P_{lm} \right) T_{lm}(\mathbf{k}) \tag{12}$$

where $P_{ij} \equiv \delta_{ij} - \hat{\mathbf{k}}_i \hat{\mathbf{k}}_j$ is a projector onto the transverse plane. The connection to isotropic turbulence is obtained via the two-point correlation function

$$\left\langle u_i(\mathbf{k}) u_j^*(\mathbf{k}') \right\rangle = (2\pi)^3 P_{ij} P(k) \delta(\mathbf{k} - \mathbf{k}') \tag{13}$$

which holds for any statistically isotropic and homogeneous velocity field. For Kolmogoroff turbulence, the power spectrum is

$$P(k) \simeq \pi^2 \bar{\varepsilon}^{2/3} k^{-11/3}. \tag{14}$$

After a lengthy calculation which combines these pieces, the resulting gravitational radiation spectrum can be estimated as [11]

$$h_c(f) \simeq 7 \times 10^{-14} \gamma^{5/3} \beta^{1/3} \left(\frac{\epsilon \rho_{\text{vac}}}{w_*} \right)^{2/3} \left(\frac{100}{g_*} \right)^{1/3} \left(\frac{1\,\text{GeV}}{T_*} \right) \left(\frac{f_S}{f} \right)^{4/3}, \tag{15}$$

with the lowest frequency f_S corresponding to the comoving scale L_S,

$$f_S = 1.6 \times 10^{-7}\,\text{Hz}\, \gamma^{-1} \left(\frac{T_*}{1\,\text{GeV}} \right) \left(\frac{g_*}{100} \right)^{1/6}, \tag{16}$$

using Eq. (4). The spectrum of the gravitational radiation extends to the highest frequency corresponding to the dissipation length scale. This can be approximated, but it suffices to note that the Reynolds number is large in the early Universe, so the range of frequencies will span at least a factor of several hundred. Turbulence can be just as efficient at producing gravitational radiation as the first-order phase transition that spawned it. In cases where the phase transition is only weakly first-order, the turbulence may be the dominant source.

Turbulence has the possibility of producing gravitational radiation via another mechanism. If any small seed magnetic field is present at the onset of turbulence, a turbulent dynamo will amplify the field exponentially until it saturates at equipartition with the turbulent kinetic energy. In Kolmogoroff turbulence, the dynamical time on a given scale is shorter at smaller scales, so the diffusion scale magnetic fields will saturate first. Typically, the smallest scale dynamical time is at least 100 times shorter than the largest scale dynamical time, so the magnetic seed field on the smallest scales will be amplified by at least 100 e-foldings. We can crudely approximate the resulting magnetic field as isotropic with a power spectrum that has the Kolmogoroff form from the smallest scale up to some saturation scale, and then an exponential drop up to the largest scale.

The magnetic fields generated by turbulence are potentially interesting because they also generate gravitational radiation. Roughly, the equipartition magnetic fields are as efficient as the turbulence at making gravitational waves, with the crucial difference that the magnetic fields last much longer. The turbulence damps out on the turbulence time scale, while the magnetic fields remain until matter-radiation equality. The resulting gravitational radiation energy density can be approximated as that from the turbulence times an additional factor of $\ln(z_*/z_{eq})$, where z_* is the redshift of the turbulence and z_{eq} is the redshift of matter-radiation equality. For the electroweak phase transition, this factor is around 25, giving a large increase in the gravitational wave amplitude. This mechanism would yield a distinctive power spectrum of gravitational waves: it peaks at a frequency corresponding to the largest scale on which the dynamo saturates, drops off like a power law at smaller scales, and drops off exponentially at larger scales. The major question about this mechanism is whether an adequate seed field exists. The plasma in the very early Universe is strongly in the MHD regime (collision time short compared to dynamical time) and has a very high conductivity, making it difficult for any seed field generation mechanism to work. It has been suggested that the expanding bubbles in the electroweak phase transition will produce adequate seed fields [14], while thermal seed fields have also been advocated [15]. We are currently investigating gravitational radiation from turbulent-dynamo magnetic fields in detail.

DETECTABILITY OF STOCHASTIC SOURCES

Phase transitions are almost sure to have occurred in the early Universe; in particular, electroweak symmetry breaking at $T \simeq 100$ GeV and the QCD phase transition at $T \simeq 1$ GeV follow from our current understanding of the standard model of particle physics. However, only phase transitions which are first order are likely to have caused large enough stress-energy fluctuations to leave behind a significant gravitational wave background. In the standard model, both the electroweak and QCD phase transitions appear to be second-order. On the other hand, no high-energy theorist believes that the standard model is the final story. Once additional symmetries and fields are allowed, as with supersymmetry, the strength of the electroweak phase transition can be greatly enhanced. Belief in a strongly first-order electroweak phase transition is motivated by electroweak baryogenesis [16], which requires significant departure from thermodynamic equilibrium and thus a strongly first-order phase transition (but see [17]). Baryogenesis at the electroweak energy scale is a natural possibility, since at higher energies, sphaleron processes which violate baryon number occur rapidly enough to erase any net baryon number. The very existence of matter in the Universe may be connected to a detectable gravitational wave background; we are currently studying this issue.

Inflationary gravitational radiation may be detectable with LISA, although it is difficult to come up with sensible models which give signals much above the anticipated LISA sensitivity [18]. The amplitude of the gravitational radiation at the present horizon scale is limited by the COBE measurement of large angular scale fluctuations in the microwave background temperature, and general considerations of slow-roll inflation argue that the power spectrum of gravitational radiation must drop as the scale decreases. Detection of such a background would be extremely important, because it would reflect the energy scale at which inflation occurred.

Cosmic defect sources, while well-motivated, offer less hopeful direct detection prospects. At this point, measurements of large-scale structure and the microwave background likely rule out any cosmic defects as the primary mechanism for structure formation in the Universe [19]. On the other hand, defects are a generic product of symmetry breaking. The gauge group of the standard model of particle physics is $SU(3)_C \times SU(2)_L \times U(1)_Y$, which has a U(1) subgroup. A topological theorem states that the breaking of any larger group to a subgroup containing a U(1) factor will result in one-dimensional defects, i.e. cosmic strings. Thus the production of cosmic strings is a generic feature of any cosmology incorporating breaking of a grand-unified (GUT) symmetry to the standard model. (although if the GUT transition is inflationary, the number density of cosmic strings will be exponentially suppressed, making them completely irrelevant). Even if strings or other defects do not drive structure formation, they could conceivably produce detectable amounts of stochastic gravitational radiation [20].

Other cosmological sources of gravitational radiation, such as an early string phase of cosmic evolution [21] or complex field dynamics during reheating after inflation [22], offer detection possibilities, but these sources are based on more speculative underlying physics. Until further theoretical advances put them on more secure footing, they are perhaps best considered examples of the flavor of theories which might give an unanticipated signal.

In summary, potential cosmological sources of a stochastic background of gravitational radiation are more speculative than the compact binary point sources towards which LIGO is largely oriented. But several sources are well motivated, and the potential scientific payoff is spectacular. In particular, a direct probe of electroweak symmetry breaking with the LISA space-based interferometer is a distinct possibility, as is the direct detection of gravitational waves from inflation. All conjectured gravitational wave sources in the early Universe are intimately connected with fundamental physics at energy scales between 100 GeV and 10^{16} GeV. For all but the lowest portion of this range, gravitational radiation represents the *only* direct probe of physics available with currently envisioned technology. It also seems likely that stochastic backgrounds offer a greater chance at serendipidous discovery than point sources, since the evolution of the very early Universe is determined by physics which is partly unknown, while point sources are probably known astrophysical objects. We strongly encourage designers of gravitational radiation detectors to make stochastic background sensitivity a primary design consideration.

This work has been supported by NASA's Astrophysics Theory Program. T.K. has been partially supported by the National Research Council's COBASE program. A.K. is a Cotrell Scholar of the Research Corporation.

REFERENCES

1. Maggiore, M., *Phys. Rep.* **331**, 283-367 (2000).
2. Misner, C.W., Thorne, K.S., and Wheeler, J.A., *Gravitation*, New York: W.H. Freeman (1973).
3. Thorne, K.S., in *300 Years of Gravitation*, ed. S. Hawking and W. Israel, Cambridge, England: Cambridge University Press (1987).
4. Witten, E., *Phys. Rev. D* **30**, 272 (1984).
5. Hogan, C.J., *Mon. Not. R. Ast. Soc.* **218**, 629 (1986).
6. Turner, M.S. and Wilczek, F., *Phys. Rev. Lett.* **65**, 3080 (1990).
7. Turner, M.S., Weinberg, E.J., and Widrow, L.M., *Phys. Rev. D* **46**, 2384 (1992).
8. Kosowsky, A., Turner, M.S., and Watkins, R., *Phys. Rev. D* **45**, 4514 (1992).
9. Kosowsky, A., and Turner, M.S., *Phys. Rev. D* **47**, 4372 (1993).
10. Kamionkowski, M., Kosowsky, A., and Turner, M.S., *Phys. Rev. D* **49**, 2837 (1994).
11. Mack, A., Kahniashvili, T., and Kosowsky, A., in preparation (2001).
12. Monin, A.S. and Yaglom, A.M., *Statistical Fluid Mechanics*, vol. 2, chapter 7, Cambridge: MIT Press (1975).
13. Durrer, R., Ferreira, P.G., and Kahniashvili, T., *Phys. Rev. D* **61**, 043001 (2000).
14. Sigl, G., Olinto, A., and Jedamzik, K., *Phys. Rev. D* **55**, 4582 (1997).
15. Tajima, T., Cable, S., Shibata, K., and Kulsrud, R.M., *Ap. J.* **390**, 309 (1992).
16. Trodden, M., *Rev. Mod. Phys.* **71**, 1463 (1999).
17. Trodden, M. and Gleiser, M., submitted to *Phys. Rev. Lett.*, preprint astro-ph/9911380 (1999).
18. Liddle, A., *Phys. Rev. D* **49**, 3805 (1994).
19. Albrecht, A., Battye, R.A., and Robinson, J., *Phys. Rev. D* **59**, 023508 (1999).
20. Vachaspati, T. and Vilenkin, A., *Phys. Rev. D* **31**, 3502 (1985).
21. Lidsey, J.E., Wands, D., and Copeland, E.J., *Phys. Rep.* **337**, 343 (2000).
22. Basset, B., *Phys. Rev. D* **56**, 3439 (1997); Khlebnikov, S. and Tkachev, I., *Phys. Rev. D* **56**, 653 (1997).

The Cosmic Microwave Background
A Gravity Wave Detector

David N. Spergel[1]

Princeton University, Princeton, NJ 08544
Institute for Advanced Study, Princeton, NJ 08540

Abstract. Long wavelength gravity waves produce a distinctive signature of polarization fluctuations in the cosmic microwave background. With very sensitive experiments with minimal systematics, this signal is potentially detectable. Since the universe was transparent to gravity waves back to the Planck time, gravity wave observations are a probe of physics in the early universe and physics at very high energy scales.

In this talk, I review the current status of microwave background experiments. I then discuss the coming generation of experiments that should be of detecting polarization fluctuations produce by scalar fluctuations. A future satellite experiment (beyond MAP and Planck) will likey be needed to detect tensor fluctuations.

INTRODUCTION: CURRENT STATE OF THE FIELD

On large scales, metric fluctuations generate CMB fluctuations. Adiabatic primordial variations in the gravitational potential: $\Delta T/T = -\Phi/3$. In inflationary models, these potential fluctuations are thought to be scale invariant. COBE's measurements on large angular scales are consistent with this basic picture [1]. Once these potential fluctuations enter the casual horizon, matter starts to move. This leads to density and velocity fluctuations that produce additional fluctuations on smaller angular scales.

---

[1] W.M. Keck Distinguished Visiting Professor of Astrophysics

TABLE 1. Parameters for three different models consistent with the MAXIMA, BOOMERANG and COBE data.

	n	h	τ	Ω_b	Ω_c
A	0.9	0.9	0.0	0.03	0.15
B	1.3	0.9	0.4	0.05	0.25
C	1.1	0.6	0.2	0.1	0.7

CP575, *Astrophysical Sources for Ground-Based Gravitational Wave Detectors,* edited by J. M. Centrella
© 2001 American Institute of Physics 0-7354-0014-8/01/$18.00

The characteristic physical scale for microwave background fluctuations is roughly set by the causal horizon at decoupling. Since the electrons and photons are tightly coupled, the actual characteristic scale is set by the acoustic horizon, the distance that a sound wave can travel from the beginning of the universe to the present time. The angular size of the characteristic fluctuations is set by the geometry of the universe. If the universe is flat, this fixed physical size corresponds to roughly 1 degree. On the other hand, if the universe is negatively curved (and the matter density today is roughly 1/3 of the critical density), then the characteristic angular size is smaller, roughly 0.5 degrees. If the universe were positively curved, then the characteristic angular size for microwave background fluctuations would be larger than 1 degree.

Current data suggests that the universe was flat and fluctuations were produced by inflation. In 1999, the TOCO experiment [9] first saw clear evidence for a peak near the one degree scale. In 2000, several experiments [8,3,4] showed every improving evidence for a clear peak in the microwave background fluctuations spectrum near the one degree scale. The amplitude of the fluctuations as a function of scale is roughly fit by inflationary models.

At the time of this article, the current best data sets appear to be in mild conflict. While all of the data sets are roughly consistent with each other and standard theory, there are several intriguing problems. The characteristic scale of

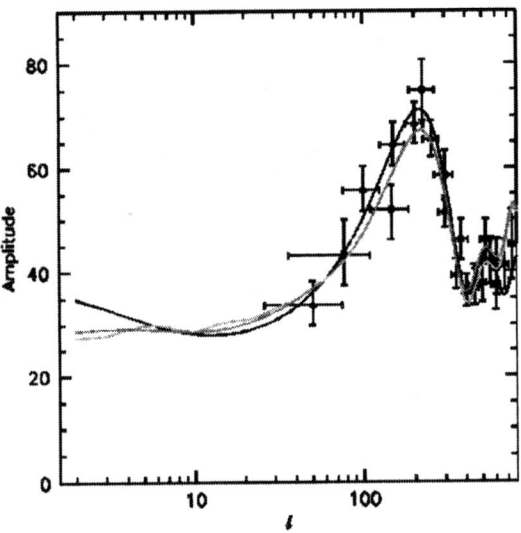

FIGURE 1. This figure shows the predictions of three different models (see table 1). The black, green and cyan lines correspond to models A,B,C. The data points are from the MAXIMA and BOOMERANG experiments. All of the models are normalized to fit the COBE data.

fluctuations in the BOOMERANG data [3] is somewhat larger than expected for an inflationary model with a cosmological constant. The amplitude of fluctuations on small angular scales in the BOOMERANG and MAXIMA [4] data is smaller than expected. One possible explanation for this "missing second peak" is that the density of baryons in the universe is 50% higher than previously thought [11]. Recently, the CBI experiment [10] reported initial results 2.7 σ higher than the BOOMERANG experiment (and 1.9σ higher than the MAXIMA experiment). The CBI results are consistent with standard values for the density of baryons.

In the past few months, there have been many papers using the current CMB data to place constraints on cosmological models. My own interpretation is that there are currently a wide range of flat universe models consistent with the current data. However, it is difficult to fit the current data with an open universe or with deffect models. As an illustration, I have listed 3 very different flat universe models in Table 1. These models are defined by six parameters: the overall normalization, the spectra index (n), the optical depth (τ), the baryon density (Ω_b) and the dark matter density Ω_c). The power spectra for these models are shown in figure 1.

The quality of experimental data will be improving dramatically in the coming months. John Carlstrom's DASI experiment will soon report results from the South Pole. Further analysis should yield improved results from the MAXIMA, BOOMERANG, and CBI experiments. The British CAT experiment should also soon report new measurements around the second peak. The TOPHAT experiment has just had a flight around the South Pole. With its good sky coverage, it should be able to better characterize the amplitude of fluctuations near the first peak.

In June 2001, NASA plans to launch the Microwave Anisotropy Probe[2]. MAP will have an angular resolution of 0.2 degrees and will image the entire sky at five different frequencies. MAP is a differential experiment that has been designed to minimize systematic errors. The MAP data should represent a significant step beyond the existing data sets and should provide a powerful test of inflation and of cosmological models [14].

POLARIZATION FLUCTUATIONS

While most of the recent attention has focused on temperature fluctuations in the microwave background, there is wealth of information in measurements of polarization fluctuations in the microwave background.

The electron-photon scattering cross-section,

$$\frac{d\sigma}{d\Omega} = \frac{3\sigma_T}{8\pi} |\hat{\epsilon}' \cdot \epsilon|^2 ,\qquad (1)$$

depends on the polarization of the incoming and outgoing waves, $\hat{\epsilon}$ and $\hat{\epsilon}'$. Here, σ_T is the Thompson cross-section. physical origin. Since scattering is anisotropic,

---

[2] http://map.gsfc.nasa.gov

measurements of the microwave polarization are sensitive to the local quadrupole seen by an electron at the surface of last scatter.

Scalar and tensor fluctuations produce distinctive signatures in the microwave sky. Scalar fluctuations produce "E-like" pattern of polarization fluctuations [7,13]: the fluctuations can be expressed as a gradiant of a scalar field. Tensor fluctuations produce both E-like and B-like fluctuations in the CMB. The B-like fluctuations have a non-zero curl associated with them on the sky. Observed polarization fluctuations can be decomposed into these two modes, so that we can detect the distinct signature of tensor modes. Martin White and Wayne Hu's CMB polarization primer (http://cfa-www.harvard.edu/ mwhite/polar/) provides a general introduction to the physics of these fluctautions.

There are a number of possible sources of these tensor fluctuations. Arthur Kosowsky will discuss in the subsequent talk how phase transitions in the early universe can generate gravity waves. Inflationary models predict a gravity wave signal whose amplitude is a sensitive probe of the details of the inflaton potential [2]. Since the universe was transparent to gravity waves back to the Planck time, compactification or other very early universe physics may be detectable through gravity waves [6].

CURRENT AND FUTURE EXPERIMENTS

Detecting these weak fluctuations is a demanding experimental challenge. This challenge is complicated by the presence of a number of galactic foregrounds. Galactic synchrotron radiation is highly polarized and is likely to be the dominant foreground at frequencies below 60 GHz. Active galactic nuclei and other point sources will be a significant foreground at small angular scales. The polarization fraction of dust emission is not well known and may be a significant foreground at shorter wavelengths.

The control of systematic errors is essential for these experiments. While the atmosphere is not significantly polarized, emission from the Earth's surface is polarized. Thus, ground and balloon-based experiments must have very accurate control of their sidelobes. A 80 dB sidelobe that picks up the Earth's 300 K signal will completely swamp the sub-micro K polarization signal. Since most experiments have different sidelobes in each polarization, this is a potentially serious experimental problem.

Suzanne Staggs' PIQUE experiment [3] has reported the best limits on CMB polarization fluctuations [5]. This experiment operated near 90 GHz where the foregrounds are thought to be lowest. It placed a limit of 14 and 13 microKelvin on E and B-like polarization fluctuations.

There are several on-going experiments that are aiming to detect the signature of CMB polarization fluctuations. The POLATRON experiment[4] at Caltech will use

---

3) http://dicke.princeton.edu/~pique/pique.html
4) http://www.astro.caltech.edu/~lgg/polatron/ppro.html

bolometers to measure polarization near 100 GHz on the scale of several arcminutes. John Ruhl's COMPOSAR experimen[5] plans to use a 5 meter diameter mirror to detect polarization fluctuations on the sub-degree scale. The CMBRope experiment at Berkeley[6] is trying to detect fluctuations on the 2 degree scale. Peter Timbie's Polar experiment[7] aims to detect fluctuations on the 10 degree scale. Later this year, a new flight of the BOOMERANG mission should achieve the most sensitive measurements of CMB polarization.

MAP will make all sky polarization maps at all five frequencies. With its two year data, MAP will have a sensitivity of roughly 45 micro Kelvin per $0.3° \times 0.3°$ pixel at its three highest frequencies. Since the expected polarization signal is only 3 micro Kelvin, MAP will need to co-add many pixels to achieve a statistically significant detection of polarization. MAP will be able to characterize the polarization signal due to various galactic foregrounds. Armed with this information, experimentalist will be able to plan the next generation of experiments.

The European science community is planning two polarization missions. SPOrt[8], an Italian built experiment that operates from 20 - 90 GHz, will measure polarization on large angular scales from the International Space Station starting in 2002. Planck Surveyor, which will launch at the end of the decade, will have much more sensitivity per pixel. This should enable to accurately map the scalar polarization signal. If galactic foregrounds or instrumental systematics do not limit the experiment, then Planck Surveyor should be capable have accurately characterizing the polarization fluctuations due to scalar fluctuations.

Many cosmologist suspect that we will need a third generation CMB experiment that is optomized to detection the "B-wave" polarization signal due to gravity waves [12]. This future experiment will have to be designed to have sensitivities of better than 100 nanoKelvin, the ability to reject various galactic foregrounds, and careful control of systematic errors.

CONCLUSION

Measurements of the microwave background polarization fluctuations can lead to the detection of ultra-long wave gravity waves. While this is a technically challenging problem, the scientific return is potentially enormous. There is a clear path from ground-based experiments through upcoming space-based experiment towards a high sensitivity polarization experiment at the beginning of the next decade. If successful, this experiment could detect the gravity waves generated at the beginning of the universe.

---

[5] http://www.physics.ucsb.edu/~ruhl_lab/composar/COMPOSAR.html
[6] http://kepler.lbl.gov
[7] http://cmb.physics.wisc.edu/polar/
[8] http://sport.tesre.bo.cnr.it

REFERENCES

1. Bennett, C.L., *et al.*, "Four-Year COBE DMR Cosmic Microwave Background Observations: Maps and Basic Results", ApJ, **464**, L1 (1996).

2. Crittenden, R. Bond, J.R. Davis, R.L., Efstathiou, G. & Steinhardt, P.J. "Imprint of gravitational waves on the cosmic microwave background", Phys. Rev. Lett. **71**, 324 (1993).

3. de Bernardis, P. *et al.*, "A flat Universe from high-resolution maps of the CMB radiation," (BOOMERANG) Nature, **404**, 955 (2000).

4. Hanany, S. *et al.*, "Maxima-1: A measurement of the cosmic microwave background anisotropy on angular scales of 10′ to 5°," astro-ph/0005123, 2000.

5. Hedman, M.M., Barkats, D., Gunderson, J.O., Staggs, S. & Winstein, G., "A Limit on the Polarized Anisotropy of the Cosmic Microwave Background at Subdegree Angular Scales", astro-ph/0010592 (2000).

6. Hogan, C., "Scales of the Extra Dimensions and their Gravitational Wave Backgrounds" Phys.Rev. **D62** 121302 (2000).

7. Kamionkowski, M., Kosowsky, A., & Stebbins, A., "Statistics of Cosmic Microwave Background Polarization", Phys. Rev. D, **D55** 7368 (1997).

8. Knox, L. & Page, L., "Characterizing the Peak in the CMB Angular Power Spectrum," PRL **85**, 1366-1369 (2000).

9. Miller, A. *et al.* "A measurement of the angular power spectrum of the CMB from $l = 100$ to 400," (TOCO) Ap.J. **524**, L1 (1999).

10. Padin, S. *et al.*, "First Intrinsic Anisotropy Observations with the Cosmic Background Imager", astro-ph/0012211 (2000).

11. Tegmark, M. & Zaldarriaga, M., "Current cosmological constraints from a 10 parameter CMB analysis", Ap.J., **544**, 30 (2000).

12. Peterson, J.B., Carlstrom, J.E., Cheng, E.S., Kamionkowski, M., Lange, A.E., Seiffert, M., Spergel, D.N. & Stebbins, A., "Cosmic Microwave Background Observations in the Post-Planck Era", astro-ph/9907286 (1999).

13. Zaldarriaga, M. & Seljak, U. 1997, "An All-Sky Analysis of Polarization in the Microwave Background", Phys. Rev. D **55** 1830 (1997).

14. Zaldarriaga, M., Spergel, D.N. & Seljak, U. "Microwave Background Constraints on Cosmological Parameters", ApJ, **488**, 1 (1997).

Stellar Collapse and Rotational Instabilities

Gravitational Waves from Stellar Collapse

Chris L. Fryer

T-6, Los Alamos National Laboratory,Los Alamos, NM 87545

Abstract. Stellar core-collapse plays an important role in nearly all facets of astronomy: cosmology (as standard candles), formation of compact objects, nucleosynthesis and energy deposition in galaxies. In addition, they release energy in powerful explosions of light over a range of energies, neutrinos, and the subject of this meeting, gravitational waves. Because of this broad range of importance, astronomers have discovered a number of constraints which can be used to help us understand the importance of stellar core-collapse as gravitational wave sources.

INTRODUCTION

The increasing reality of gravitational wave (GW) detectors sufficiently sensitive to observe a host of astrophysical GW sources has led to a flurry of activity among astrophysicists to estimate these sources. One of the most studied class of GW sources involves the collapse of massive stars to form compact remnants (either neutron stars or black holes). This class includes a variety of astrophysical objects with a range of masses from the collapse of a Chandrasekhar-massed white dwarf to the collapse of very massive stars with masses in excess of $250\,M_\odot$.

Although there is little doubt that stellar collapse produce gravitational waves, it is difficult to accurately estimate the characteristics of the signal produced. These difficulties arise not just from uncertainties in the collapse itself, but also in the evolution of the *progenitors* of these objects. Fortunately, observational evidence of stellar collapse is not limited to gravitational waves. Stellar collapses produce compact remnants, neutron rich isotopes, neutrino and photon outbursts. All of these "observables" can be used to place some constraints on these events. By tapping into this store of astrophysical knowledge, we can get some understanding of GW emission from stellar collapse.

Here we review the current understanding of 3 distinct collapse events: the collapse of an accreting white dwarf pushed beyond the Chandrasekhar limit (Accretion Induced Collapse), the standard core-collapse supernova model, and the collapse of very massive stars ($\sim$250-500$M_\odot$). The rate that each of these collapse

CP575, *Astrophysical Sources for Ground-Based Gravitational Wave Detectors*, edited by J. M. Centrella events occurs is known to some degree (although except for core-collapse supernovae, we can only place upper limits on the rate). Also, at varying levels of accuracy, we know the photon and neutrino outbursts that accompany the collapse (and the GW emission). Predicting the GWs themselves is much more difficult and a complete discussion of the possibilities is beyond the scope of this paper. However, for many instabilities which drive the emission of GWs, the strength of the gravitational wave emission depends upon the mass and angular momentum in the emitting region. Here we will review the constraints astronomers can place on the collapse rates, neutrino and photon outbursts, and the mass and angular momentum distributions that arise from stellar collapse.

ACCRETION INDUCED COLLAPSE

When a white dwarf's mass exceeds the Chandrasekhar limit, it begins to collapse. As it contracts, its temperature increases adiabatically. Neutrino cooling (via Urca processes) limits the rise in temperature. If neutrino cooling does not reduce the adiabatic heating significantly, the collapsing white dwarf will reach temperatures hot enough to ignite nuclear burning. The entire white dwarf explodes in a thermonuclear explosion known as a Type Ia supernova. If, on the other hand, cooling initially prevents nuclear ignition, the white dwarf will collapse more and more quickly as electrons capture onto protons and the white dwarf will ultimately form a neutron star.

This "Accretion-Induced Collapse" (AIC) of a white dwarf is very similar to core-collapse supernovae. The collapse of white dwarfs has been studied in some detail over the past few decades (Hillebrandt, Nomoto, & Wolff 1984, Woosley & Baron 1992, Fryer et al. 1998) and we have some understanding of the collapse process and the resultant explosion. Since the white dwarf is pushed over the Chandrasekhar mass limit through disk accretion, it is likely that the collapsing white dwarfs will contain a considerable amount of rotation, allowing the possibility of a variety of instabilities and the emission of GWs. For this paper, we rely upon the rotating core collapse models from Fryer et al. (1998).

Formation Rate

Calculating the formation rate of AICs from first principles is fraught with a number of difficulties from understanding binary star evolution to uncertainties in the accretion process itself. We have already mentioned one such uncertainty: Does the star ignite in a thermonuclear explosion or collapse to form an AIC? However, we currently don't even understand what conditions are necessary for a white dwarf to actually accrete matter instead of losing it via a series of nova explosions. To calculate the rate of AICs, we must then rely upon indirect methods.

First, the thermonuclear explosion of a Chandrasekhar-massed white dwarf seems to match supernova observations well (see Pinto & Eastman 2000 for example) and it is almost certainly the mechanism which produces Type Ia supernovae. Hence, we know that, roughly every few hundred years in the Galaxy, a white dwarf does accrete enough mass to exceed the Chandrasekhar limit (Cappellaro et al. 2000). Some fraction of these white dwarfs will collapse to form a neutron star. Which fate befalls the white dwarf depends sensitively upon the initial mass of the white dwarf, its chemical composition, and the rate at which it accretes matter (see Nomoto & Kondo 1991 for review), making an accurate estimate nearly impossible.

However, by modeling the collapse, we can place constraints on the AIC rate. During the collapse, the white dwarf ejects the outer $\sim 0.1\,M_\odot$ of its envelope, some of which became very neutron rich due to electron capture. As this material is ejected, it forms some extremely rare, neutron-rich isotopes, which "pollutes" the Galaxy. By measuring the total amount of these isotopes in the Galaxy, and assuming these isotopes are formed solely in AICs, we can place an upper limit on the rate of AICs in the Galaxy at about 10^{-5} per year (Fryer et al. 1998).

Neutrino and Photon Outbursts

During the collapse, electrons capture onto protons ($e^- + p \rightarrow n + \nu_e$) and the collapsing object produces a burst of electron neutrinos. As the core collapses, it gradually becomes optically thick to neutrinos and the neutrino luminosity is cut off. Slowly, neutrinos leak out of the core, but as the core temperature increases, pair annihilation produces neutrinos of all flavors and these neutrinos cool the proto-neutron star (just as the electron neutrinos reduce the electron fraction) and allow the core to eventually become a young neutron star.

At the same time as this neutron burst, the matter in the core collapses down to nuclear densities where nuclear forces and neutron degeneracy pressure abruptly halt the infall and the core bounces. The inner core has become a "proto-neutron star". The bounce causes the outer layers of the white dwarf to expand out of the potential well of the proto-neutron star. Neutrinos leaking out of the core heat the outer $\sim 0.1\text{-}0.15\,M_\odot$ of the white dwarf and eject it in a mini-supernova explosion. As the ^{56}Ni in this ejecta decays, it powers a weak supernova outburst similar to Type Ia supernovae, but ~ 10 times dimmer. To date, there are no convincing observations of AIC outbursts. Remember, however, that the rate of AICs is less than 1% of the Type Ia supernovae rate and it is not surprising that we have not yet observed an AIC.

Gravitational Wave Emission

Typically, estimates of the GW emission from core-collapse concentrate on the initial collapse and bounce phase of the star. It is at this early stage that the core is moving quickly and a rapidly varying quadrupole moment can be produced *if* some instability occurs. The strength of the gravitational waves depends on the amount of mass in the core and the rotation rate (which determines the likelihood of an instability to develop).

First, lets discuss the density distribution shortly after bounce. Because the neutrinos are trapped in the core, both the temperature and the electron fraction of the proto-neutron star remain high for nearly 10 s. Hence, the proto-neutron star remains extended during this time. To estimate gravitational waves, we must use the density distributions of these extended neutron stars. The density profile of an AIC 0.18s after bounce is given in figure 1.

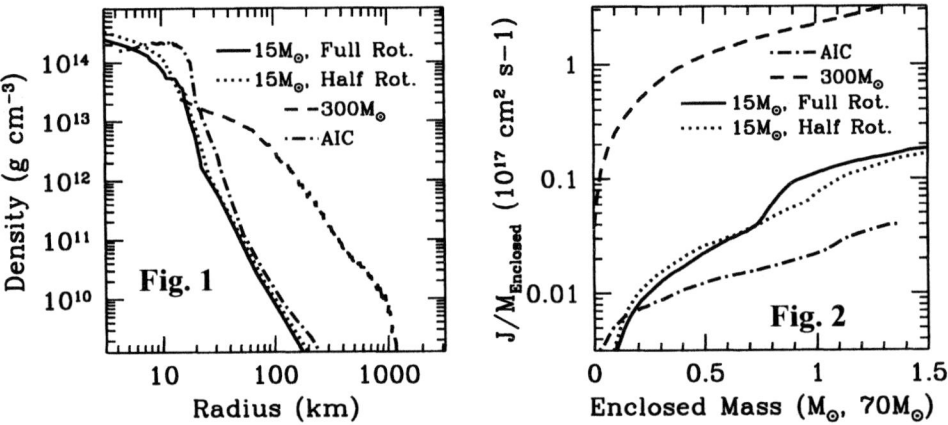

FIGURE 1. Density (Fig. 1), Angular Momentum (Fig. 2) versus radius for collapsing stars: AIC-0.18 s after collapse, rotating core-collapse (full rotation)-1.6 s after bounce, core-collapse (half rotation)-1.4 s after bounce, 300 $M_\odot$ direct collapse-1.9 s after collapse. For the core-collapse simulations, the slower rotator is more dense. Although the maximum density of the 300 $M_\odot$ direct collapse is much lower than other core-collapse, its mass out to 1000 km is 50 times that of the other collapsed objects. Note in Fig. 2 that the 300 $M_\odot$ has, by far, the highest angular momentum.

Even more important is the distribution of angular momentum in the proto-neutron star. The moment of inertia of a typical white dwarf is small ($IM^{-1}R^{-2} < 0.1$) and the accretion of only a few tenths of a solar mass through an accretion disk can cause the the white dwarf to spin up nearly to break-up. For a 10,000 km, Chandrasekhar-massed white dwarf, this corresponds to a spin period of 14.5 s (for a 2,500 km white dwarf, the corresponding break-up spin period is 1.8 s). Such high spin rates are not seen in white dwarfs. Indeed the fastest spinning white dwarfs observed thus far have periods in excess of 100 s. Fryer et al. (1998) modeled AICs assuming a maximum total angular momentum of 10^{49} g cm² s⁻¹. This corresponds to a 100 s spin period for a 10,000 km white dwarf or a 12.5 s period for a 2,500 km white dwarf. 0.18 s after bounce, half of the angular momentum is in the $0.8 \, M_\odot$ proto-neutron star (Fig. 2).

Let's discuss two types of instabilities which might drive the generation of gravitational waves: bar modes and rossby modes. Bar mode instabilities occur in objects whose rotational energy exceeds some fraction of its potential energy. This fraction is generally written as $\beta \equiv T/|W|$. The standard lore is that an object is unstable on a secular time scale if $\beta \gtrsim 0.14$ and it is dynamically unstable if $\beta \gtrsim 0.27$. Unfortunately, for AICs, our $J=10^{49} \, \mathrm{g \, cm^2 \, s^{-1}}$ is not unstable 0.18 s after bounce (Fig. 3). As the proto-neutron star cools, β may increase, but it is unlikely that such instabilities will affect much of the mass in the nascent neutron star and it is unlikely that they will produce strong gravitational waves. One might imagine a faster rotating white dwarf, but it is not clear that nature produces them and the one should not rely upon bar-modes producing any detectable GW signal.

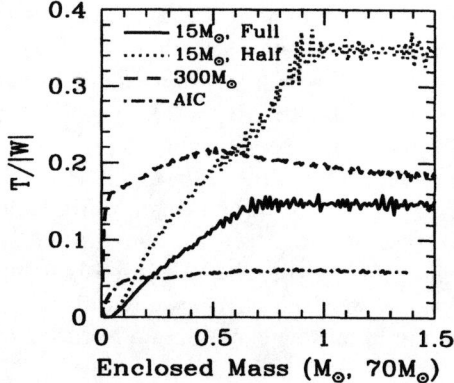

FIGURE 2. Rotational energy divided by gravitational energy ($T/|W|$) versus mass for the 4 collapse progenitors from Figs. 1,2. For the core-collapse stars, $T/|W|$ is actually higher for the star initially spinning at half the rotation rate. This is because it is more compact.

However, as the neutron star cools and contracts, the angular momentum in this object will produce a sub-ms neutron star. If r-modes exist, AICs will produce a signal as strong as any young neutron star formation scenario. However, given that the rate is 1000 times lower than standard core-collapse supernovae, it appears that core-collapse supernovae, not AICs are a better GW source.

CORE-COLLAPSE SUPERNOVAE

Stars more massive than $\sim 8\,M_\odot$ also end their lives in a core-collapse. During their lives, successive stages of nuclear burning build up a massive iron core in the stellar center. This iron core is supported by electron degeneracy and thermal pressures. When the density and temperature in the core become so high that a) iron is dissociated into alpha particles and b) electron capture occurs, the support pressure is suddenly removed and the core collapses. As it collapses, the core density and temperature increases, causing more iron dissociation and electron capture which leads to a runaway infall of the core. Just as with AICs, the core collapses until it reaches nuclear densities where nuclear forces and neutron degeneracy pressure abruptly halt the collapse.

Astronomers have long understood that the potential energy released as a star collapses down to a neutron star could power a supernova explosion (Baade & Zwicky 1934). However, it was not until 1966 that Colgate & White realized that neutrinos could be the medium which transported energy released during the collapse of the core into the outer layers of the star which could then explode and drive the supernova explosion. Since this time, astronomers have continued to refine the neutrino-driven model. Indeed, core-collapse supernovae are one of the few objects in astronomy that we do not invoke fudge factors to explain (albeit, this means that we do not yet match the observations well either).

The basic mechanism behind core-collapse supernovae has developed from 3 decades of study and is very similar to AICs. The main difference arises from the fact that the proto-neutron star must somehow eject $\gtrsim 10 - 15\,M_\odot$ of material instead of $0.1\,M_\odot$ in the case of AICs. After bounce, the inner portion of the star rains down upon the proto-neutron star, preventing a quick explosion that occurs in AICs. A convective layer above the proto-neutron star and below the pressure cap of the infalling material converts the heat deposited by neutrinos into kinetic energy, aiding the explosion. As we shall see, convection plays an important role in the supernova mechanism and in our understanding of rotating supernova collapse models.

A great deal of work has been devoted to studying gravitational waves from core-collapse. Generally, these simulations have simplified the physics in an effort to concentrate on the gravitational wave emission (see Rampp, Müller, & Ruffert 1998 for a review). The advantage of these simplifications is that some of the collapse simulations can actually be modeled in 3 dimensions. The disadvantage is that these simulations do not include enough physics to accurately model the structure of the star at collapse and this limits the reliability our models of the gravitational wave signals from core-collapse supernovae. In addition, until recently, no massive stellar models existed which evolved rotating stars to collapse, and the angular momentum profiles used in GW calculations of core-collapse have all been artificially put in (generally at spin rates which are much higher than we now expect in nature).

Formation Rate

The formation rate of core-collapse supernovae is fairly well known and lies somewhere between 1 per 50-140 years in the Galaxy (Cappellaro et al 1997). What we do not know is what fraction of these core-collapse supernovae (if any) are rotating rapidly enough to emit detectable amounts of gravitational waves. From measurements of young pulsars, we know that at least some neutron stars are born with periods faster than 20 ms. But whether or not any neutron stars are born with millisecond periods is hard to ascertain. The problem is that pulsars spin down as they emit radiation, but we don't know exactly how fast the spin-down occurs. The most recent analysis by Chernoff & Cordes (pvt. communication) found that they could fit the initial spin periods with a Gaussian distribution peaking at 7 ms with sub-ms pulsars lying beyond the 2-sigma tail. Does this mean that less than 10% of pulsars are born spinning with millisecond periods, or does it mean that many pulsars are born spinning rapidly and GW emission removes a considerable amount of their angular momentum? In addition, the analysis of Chernoff & Cordes is very sensitive to their choice of spin down rates and other uncertainties in their population study and such results should be taken with a great deal of caution.

Hence, although we know the rate of core-collapse supernovae to high accuracy (for astronomical standards), we do not know the rate of core-collapse supernovae which occur with sufficiently high spins to be interesting to observations of GWs. But perhaps, the explosion itself can provide us with clues.

Neutrino and Photon Outbursts

The neutrino burst from core-collapse supernovae is very similar to that of AICs (Fig. 4). However, because it takes more time after bounce for neutrino heating to drive an explosion in core-collapse supernovae, the time variation of the neutrino spectrum differs from core-collapse supernovae to AICs. Technically, one could differentiate core-collapse supernovae from AICs simply by this spectral resolution.

It is easier to differentiate core-collapse supernovae from AICs simply by their optical output. Supernovae are classified by their spectra (based on what lines are visible). The collapse of massive stars have spectra that match Type Ib, Ic, II supernovae, whereas AICs have spectra which are very similar to Type Ia supernovae.

We may also be able to distinguish rapidly rotating core-collapse supernovae from slowly rotating supernovae based simply upon their polarization. Using the first stellar models including rotating, Fryer & Heger (2000) found that rotation produced asymmetric supernova explosions, which may explain the polarization measurements of supernovae. This effect arises from the fact that the positive angular momentum gradient in the spinning core stabilizes against convection in the equator. Since convection is limited to the polar region, the supernova explosion is strongest there, and the resultant explosion is highly asymmetric (Fig. 5). It is currently believed that such asymmetries are necessary to produce the observed

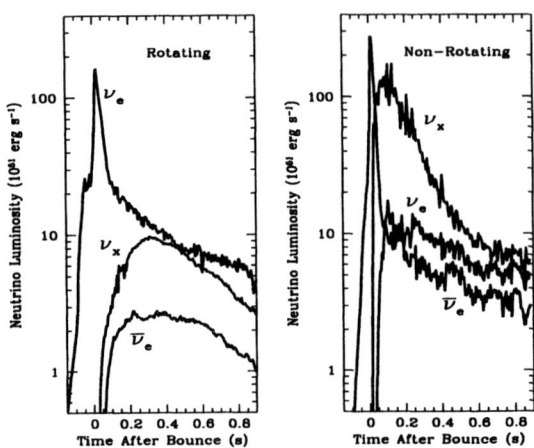

FIGURE 3. Neutrino luminosities versus time for a rotating and non-rotating progenitor. The non-rotating core has a much larger μ and τ (ν_x) neutrino luminosity, especially just after bounce. This is because the non-rotating core compresses more and, at the μ and τ neutrinosphere, the temperature is over a factor of 1.5 higher than that of the rotating core. Because of the large dependence of neutrino emission on temperature (the luminosity from pair annihilation $\propto T^9$), this small change in temperature has large effects on the neutrino luminosity.

polarization (Höflich 1991). With more accurate calculations and careful observations, we may be able to distinguish quickly and slowly rotating core-collapses.

Gravitational Wave Emission

In figure 1, we show the density profiles of 2 separate core-collapse simulations: one using the rotation profile calculated by Heger (1998), and the other using half that amount of angular momentum (see Fryer & Heger 2000 for details). The simulations by Heger assumed a near-maximally rotating main-sequence star and did not include any angular momentum transport caused by magnetic fields. Since this time, Heger has included a prescription for magnetic-field induced angular momentum transport, and the total angular momentum in the core has decreased somewhat. Note first that the lower angular momentum simulation is much denser than the full rotation simulation. Even though this simulation has less angular momentum (Fig. 2), its compactness leads to a proto-neutron star which is much more unstable (Fig. 3). In all cases, the angular momentum in these stars is initially much less than what is used in most GW core-collapse simulations, but as the strong polar explosion removes the matter along the poles (which had little angular momentum) the specific angular momentum of the remnant proto-neutron star can get very large. Unfortunately, at these times, the density is not high enough to produce very large GW signals from bar-modes and, if we are to detect GW emission from core-collapse, it will likely be from r-modes. In both cases, the resultant neutron star has millisecond spin periods, making these objects ideal r-mode candidates. At a rate up to 1 per 50 years in the Galaxy, r-mode driven GW emission from core-collapse supernovae remains a promising source of GW waves.

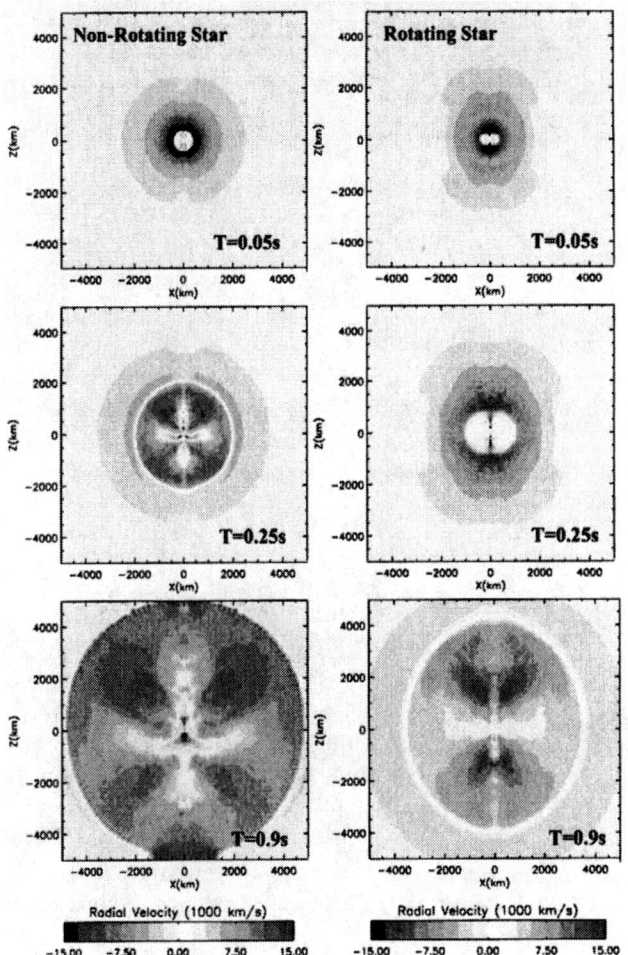

FIGURE 4. Radial velocity distribution of non-rotating and rotating models 0.05, 0.25, 0.5, and 0.9 s after bounce. At 0.9 s, the non-rotating model remains essentially spherical. The asymmetries in the velocities are caused by the buoyant convective bubbles which are driving the explosion. In contrast, the rotating model already shows strong asymmetries in the shock position and velocities.

COLLAPSE OF VERY MASSIVE STARS

As the mass of the collapsing star increases, the basic picture described above on core-collapse supernovae begins to change. Above 20-25 $M_\odot$, the supernova explosion is too weak to eject the entire star, and some of the star will fall back onto the neutron star 100-100,000 s after the supernova explosion (and after the GW emission). This fallback matter will push the remnant mass above the maximum neutron star mass, and it will collapse to form a black hole. Beyond ~40-50 $M_\odot$, neutrino heating is unable to drive a supernova explosion and the star collapses to form a black hole. These direct-collapse stars will definitely have different GW emission than normal core-collapse simulations and also different optical output (they are known as "Collapsars" or "Hypernovae" and constitute one of the favored models for gamma-ray bursts). Unfortunately, beyond ~40-50 $M_\odot$, mass-loss from stellar winds can dramatically change the mass of the star before collapse and it may be that nature does not produce any high-metallicity collapsar progenitors.

Stellar winds are driven by the opacity of metals in the stellar envelope. It is likely that as we reduce the fraction of metals in the star, mass-loss from winds will decrease. Population III stars are the first generation of stars formed in the early universe, when no metals existed (stars produce all of the metals we see today). In this section, we review the collapse of very massive, population III stars (100-500 $M_\odot$). If these stars are rotating, rotational (plus thermal) support prevents the star from immediately collapsing into a black hole. Just like core-collapse supernovae, rotating, very massive stars collapse and bounce, forming a proto-black hole (50-70 $M_\odot$ within 1000-2000 km). This rotating proto-black hole is susceptible to bar instabilities and may produce a strong GW signal.

Formation Rate

Estimating an accurate rate of core-collapse from very massive stars depends on two major uncertainties: the fraction of stars which form with masses above 100 $M_\odot$ and the number of these stars which actually collapse to form black holes. The mass distribution of stars at birth is known as the initial mass function (IMF). Today, the IMF is peaked toward low mass stars such that 90% of stellar core-collapse occurs in stars between 8 and ~20 $M_\odot$ and only 1% of core-collapse occurs in stars more massive than 40 $M_\odot$. However, recent simulations by Abel, Bryan, & Norman (2000) suggest that the typical mass of first generation stars may be peaked towards ~ 100 $M_\odot$ and it could be that a majority of Population III stars had masses in excess of 100 $M_\odot$.

The light from these very massive stars re-ionizes the early universe, and from this, we can derive a constraint on the formation rate of these stars. Although we expect that these photons ionized a significant fraction of the early universe, there should not be so many stars that they ionize the universe several times over. Using our best estimates of the re-ionization fraction, the amount of ultraviolet photons produced by these massive stars, and the ionization efficiency of massive stars, one estimates that roughly 0.01%-1% of the baryonic matter in the universe was incorporated into very massive stars. This calculation corresponds to roughly $10^4 - 10^7$ very massive stars produced in a $10^{11} M_\odot$ galaxy, or a rate of massive stellar collapse as high as 1 every few thousand years! However, stars less massive than $\sim 260 M_\odot$ do not collapse, but explode in a giant thermonuclear explosion known as a pair-instability supernovae. Unfortunately, although we might believe our formation rate of very massive stars (within a few orders of magnitude), it is currently impossible to determine how many very massive stars are produced with masses beyond $\sim 260 M_\odot$. The Galaxy could produce a million of these objects, or maybe just a few hundred. 1-10 million very massive stars is a secure upper limit.

Neutrino and Photon Outbursts

As a rotating massive star collapses, rotational support (plus thermal) support can actually halt the collapse and produce a weak bounce. What remains behind is a massive (but not dense in core-collapse standards) proto-black hole (Fig. 1). It takes a few seconds for neutrinos to cool this proto-black hole, allowing it to collapse to a black hole. The neutrino luminosity is nearly an order of magnitude higher than that of core-collapse supernovae (Fig. 6), but only lasts until the star collapses to form a black hole. When the star collapses, the μ and τ neutrino flux drops off first, and later, the electron neutrino flux. This occurs because the μ and τ neutrinos probe the interior of the proto-black hole, and as soon as an event horizon is formed, these neutrinos become trapped in the black hole. The electron neutrinos do not decrease significantly until the black hole expands enough to produce a cold accretion disk.

Fryer et al. (2001) have suggested that, if a magnetic jet mechanism works, the black hole accretion disk system produced during the collapse may produce a gamma-ray burst. The burst would likely have a longer duration than typical gamma-ray bursts and would, like the collapsar, be accompanied by a supernova-like explosion. Even if such a jet is not produced, the further accretion of material onto this black hole would produce an X-ray transient (at the Eddington flux) which would persist for about a day.

Gravitational Wave Emission

Although the density of the proto-black hole is much less than the density found in core-collapse supernovae (Fig. 2), the amount of mass in the proto black hole (up to $80 M_\odot$) is nearly 70 times that of the proto-neutron star. The GW signal is very sensitive to the mass, and the collapse of these massive stars can produce very strong GW emission. The angular momentum in the the proto-black hole is high (Fig. 3) and these stars will certainly be unstable to secular instabilities and possibly dynamical bar-mode instabilities (Fig. 4). However, the temperature is too high to form r-mode instabilities. A final GW source could arise from ringing in the nascent black hole and we are actively studying its potential now.

REFERENCES

1. Abel, T., Bryan, G. L., & Norman, M. L., *ApJ*, **540**, 39 (2000).
2. Baade, W., & Zwicky, F., *Phys. Rev.*, **45**, 138 (1934).
3. Cappellaro, E., Turatto, M., Tsvetkov, D. Yu., Bartunov, O.S., Pollas, C., Evans, R., Hamuy, M., *A&A*, **322**, 431 (1997).
4. Colgate, S. A., White, R. H., *ApJ*, **143**, 626, (1966).
5. Fryer, C.L., Benz, W., Herant, M., Colgate, S.A., *ApJ*, **516**, 892 (1998).
6. Fryer, C.L., Heger, A., *ApJ*, **541**, 1033 (2000).
7. Fryer, C.L., Woosley, S.E., Heger, A., *accepted by ApJ*, astro-ph/0007176 (2001).
8. Heger, A., *Ph.D. thesis*, Technische Univ. München (1998).
9. Hillebrandt, W., Nomoto, K., & Wolff, R. G., *A&A*, **133**, 175, (1984).
10. Höflich, P.*A&A*, **246**, 481, (1991).
11. Nomoto, K., Kondo, Y., *ApJ*, **367**, L19, (1991).
12. Pinto, P.A., Eastman, R.G., *accepted in New Astronomy*, astro-ph/0006171 (2000).
13. Rampp, M., Müller, E., & Ruffert, M., *A&A*, **332**, 969 (1998).
14. Woosley, S. E., & Baron, E., *ApJ*, **391**, 228, (1992).

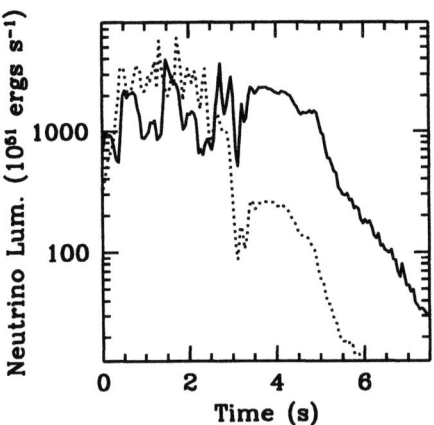

FIGURE 5. Neutrino Luminosity as a function of time from Model B. The μ and τ neutrinos (dotted line) dominate the neutrino emission until black hole formation. Shortly after the black hole forms, the event horizon grows beyond the μ and τ neutrinosphere (at 2.5 s) and drastically diminishes the neutrino luminosity. The electron neutrinos (solid line) do not decrease significantly until the black hole expands enough to produce a cool accretion disk.

Rotational Instabilities and Centrifugal Hangup

Kimberly C. B. New* and Joan M. Centrella†

*Los Alamos National Laboratory, Los Alamos, New Mexico 87545[1]
†Drexel University, Philadelphia, Pennsylvania, 19104

Abstract. One interesting class of gravitational radiation sources includes rapidly rotating astrophysical objects that encounter dynamical instabilities. We have carried out a set of simulations of rotationally induced instabilities in differentially rotating polytropes. An $n=1.5$ polytrope with the Maclaurin rotation law will encounter the $m=2$ bar instability at $T/|W| \gtrsim 0.27$. Our results indicate that the remnant of this instability is a persistent bar-like structure that emits a long-lived gravitational radiation signal. Furthermore, dynamical instability is shown to occur in $n=3.33$ polytropes with the j-constant rotation law at $T/|W| \gtrsim 0.14$. In this case, the dominant mode of instability is $m=1$. Such instability may allow a centrifugally-hung core to begin collapsing to neutron star densities on a dynamical timescale. If it occurs in a supermassive star, it may produce gravitational radiation detectable by LISA.

INTRODUCTION

One interesting class of gravitational radiation sources includes rapidly rotating astrophysical objects that encounter dynamical instabilities. Linear stability analysis has shown that rapidly rotating bodies may experience global deformations due to the growth of unstable azimuthal modes $e^{\pm im\phi}$ [1,2]. The mode numbers m describe the shape of the induced deformation. For example, an $m=1$ mode could result in the development of a one-armed spiral or a simple translation; an $m=2$ mode may produce a bar-shaped distortion; an $m=3$ mode, a triangular distortion; and so on. The onset of such an instability occurs when an object's ratio of rotational kinetic energy T to gravitational potential energy W, $\beta = T/|W|$, exceeds a critical value β_{crit}.

Both secular and dynamical varieties of these instabilities may exist. A dynamical instability is driven by gravitational and hydrodynamical forces and develops on the order of the rotation period of the system. A secular instability is driven by a dissipative mechanism, such as viscosity or gravitational radiation, and develops

---

[1] A portion of this work was performed under the auspices of the U.S. Department of Energy by Los Alamos National Laboratory under contract W-7504-ENG-36.

on the timescale of that mechanism (which can be many rotation periods). In this paper, we focus on the numerical study of dynamical instabilities, since their development can be followed in a reasonable amount of computational time with explicit hydrodynamical simulations.

There are several types of astrophysical objects that may encounter these rotational instabilities. A star that accretes matter and angular momentum from a binary companion may be spun up to rapid rotation [3]. A second example is a centrifugally hung stellar core or "fizzler" [4–6]. The formation of a fizzler begins when the core of a massive star depletes its nuclear fuel and begins to collapse to neutron star densities. If the core was rotating initially, conservation of angular momentum requires that the core spin up as it collapses. This spin up could actually halt the collapse if the centrifugal force increases to the point where it overcomes the inward gravitational push. The results would be a rapidly rotating, partially collapsed stellar core. Another example is a cooling supermassive star (mass $> 10^6 M_\odot$) that also spins up as it contracts [7,8]. Finally, if the merger of a compact binary does not result in an immediate collapse to a black hole, the remnant will be a rapidly rotating compact object [9]. The nonaxisymmetric deformations induced by rotational instabilities could result in relatively strong gravitational radiation emission from these rapidly rotating objects.

We have performed Newtonian hydrodynamics simulations of dynamical rotational instabilities in two different types of objects [10,11]. The much studied $m=2$ bar mode is the strongest of the set of global instabilities encountered by an $n=1.5$ polytrope (for which the equation of state gives the pressure P in terms of the density ρ as $P = K\rho^{(1+1/n)}$, where K is the polytropic constant) with the Maclaurin differential rotation law. This instability sets in when $\beta \gtrsim 0.27$. Our results indicate that dynamical instabilities also occur in objects with much lower values of β. In fact, $n=3.33$ polytropes with the j-constant rotation law encounter a dominant $m=1$ instability when β reaches 0.14. Simulations of the bar mode instability and the $m=1$ instability will be discussed in the following sections.

THE BAR MODE INSTABILITY

There has been a discrepancy in the outcome of previous hydrodynamical simulations of the dynamical bar mode. Namely, simulations carried out by Centrella's group showed that an object deformed by the bar mode would become axisymmetric again after a short interval [12,13]. This is in contrast to the simulations performed by several other groups, which resulted in bar-shaped final configurations (e.g., [14–16]).

The degree of asymmetry in the final configuration is important because an axisymmetric object (rotating about its short symmetry axis) will not emit gravitational radiation. If the simulations performed by Centrella's groups's simulations are correct, the instability would produce only a short burst of radiation. However, if the object retains a bar-like structure, as the other simulations seem to indicate,

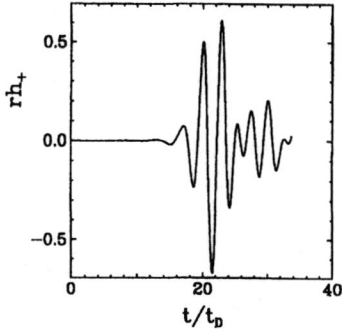

FIGURE 1. The h_+ polarization of the gravitational waveform from a $\mathcal{D}$ code bar mode simulation, as viewed by an observer located along the rotation axis at a distance r from the system. The value rh_+ has been normalized to $R^{-1}c^{-4}M^2G^2$; the time is normalized to the dynamical time $t_D = [R^3/(GM)]^{1/2}$.

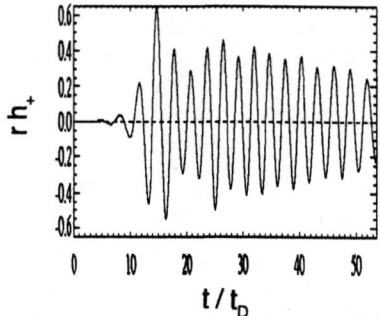

FIGURE 2. Same as Figure 1, for a simulation performed with the $\mathcal{L}$ code.

it will go on emitting gravitational radiation, producing a longer-lived signal that would be easier to detect.

This discrepancy is illustrated by the gravitational waveforms shown in Figures 1 and 2. The waveform in Figure 1 is from a simulation performed by Centrella's group [12], with a finite-difference hydrodynamics (FDH) code developed at Drexel University (hereafter called the $\mathcal{D}$ code). This waveform is burstlike; the amplitude dies off as the object loses its bar-like shape. The waveform in Figure 2 is from a simulation performed by New [14], with a FDH code developed at Louisiana State University (hereafter called the $\mathcal{L}$ code). This waveform rings for the duration of the simulation because the object retains its bar-shaped structure.

Again, the resolution of this discrepancy is important because if the gravitational radiation signal persists, the total accumulated amplitude could make it easier to detect. The characteristics of New's waveform are such that if the star's initial mass and radius are $M = 1.4M_\odot$ and $R = 36\,$km, and it is located at a distance of 20 Mpc, the maximum amplitude is $h_{max} \sim 10^{-23}$ and the frequency f_{gw} ranges from $\sim 600 - 670\,\mathrm{Hz}$. The scaling is such that $h_{max} \propto R^{-1}$ and $f_{gw} \propto R^{-3/2}$. This signal could possibly be detected with advanced interferometers like LIGO II.

In a recent paper [11], we demonstrate that a good deal of progress has been made in resolving the discrepancy among the outcomes of the previous simulations of the bar instability. In an attempt to get at the root of the problem, we examined the differences between the simulations performed with the $\mathcal{D}$ and $\mathcal{L}$ codes. One of the main differences is that New's simulation with the $\mathcal{L}$ code [14] used a condition called π-symmetry. The π-symmetry condition enforces periodic azimuthal symmetry and thus allows the azimuthal resolution to be doubled as the solution only needs to be evolved from 0 to π. This symmetry condition does not allow the growth of odd modes in a simulation. However, the use of π-symmetry in a bar mode simulation seems justified by analytic perturbation anaylsis, which indicates that odd modes will not grow if $\beta < 0.32$ [17] ($\beta=0.30$ in the initial models used by [12–14]).

In order to investigate if the use of π-symmetry was responsible for differences in the outcomes of the $\mathcal{D}$ and $\mathcal{L}$ codes' bar mode simulations, we reran the $\mathcal{L}$ code simulation with the π-symmetry condition turned off (hereafter referred to as simulation $\mathcal{L}1$). The $\mathcal{L}1$ simulation started with the same axisymmetric initial model used in the previous simulations [12,14]. It is constructed in hydrostatic equilibrium with Hachisu's Self-Consistent Field (SCF) method [18]. It has an $n=1.5$ polytropic equation of state and is differentially rotating, with an angular momentum distribution identical to that of a Maclaurin spheroid [19]. It is a highly flattened, rapidly rotating object with $\beta \sim 0.3$ (recall $\beta_{crit} \approx 0.27$).

Equatorial density contours from the last half of the $\mathcal{L}1$ simulation are shown in Figure 3. As the final frame indicates, the object appears rather symmetric at the end of the run. The resulting waveform (Figure 4) reflects the object's return to symmetry. This waveform looks much more like the burstlike waveforms of Centrella's group [12,13] (see, e.g., Figure 1).

Thus the $\mathcal{L}1$ run appears to be in better agreement with the $\mathcal{D}$ code simulations. But the important question is whether the result on which they agree is the "physical" result. That is, can we with confidence tell the gravitational wave detection community that a star that encounters the bar instability will emit only a burst of radiation?

The answer to that question is *no*. This is because of an effect that appears late in the $\mathcal{L}1$ simulation. At about the time that the object loses its bar-like structure, the center of mass of the object moves off the center of the grid. This can be seen in the final frame of Figure 3. In the absence of external forces, the center of mass should not move. This is thus a nonphysical result and must be due to a shortcoming in the numerics of the $\mathcal{L}$ code.

Before we could say for certain that an axisymmetric configuration is the correct remnant of the bar instability, we needed to find a way to minimize the center of mass motion and see how this affected the outcome of the evolution. Further testing indicated that increasing the radial resolution of the simulation does indeed

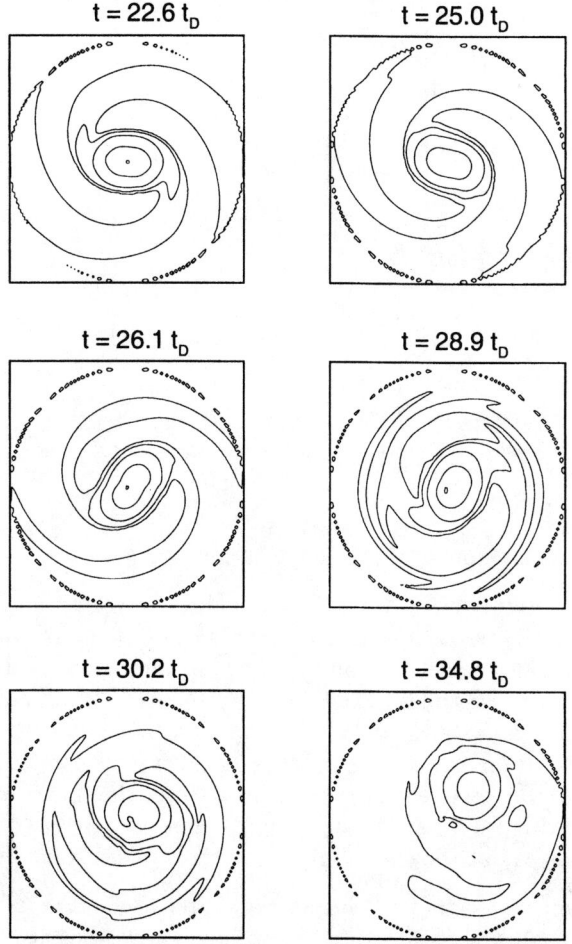

FIGURE 3. Density contours in the equatorial plane for the later stages of the $\mathcal{L}1$ simulation are shown. The maximum density has been normalized to unity at $t=0$ and the contour levels are at 0.5, 0.05, 0.005, and 0.0005.

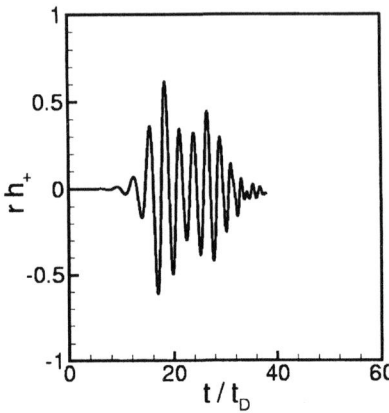

FIGURE 4. The gravitational waveform from the $\mathcal{L}1$ simulation is shown. The normalizations are identical to those used in Figure 1.

delay the onset of the motion of the center of mass.

We ran the bar mode simulation with the $\mathcal{L}$ code once again, this time with double the radial (and axial) resolution used in the $\mathcal{L}1$ run (this high resolution run will be referred to as $\mathcal{L}2$). In the $\mathcal{L}2$ evolution, the bar-like structure is maintained for about 15 more dynamical times (where $t_D = (R^3/GM)^{1/2}$) than in the $\mathcal{L}1$ run. The waveform from this run, shown in Figure 6, has a correspondingly longer duration. Doubling the resolution delayed the onset of the center of mass motion, but it did not prevent it. Equatorial density contour plots from the latter portion of the $\mathcal{L}2$ run are shown in Figure 5. As is evident in the final frame, the bar-like structure (and waveform) decay at the time that the center of mass starts to move.

We have demonstrated that the longer the center of mass motion is suppressed, the longer the bar-like structure is maintained. Since the center of mass motion is unphysical, the decay of the bar is as well. Thus a simulation with very high resolution, or better finite-differencing algorithms that did not develop center of mass motion, should produce a long-lived nonaxisymmetric structure. A recent paper by Brown confirms this [20]. The FDH PPM hydrocode Brown used to simulate the bar instability explicitly conserves linear momentum, thus preventing center of mass motion from developing. The outcome of Brown's simulation was indeed a persistent bar.

Recall that this is the outcome of the $\mathcal{L}$ code's simulation with π-symmetry [14]. This symmetry condition prevents the growth of any center of mass motion because it will not allow an $m=1$ mode to grow. Our results and the recent paper by Brown [11,20] indicate that the physically accurate outcome of the bar instability in this object is a persistent bar-like configuration that emits gravitational radiation over many cycles and thus may be capable of producing a detectable signal.

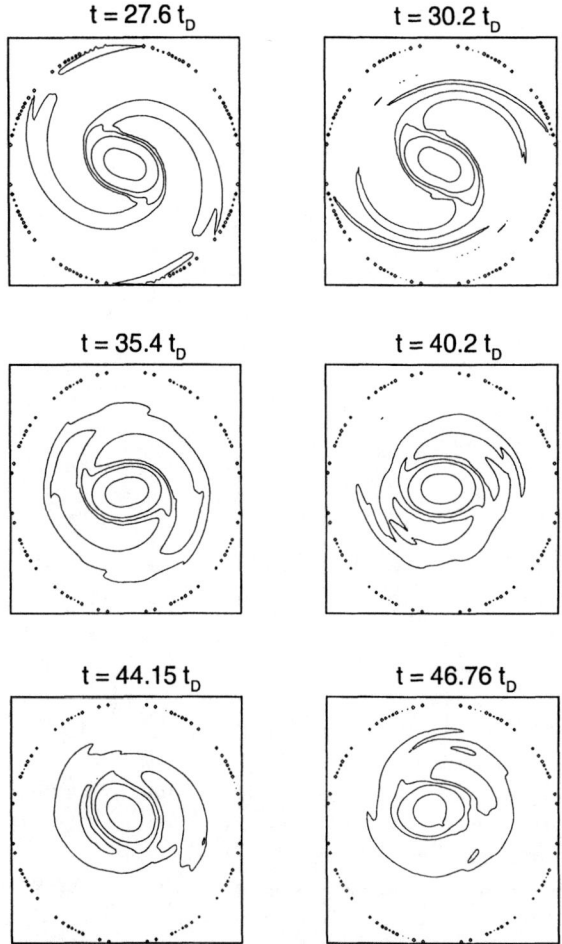

FIGURE 5. Density contours in the equatorial plane for the later stages of the $\mathcal{L}2$ simulation are shown. The contour levels are the same as those shown in Figure 3.

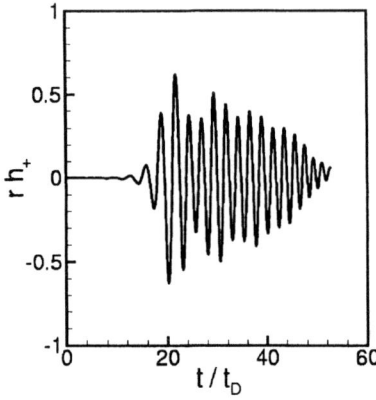

FIGURE 6. The gravitational waveform from the $\mathcal{L}2$ simulation is shown. The normalizations are identical to those used in Figure 1.

INSTABILITY AT LOW β

We have recently also investigated dynamical rotational instabilities in objects with lower values of β [10]. This study was motivated by the fact that centrifugally hung stellar cores, or fizzlers, are likely to have $\beta < 0.27$ and thus will not encounter the particular bar mode instability that was discussed in the preceeding section [5,21,22].

Previous authors have identified dynamical instabilities in toroidal configurations at values of $\beta \gtrsim 0.1$ [23,24]. Thus we decided to investigate the stability properties of spheroidal configurations with off-center density maxima (thinking they may behave more like tori). Our goal was to determine whether spheroidal configurations, representative of fizzlers, could encounter instabilities at lower values of β.

The initial models for this stability study were constructed with Hachisu's SCF method. An $n=3.33$ polytropic equation of state was used and is representative of a partially collapsed stellar core. Because we wanted to study models with off-center density maxima, we needed to use a rotation law other than the Maclaurin law. The j-constant law does produce off-center density maxima in $n=3.33$ polytropes. The angular velocity distribution $\Omega(\varpi)$ given by this law is

$$\Omega^2 = \frac{j_0^2}{(d^2 + \varpi^2)^2},\tag{1}$$

where ϖ is the cylindrical radius and d is an arbitrary constant. As d approaches zero, the specific angular momentum approaches the constant j_0. We constructed models with $d=0.2$. Figure 7 displays meridional density contour plots of four of the equilibrium models we constructed. Off-center density maxima appear as β is increased.

We performed simulations of these models with the $\mathcal{L}$ code and with Brown's PPM hydrocode. The $\mathcal{L}$ code uses a cylindrical grid and Brown's code uses a

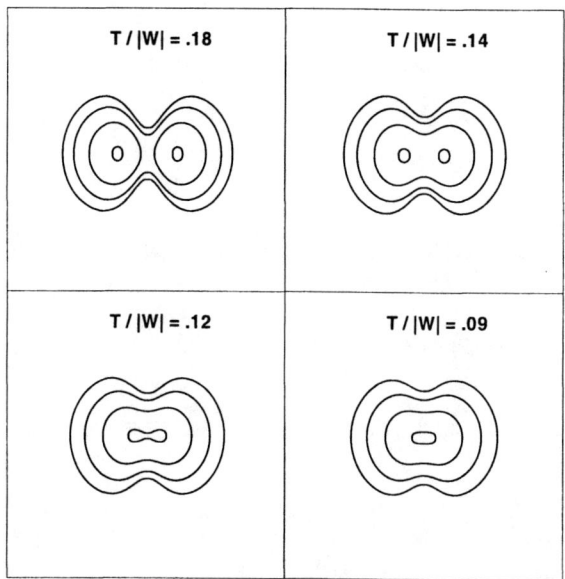

FIGURE 7. Density contours in the meridional plane are shown for 4 equilibrium models with $d=0.2$ and $n=3.33$. The models are labeled with their values of $\beta = T/|W|$. The initial maximum density has been normalized to unity and the contour levels are at 0.9, 0.1, 0.01, and 0.001.

Cartesian grid. We have evolved all four of the models shown in Figure 7. The onset of instability is near $\beta=0.14$. Both the $\beta=0.14$ and 0.18 models are clearly unstable (see below). We have also run simulations of the models with $\beta=0.12$ and 0.09. The $\beta=0.12$ model was run for about 40 t_D. At the end of the simulations, the $m=1$ mode was just starting to grow. The $\beta=0.09$ model was run for 35 t_D and showed no mode growth. We plan to run longer, higher resolution simulations to determine the value of β at which the instability sets in. Note that no significant center of mass motion was seen in these simulations.

The evolution of the model with $\beta=0.14$ exhibits a dynamical instability that is very similar in simulations performed with both codes. The development of the instability is shown in the equatorial density contour plots of Figure 8 (from the $\mathcal{L}$ code run). The torus pinches off to produce a single high density region, indicative of a $m=1$ mode. This dense region starts to collapse at late times, as is evident from the appearance of higher density contours. Figures 9 and 10 show the amplitude of the modes that grow during the evolutions performed with both codes. The $m=2$, 3, and 4 modes grow in sequence after the $m=1$ mode. The pattern speeds for the $m=1$ and $m=2$ modes are identical. Thus the $m=2$ mode is an harmonic of the $m=1$ mode. The constant amplitude of the $m=4$ mode in Brown's simulation is a result of the symmetry of his Cartesian grid.

The evolutions of the $\beta=0.18$ model also exhibit dynamical instability. The mode amplitude plots from these runs are shown in Figures 11 and 12. The instability

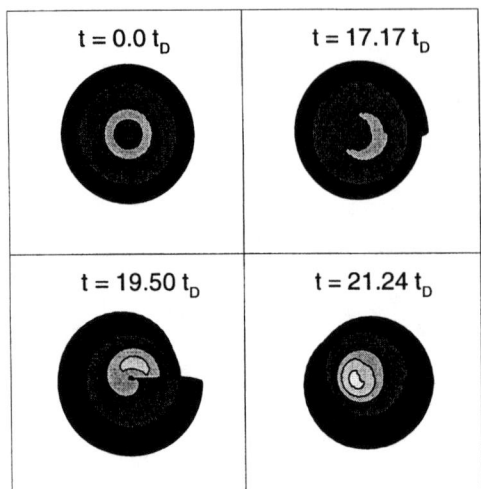

FIGURE 8. Density contours in the equatorial plane are shown for the model with $\beta=0.14$. The darkness of the shading increases as the density decreases. The contour levels are 0.01, 0.1, 0.9, 2, and 4 times the maximum density at the initial time.

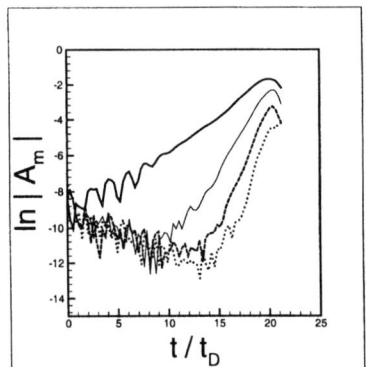

FIGURE 9. The growth of the amplitudes $|A_m|$ for $m=1$ (thick solid line), $m=2$ (thin solid line), $m=3$ (dashed line), and $m=4$ (dotted line) is shown for the $\mathcal{L}$ code simulation of the $\beta=0.14$ model. These amplitudes were calculated in the equatorial plane for a ring with radius $\varpi=0.32$.

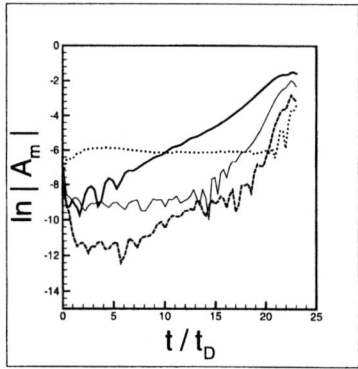

FIGURE 10. Same as Figure 9, for Brown's simulation.

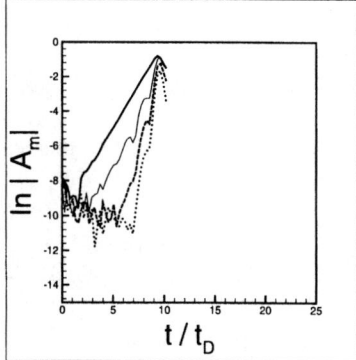

FIGURE 11. Same as Figure 9, except that the model has $\beta=0.18$.

grows more quickly in this model than in the $\beta=0.14$ model, as expected. The $m=1$ mode grows at about the same rate in the simulations of both codes. The $\mathcal{L}$ code run shows the $m=2$, 3, and 4 modes growing in sequence, just like in the $\beta=0.14$ simulation; the $m=2$ mode is again an harmonic of the $m=1$ mode. Brown's run, however, shows the $m=1$ and $m=2$ modes growing at the same rate. These modes are not harmonics in his simulation. We plan to investigate the differences in these results with higher resolution simulations.

Our study demonstrates that dynamical instabilities can occur in differentially rotating polytropes with relatively low values of β. Note that Pickett, Durisen, and Davis [15] also found a dominant $m=1$ instability in a centrally condensed configuration. The instability in their model set in at $\beta=0.2$. The model had an $n=1.5$ polytropic equation of state and was constructed with the so-called $n'=2$ rotation law. This rotation law places more angular momentum in the outer regions of the model than does the j-constant law. Their model did not have off-center density maxima.

We plan to carry out more detailed studies to investigate the character of these

231

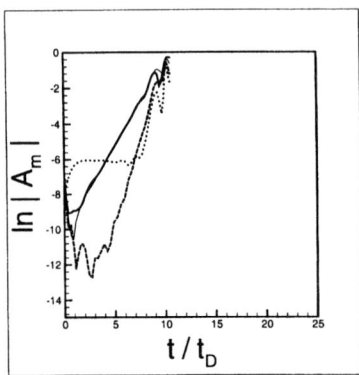

FIGURE 12. Same as Figure 10, except that the model has β=0.18.

unstable modes and their properties for various values of the polytropic index n and the rotation law parameter d.

If this instability occurs in centrifugally hung stellar cores, collapse to neutron star densities may result. Our simulations do show that the density is increasing at the end of the unstable runs. Further studies are needed to determine how dense the remnant becomes and to see if the m=1 mode will result in the star moving with velocities comparable to those of actual neutron stars.

Longer runs on larger grids are needed to obtain the full gravitational radiation waveforms. However, we can estimate the properties of the emission during the initial stages of the development of the instability. For a fizzler that starts out with $M \sim 1.4 M_\odot$ and $R \sim 200$km, the peak emission will occur at $f_{gw} \sim 200$Hz. The peak amplitude h_{max} will be $\sim 10^{-24} r_{20}^{-1}$ for $\beta = 0.14$, and $\sim 10^{-23} r_{20}^{-1}$ for $\beta = 0.18$. Here, r_{20} is the distance to the source in units of 20 Mpc. Emission from such unstable cores may be detectable with advanced ground-based interferometers like LIGO II.

An even more optimistic scenario for the detection of gravitational radiation occurs if this type of instability is encountered by a cooling and contracting super-massive star [7,8]. If the star's ratio GM/Rc^2 is ~ 15 when the instability develops (a value that is approximately appropriate for an uniformly rotating star), f_{gw} will be $\sim 10^{-3}$ Hz and h_{max} will be $\sim 10^{-18} r_{20}^{-1}$ for $\beta = 0.14$ and $\sim 10^{-17} r_{20}^{-1}$ for $\beta = 0.18$. Such signals would be easily detectable by the space-based LISA detector.

REFERENCES

1. Chandrasekhar, S., *Ellipsoidal Figures of Equilibrium*, Yale University Press, New Haven, 1969.
2. Tassoul, J., *Theory of Rotating Stars*, Princeton University Press, Princeton, 1978.
3. Schutz, B., *Class. Quantum Gravity* **6**, 1761 (1989).

4. Thorne, K., in *Proceedings of the Snowmass 95 Summer Study on Particle and Nuclear Astrophysics and Cosmology*, edited by E. W. Colb and R. Peccei, World Scientific, Singapore, 1996.

5. Tohline, J. E., *Astrophys. J.* **285**, 721 (1984).

6. Shapiro, S. and Lightman, A., *Astrophys. J.* **207** 263 (1976).

7. New, K. C. B. and Shapiro, S. L., *Astrophys. J.*, in press (2000) (astro-ph/0010574).

8. Baumgarte, T. and Shapiro, S. L., *Astrophys. J.* **526**, 941 (1999).

9. Lai, D. and Shapiro, S. L., *Astrophys. J.* **442**, 259 (1995).

10. Centrella, J. M., New, K. C. B., Lowe, L. L., and Brown, D., *Astrophys. J. Lett*, submitted (2000).

11. New, K. C. B., Centrella, J. M., and Tohline, J. E., *Phys. Rev. D* **62**, 064019 (2000).

12. Smith, S., Houser, J., and Centrella, J., *Astrophys. J.* **458**, 236 (1996).

13. Houser, J. and Centrella, J., *Phys. Rev. D* **54**, 7278 (1994).

14. New, K. C. B., Ph.D. Thesis, Louisiana State University (1996).

15. Pickett, B., Durisen, R., and Davis G., *Astrophys. J.* **458**, 714 (1996).

16. Durisen, R. Gingold, R., Tohline, J., and Boss, A., *Astrophys. J.* **305**, 281 (1986).

17. Toman, J., Imamura, J. N., Pickett, B. K., and Durisen, R. H., *Astrophys. J.* **497**, 370 (1998).

18. Hachisu, I., *Astrophys. J. Suppl.* **61**, 479 (1986).

19. Bodenheimer, P. and Ostriker, J., *Astrophys. J.* **180**, 159 (1973).

20. Brown, J. D., *Phys. Rev. D*, in press (2000) (gr-qc/0004002).

21. Zwerger, T. & Müller, E., *Astronomy & Astrophysics* **320**, 209 (1997).

22. Eriguchi, Y. & Müller, E., *Astronomy & Astrophysics* **147**, 161 (1985).

23. Woodward, J., Tohline, J., & Hachisu, I., *Astrophys. J.* **420**, 247 (1994).

24. Tohline, J. & Hachisu, I. 1990, *Astrophys. J.* **361**, 394 (1990).

Rotational Instabilities
in Post–Collapse Stellar Cores

J. David Brown

Department of Physics, North Carolina State University, Raleigh, NC 27695-8202

Abstract. A core–collapse supernova might produce large amplitude gravitational waves if, through the collapse process, the inner core can aquire enough rotational energy to become dynamically unstable. In this report I present the results of 3-D numerical simulations of core collapse supernovae. These simulations indicate that for some initial conditions the post–collapse inner core is indeed unstable. However, for the cases considered, the instability does not produce a large gravitational–wave signal.

INTRODUCTION

Core–collapse supernovae are potentially rich sources of gravitational waves. When a rotating stellar core exhausts its nuclear fuel the matter in the polar regions collapses more rapidly than the matter in the equatorial plane, since the latter must fight harder against centrifugal forces as it spirals inward. The changing oblateness of the core during this infall phase, and the subsequent changes due to core bounce, will generate gravitational waves (see, for example, Refs. [1,2]). After core bounce convective instabilities will cause the hot neutron star to boil [3], and the resulting convective motions will give rise to a gravitational–wave signal [4]. To a lesser extent, the same boiling process and release of gravitational waves can occur in the neutrino–heated material interior to the shock wave [4]. The supernova explosion itself may have a preferred direction, as evidenced by the large kick velocities of many neutron stars [5]. Such an asymmetry will generate a gravitational–wave signal with memory [6]. Perhaps the most interesting possibility, and the one discussed in this paper, is that the stellar core will spin up as it collapses and produce a very rapidly rotating neutron star. The neutron star might be subject to dynamical instabilities that act to deform, or even fragment the star, and in the process produce large amplitude gravitational waves.

In this paper I present the results of numerical simulations aimed at determining the types of initial conditions for a pre–collapse stellar core that lead to a dynamically unstable post–collapse inner core. The only previous investigations along these lines is found in the work of Rampp, Müller, and Ruffert [7].

CP575, *Astrophysical Sources for Ground-Based Gravitational Wave Detectors*, edited by J. M. Centrella
© 2001 American Institute of Physics 0-7354-0014-8/01/$18.00

Analytical and numerical work on rapidly rotating fluid stars with Maclaurin-like rotational laws has shown that dynamical instabilities, in particular the $m = 2$ bar–mode instability, will grow when the stability parameter $T/|W|$ (the ratio of rotational kinetic energy to gravitational potential energy) exceeds about 0.27 [8]. For certain angular velocity profiles, a dynamical $m = 1$ instability can grow for $T/|W|$ as low as ~ 0.14 [9]. A back–of–the–envelope calculation frequently quoted in the literature suggests that the stability parameter will scale as $T/|W| \sim 1/R$ during collapse, where R is the radius of the core. For a 10 solar mass star whose core collapses completely to neutron star size, this implies a factor ~ 100 increase in $T/|W|$. Based on this argument, one would expect that even the most slowly rotating cores will be dynamically unstable after collapse.

The factor ~ 100 increase in $T/|W|$ is overly optimistic for two reasons. First, centrifugal forces can halt the collapse at subnuclear density, producing a bloated inner core with $T/|W|$ below the threshold for dynamical instability. Second, the core actually "implodes", rather than collapses, and some of the stellar core's matter and angular momentum remain outside the inner core. Since only a percentage of the core's angular momentum is drawn into the inner core, the increase in rotational kinetic energy is less than one would predict by assuming a more complete collapse. On the other hand, the simulations described here provide evidence that the entire core need not have $T/|W| \gtrsim 0.27$ for the bar mode to grow on a relatively short timescale. It might be possible for the bar instability to grown as a secular process on a timescale of a few rotation periods, due to coupling between the inner and outer core regions.

Two of the most rapidly rotating pre–collapse models considered here lead to dynamically unstable post–collapse inner cores. These models are unstable to the $m = 2$ bar mode, but the resulting bars have insufficient density and spatial extent to generate large gravitational–wave signals. In both cases, the contribution to the gravitational–wave signal from the bar deformation is smaller in amplitude, by a factor of ~ 2–5, than the purely axisymmetric signal generated by the core's changing oblateness during collapse and bounce. Overall, for sources in the Virgo cluster, these signals are too weak to be detected by laser interferometers operating in broadband mode. For galactic sources, these signals should be detectable.

NUMERICAL CODE AND INITIAL MODELS

The numerical code used for these investigations was written in collaboration with John Blondin. It uses Newtonian hydrodynamics and Newtonian gravity, ignores neutrino heating and cooling, and uses the relatively simple analytical equation of state discussed by Zwerger and Müller [2,7]. The gravitational–wave signal is computed in the quadrupole approximation. The hydrodynamical equations are solved with VH-1, written by Blondin, J. Hawley, G. Lindahl, and E. Lufkin. VH-1 uses the piecewise parabolic method [10]. The Poisson equation for the gravitational potential is solved with multigrid techniques.

Our code models the supernova in a minimal way, retaining just enough physics to capture the gravitational–wave signal in the leading–order quadrupole approximation. Nevertheless, the computation is a challenge due to the discrepancy in length scales involved. While the stellar core has a radius of 1000's of kilometers, the computational grid must have zone sizes of no more than ~ 1 km to support the steep density and velocity gradients in the inner core. A uniform 3-D Cartesian grid would require $\sim 10^{10}$ zones for these simulations, which is not feasible with current technology. One possible solution to this problem is to use a spherical–coordinate grid with non–uniform radial spacing, so the zones are smallest in the inner regions of the grid. The difficulty with this approach is that the zones become too narrow in the angular directions near the coordinate origin, and this drives the Courant–limited timestep to zero. The solution we have adopted is a nested grid scheme, in which the computational domain is covered by a sequence of Cartesian grids with increasing resolution and decreasing size. In this way only the inner–most region is covered with high resolution. This approach was also used by Rampp, Müller, and Ruffert [7]. The simulations described here use 7 grids, each with 64^3 zones.

In addition to the 3-D code just described, I also work with a 2-D code that assumes axisymmetry. The 2-D code uses a system of nested Cartesian grids in the r–z plane to achieve the necessary resolution of the inner core. The 2-D simulations reported here use 6 grids, each with 128^2 zones.

In this paper I consider a sequence of initial models of the pre–collapse stellar core. The goal is to determine which of these models lead to dynamically unstable post–collapse inner cores. Each initial model is a rotating, equilibrium polytrope with $\Gamma = 4/3$ and central density $\rho_c = 10^{10}$ g/cm^3. The rotation laws for the sequence are given by

$$\Omega(\varpi) = \Omega_0 e^{-(\varpi/1500\,\text{km})^2}, \qquad \Omega_0 = 8,\ 12,\ 16,\ 20,\ 24\,\text{rad/s}, \qquad (1)$$

where Ω is the angular velocity and ϖ is the distance from the rotation axis. The Gaussian form for $\Omega(\varpi)$ was motivated in part by the results of Heger, Langer, and Woosley [11]. Their simulations of stellar evolution with rotation yield pre–collapse cores with relatively broad, Gaussian–like angular velocity profiles, similar in shape to the profile (1). Note, however, that even their most rapidly rotating model has an overall scale of $\Omega_0 \approx 10$, somewhat smaller than most of the models (1).

In their previous work Rampp, Müller, and Ruffert [7] used an initial data set with angular velocity profile $\Omega(\varpi) = \Omega_0/[1 + (\varpi/100\,\text{km})^2]$ and $\Omega_0 \approx 140$ rad/s. The resulting model is highly differentially rotating. The angular velocity drops from its central value of about 140 rad/s to less than 4 rad/s at a distance of $R_{eq}/2$, where $R_{eq} \approx 1260$ km is the equatorial radius. By contrast, the angular velocity of the most rapidly rotating model (1) drops from 24 rad/s to about 13 rad/s at $R_{eq}/2$, where $R_{eq} \approx 2500$ km.

The equation of state [2,7] contains a polytropic part whose stiffness depends on whether the matter is below or above nuclear density, and a thermal part that models the thermal pressure of matter heated by shock waves. Collapse of the initial models is induced by choosing the polytropic index in the sub–nuclear density regime to be $\Gamma = 1.28$, significantly below the value of $4/3$ required for the model to remain in equilibrium. Random density perturbations at the 1% level were imposed at the beginning of the 3-D simulations.

RESULTS

The stability parameter is plotted as a function of time in Figure 1 for the sequence of models (1). These simulations were run with the 2-D code. Observe that $T/|W|$ increases by a factor of 2 or less as a result of core collapse. As shown in Figure 2, for the initial models with angular velocity $\Omega_0 = 16$, 20 and 24 rad/s, the core experiences a centrifugal bounce and never reaches nuclear density $\rho_{\mathrm{nuc}} = 2.0 \times 10^{14}\,\mathrm{g/cm^3}$. For the model with $\Omega_0 = 12\,\mathrm{rad/s}$, the inner core reaches nuclear density at core bounce then relaxes to a central value slightly below nuclear density. Only the most slowly rotating initial data, with $\Omega_0 = 8\,\mathrm{rad/s}$, forms a stiff inner core with density greater than ρ_{nuc}. These results show that centrifugal forces can severely inhibit core collapse, and prevent the stability parameter from experiencing the kind of growth suggested by the back–of–the–envelope arguments discussed in the introduction.

The stability parameter for the two most rapidly rotating models exceeds 0.27 after core bounce. I will use the labels $\Omega20$ and $\Omega24$ to denote the models with $\Omega_0 = 20\,\mathrm{rad/s}$ and $\Omega_0 = 24\,\mathrm{rad/s}$, respectively. For these models one might expect the post–bounce inner core, although centrifugally hung at sub–nuclear densities, to be dynamically unstable to growth of the $m = 2$ bar mode. The inner cores are not likely to be unstable to the $m = 1$ mode discussed in Reference [9], since their

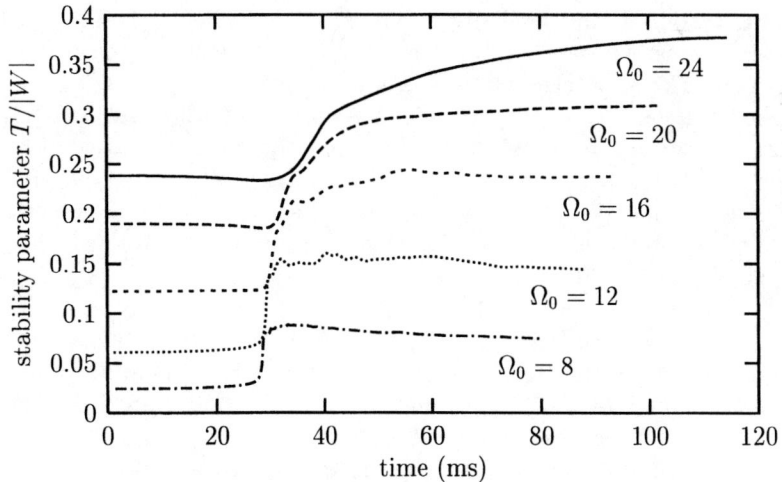

FIGURE 1. $T/|W|$ vs. t for the sequence of models (1).

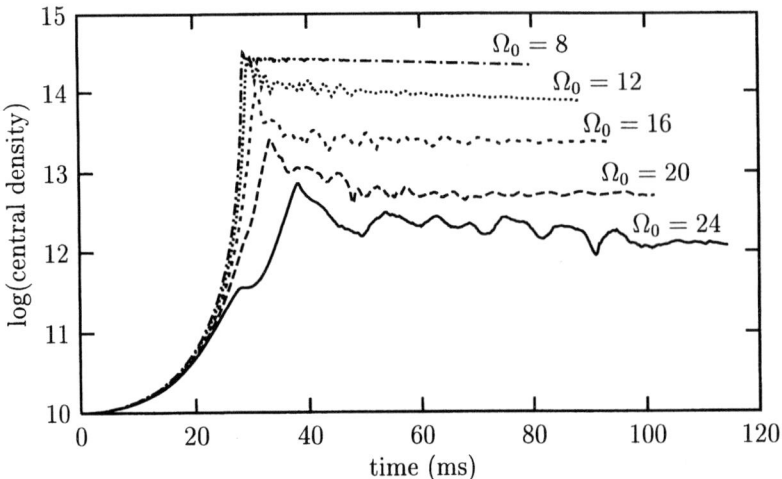

FIGURE 2. $\ln(\rho_c)$ vs. t for the sequence of models (1).

density profiles are centrally peaked. Note, however, that prior to collapse both of these models exceed the nominal threshold $T/|W| \approx 0.14$ for growth of secular instabilities. Thus, these models are unrealistic—a realistic stellar core with such a high rate of rotation would lose its axisymmetry prior to collapse. With this caveat in mind, I have forged ahead and evolved these models with the 3-D code to check for dynamical instabilities in the post–collapse inner cores. These simulations show that model $\Omega 20$ remains dynamically stable after collapse. Model $\Omega 24$ is unstable after collapse, and its inner core deforms into a bar shape.

For the model considered by Rampp, Müller, and Ruffert (RMR) [7], the pre–collapse core has an initial value for $T/|W|$ of about 0.04, well below the threshold for growth of secular instabilities. Due to the high degree of differential rotation, the material in the outer layers rotates relatively slowly. As collapse proceeds, the lack of substantial centrifugal support allows the matter in the outer layers to strongly compress the inner core. The result is that the stability parameter increases by a larger fraction than that obtained with the data sets (1). The peak value of $T/|W|$ is about 0.35, although $T/|W|$ stays above 0.27 for less than 2 ms. After bounce, the inner core relaxes and $T/|W|$ quickly settles to a value of about 0.19.

In their 3-D simulations, RMR imposed 10% random and 5% $m = 3$ density perturbations at a time of 2.5 ms before core bounce. The inner core showed $m = 2$, 3, and 4 asymmetries after bounce, but no significant enhancement in the gravitational–wave signal. Their simulations were halted at about 45 ms, approximately 15 ms after core bounce. The following question naturally arises: are the post–bounce asymmetries seen in the RMR simulation merely a transient effect caused by the asymmetrical bounce? I have carried out a 3-D simulation with the RMR initial data using 1% random density perturbations imposed at the onset of collapse, to see if asymmetries will grow from a nearly axisymmetric bounce. I will refer to this simulation as model RMR. A second motivation for taking another look at the RMR initial data comes from the recent results in Reference [9], which suggest that a fluid body with toroidal density maximum (as occurs for model RMR both before and after collapse) can be dynamically unstable to an $m = 1$ mode for $T/|W|$ as low as ~ 0.14. The results of my simulation show that, as expected, the inner core is nearly axisymmetric immediately after core bounce. It remains axisymmetric until about 45 ms, which is the time at which RMR stopped their simulations. The dominant unstable mode that begins to grow at that time is the $m = 2$ bar mode, not the $m = 1$ mode. This occurs in spite of the fact that the stability parameter has a value of around 0.19.

Before presenting the detailed results of the 3-D simulations, I need to establish some notation. The shape of the core can be described by expanding the matter density ρ in spherical harmonics:

$$\rho(t, r, \theta, \phi) = \sum_{\ell=0}^{\infty} \sum_{m=-\ell}^{m=\ell} A_{\ell m}(t, r) Y_{\ell m}(\theta, \phi) \, . \tag{2}$$

The quadrupole formula relates the gravitational-wave amplitude to the second time derivative of the quadrupole moment of the mass distribution [12]. Inserting the expansion (2) into the quadrupole formula, we find

$$\frac{c^4 R}{2G} h_+^{TT} = \sqrt{\frac{\pi}{5}} \sin^2 \Theta \langle \ddot{A}_{20} \rangle + \sqrt{\frac{2\pi}{5}} (1 + \cos^2 \Theta) \, \Re(\langle \ddot{A}_{22} \rangle e^{2i\Phi})$$
$$+ \sqrt{\frac{8\pi}{15}} \sin \Theta \cos \Theta \, \Re(\langle \ddot{A}_{21} \rangle e^{i\Phi}) \, , \tag{3a}$$

$$\frac{c^4 R}{2G} h_\times^{TT} = -\sqrt{\frac{8\pi}{15}} \cos \Theta \, \Im(\langle \ddot{A}_{22} \rangle e^{2i\Phi}) - \sqrt{\frac{8\pi}{15}} \sin \Theta \, \Im(\langle \ddot{A}_{21} \rangle e^{i\Phi}) \, , \tag{3b}$$

for the $+$ and $\times$ components of the gravitational-wave amplitude in the transverse-traceless (TT) gauge. In these formulas R, Θ, and Φ specify the distance and angular direction from the source to the observation point and $\Re$ and $\Im$ denote real and imaginary parts. The angle brackets that appear in equations (3) are defined by $\langle \ddot{A}_{\ell m} \rangle = \int dr \, r^4 \ddot{A}_{\ell m}$, which is the spatial average of the second time derivative of $A_{\ell m}$, weighted with r^2.

From equations (3) we see that only the $\ell = 2$ spherical harmonics contribute to the gravitational-wave signal in the quadrupole approximation. The coefficient A_{20} determines the oblateness of the mass distribution. Note that the space average $\langle \ddot{A}_{20} \rangle$ appears in h_+^{TT}, but not in $h_\times^{TT}$. The coefficient A_{22} corresponds to a bar-shaped deformation in the equatorial plane. Growth of this coefficient implies growth of the usual Fourier $m = 2$ bar mode. The coefficient A_{21} describes a bar-shaped deformation that is tilted out of the equatorial plane. This coefficient remains zero (apart from numerical noise) throughout the simulations, as one might expect from symmetry considerations. Note that the coefficient A_{11}, which corresponds to the $m = 1$ mode of Reference [9], does not appear in the approximate formulas (3) for the gravitational–wave amplitude.

Figures 3 and 4 show the ratios A_{20}/ρ_0 and $|A_{22}|/\rho_0$ for model $\Omega24$, at various radii. Here, ρ_0 is the average density at the given radius. For clarity of presentation, the results for radius $r = 60$ km are shown with heavy curves. From Figure 3 we see that, initially, the oblateness of the core increases rapidly as matter rushes inward

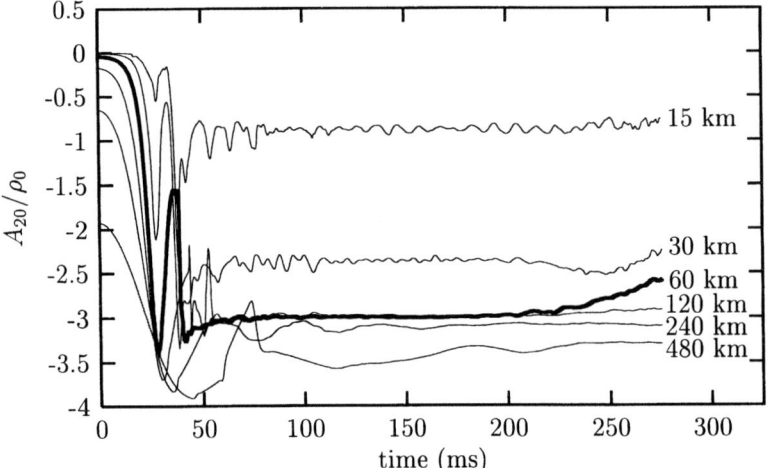

FIGURE 3. A_{20}/ρ_0 vs. t for model $\Omega24$.

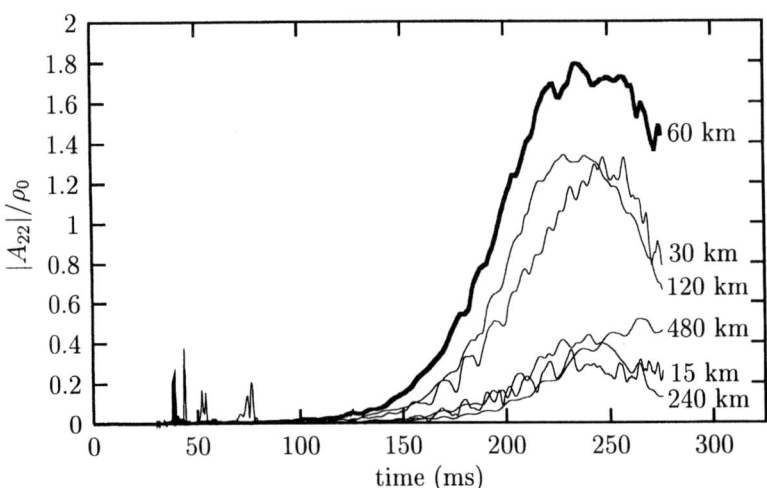

FIGURE 4. $|A_{22}|/\rho_0$ vs. t for model $\Omega24$.

240

from all directions, but most rapidly along the polar regions. (A_{20} is negative for an oblate spheroid, positive for a prolate spheroid.) At ~ 28 ms the core experiences a "polar bounce" in which the matter in the polar regions is reflected off the inner core, reverses direction, and forms shock waves that propagate outward away from the equatorial plane. During the next ~ 10 ms, the oblateness of the inner core decreases as matter continues to rush in from the equator and out from the poles. At about 38 ms the core experiences an "equatorial bounce" in which the matter in the equatorial plane is reflected and forms an outwardly propagating shock wave. As the inner core relaxes, it spreads in the equatorial direction and again assumes a highly oblate shape. The oblateness remains fairly constant until near the end of the simulation, when the bar deformation becomes strong. Figure 4 shows growth of the bar mode, which begins around 100 ms. The spikes between 40 ms and 80 ms are caused by shock waves passing through the various radii. Since the shocks are not perfectly axisymmetric they produce relatively large but short–lived distortions that show up in the A_{22} coefficient. The growth rate of the bar mode for model $\Omega 24$ is $d\ln(|A_{22}|/\rho_0)/dt \approx 46/s$.

A graph of the ratio $|A_{22}|/\rho_0$ for model RMR shows that the bar mode begins to grow at about 45 ms, at a rate of $d\ln(|A_{22}|/\rho_0)/dt \approx 180/s$. $|A_{22}|/\rho_0$ reaches a peak value of 1.6 at ~ 70 ms. The strongest bar deformation occurs within a radius of ~ 100 km.

The graphs in Figures 5 and 6 show the gravitational–wave signals for models $\Omega 24$ and RMR, respectively. The solid curves show the + polarization amplitude (3a) as measured in the equatorial plane $\Theta = \pi/2$ at a distance of $R = 20$ Mpc (the approximate distance to the Virgo cluster). The dashed curves show the + polarization amplitude (3a) as measured along the rotation axis $\Theta = 0$ at $R = 20$ Mpc.

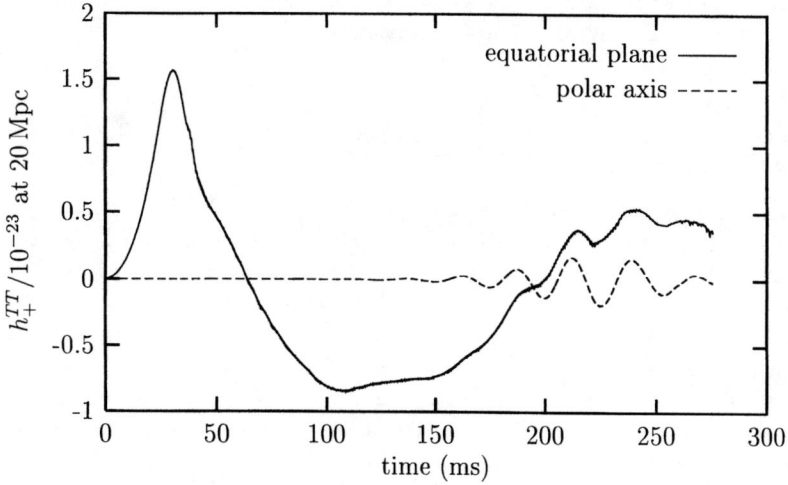

FIGURE 5. The gravitational wave amplitude for model $\Omega 24$.

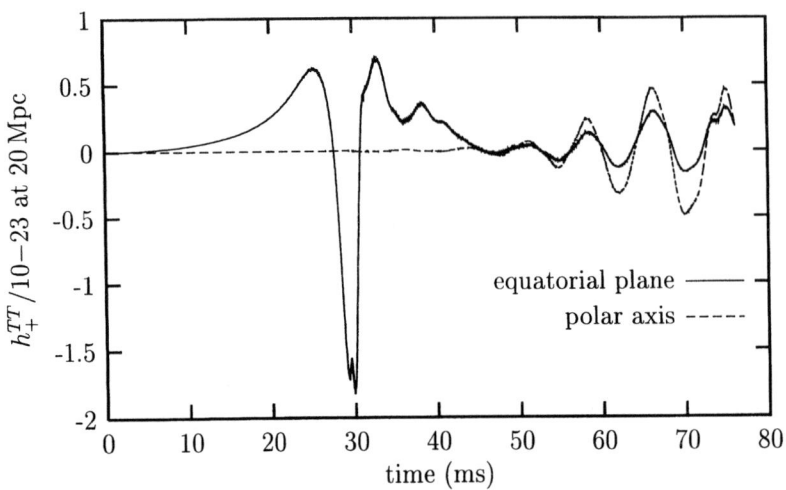

FIGURE 6. The gravitational wave amplitude for model RMR.

Since the coefficient A_{21} remains essentially zero, the $\times$ polarization amplitude (3b) at any angle is proportional to the $+$ amplitude at $\Theta = 0$. Observe that for both models, the gravitational waves produced by the bar—the wiggles on the graphs at late times—are relatively small in amplitude compared to the gravitational waves produced by the core's (nearly axisymmetric) collapse and bounce. Fourier analysis of the gravitational–wave signal for model $\Omega24$ shows two peaks, one around 5 Hz due to the collapse and bounce motion of the core, and the other around 40 Hz due to the rotation of the bar–shaped inner core. The total energy radiated in gravitational waves for model $\Omega24$, up to the end of the simulation, is only a few $\times\ 10^{-9}\ M_\odot c^2$. For model RMR, the gravitational–wave signal is spread across the frequency range 25–250 Hz. The total energy radiated for this model, up to the end of the simulation, is a few $\times\ 10^{-7}\ M_\odot c^2$.

DISCUSSION

After collapse, models $\Omega24$ and RMR are unstable to growth of the bar mode. On the other hand, model $\Omega20$ is stable. The 3-D simulation of model $\Omega20$ was carried out to ~ 200 ms beyond core bounce, and the coefficient A_{22} showed no signs of growth. At first sight these results might seem surprising: As shown in Figure 1, the stability parameter for models $\Omega24$ and $\Omega20$ after core bounce exceed the nominal threshold ~ 0.27 while the stability parameter for model RMR has a sustained, post–bounce value of less than 0.20. Of course, our understanding of the bar mode instability is based primarily on studies of isolated, equilibrium polytropes with Maclaurin–like rotation laws. There is little reason to believe that such a body would be a good approximation to a post-collapse stellar core and,

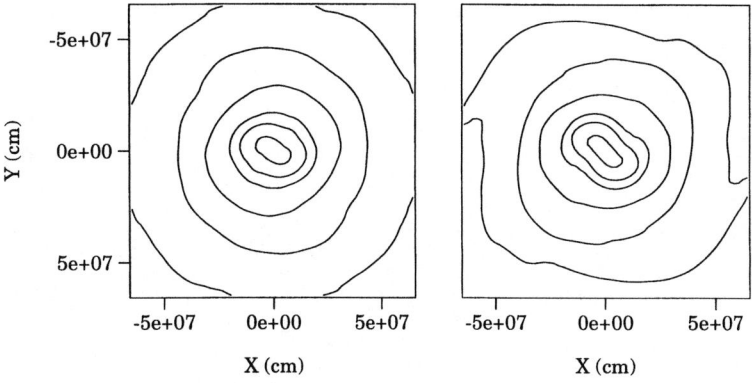

FIGURE 7. Density contours for model $\Omega 24$ at 199 ms (left) and 229 ms (right). The contour levels, in g/cm^3, are 5.0×10^9, 1.0×10^{10}, 2.0×10^{10}, 5.0×10^{10}, 1.0×10^{11}, and 5.0×10^{11}.

indeed, for the cases studied here, it is not. The value of the stability parameter for the post–collapse core is not a good diagnostic for the presence of the bar instability.

Inspection of the data for models $\Omega 24$ and RMR shows that only the most dense regions of the post–collapse core participate in growth of the bar mode. For the model $\Omega 24$ in particular, the bar deformation is contained within the region with density $\rho \gtrsim 10^{10}$ g/cm^3. Figure 7 shows contour plots of the density in the equatorial plane at 199 ms, when the bar has moderate strength, and at 229 ms, when the bar is near full strength. In the first plot, the contours below $\sim 10^{10}$ g/cm^3 are nearly circular apart from some $m = 4$ noise caused by the Cartesian grid. By the time of the second plot, the bar–shaped region with $\rho \gtrsim 10^{10}$ g/cm^3 has created "wakes" in the surrounding matter. These wakes form spiral arms that trail from the ends of the bar, and give rise to the spikes seen in the contour at 5.0×10^9 g/cm^3. Figure 8 shows the density as a function of radius in the equatorial plane for models $\Omega 20$ and $\Omega 24$. For model $\Omega 24$ the density data is taken at 161 ms, near the beginning of the bar mode growth. Note that for both models, the density has a peak in the center then levels off to a value of about 10^{10} g/cm^3. Beyond ~ 700 km, the density drops sharply.

Motivated by the observations above, I will define the inner core for models $\Omega 20$ and $\Omega 24$ to be the region interior to $\sim 10^{10}$ g/cm^3. Thus, we can view the post–collapse configuration as consisting of a dense inner core with $\rho \gtrsim 10^{10}$ g/cm^3 surrounded by relatively low density material. It is this inner core region that is unstable for model $\Omega 24$ and stable for model $\Omega 20$. The inner core for model RMR can be defined roughly by $\rho \gtrsim 10^{10}$ g/cm^3 as well.

Some insights into the behavior of the three models $\Omega 20$, $\Omega 24$, and RMR can be gained by defining a stability parameter for the inner core, $T_{ic}/|W_{ic}|$. The rotational kinetic energy of the inner core T_{ic} is straightforward to compute. The

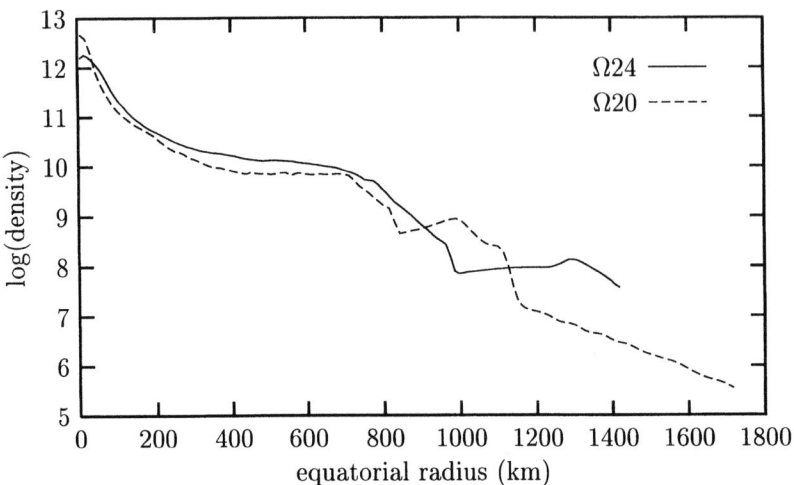

FIGURE 8. Density profile in the equatorial plane for model $\Omega24$ at 161 ms and for model $\Omega20$ at 182 ms.

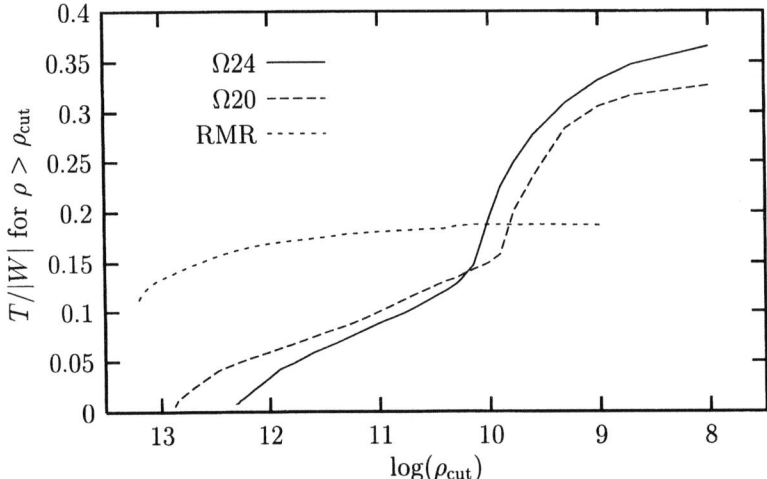

FIGURE 9. The stability parameter as a function of density cut–off for models $\Omega24$ at 161 ms, $\Omega20$ at 182 ms, and RMR at 52.9 ms.

gravitational potential energy W_{ic} must include the binding energy of the inner core material with itself, as well as the binding energy between the inner core material and the outer core material. The graph in Figure 9 shows the stability parameter of the region with density $\rho > \rho_{cut}$, as a function of the cut–off value ρ_{cut}, for the three models. The stability parameter for the inner core is obtained by setting $\rho_{cut} \approx 10^{10} \, \mathrm{g/cm^3}$. The graph shows that for model $\Omega20$, $T_{ic}/|W_{ic}| \approx 0.15$ while for models $\Omega24$ and RMR, $T_{ic}/|W_{ic}| \approx 0.19$. Based on these results, we might expect that model RMR is less stable than model $\Omega20$. Indeed, as the 3-D simulations show, RMR is unstable and $\Omega20$ is stable. Perhaps the more interesting question is this: Why are the inner cores for $\Omega24$ and RMR unstable, given the fact that the values for their stability parameters are well below 0.27? The explanation might simply be that the post–collapse inner cores are not equilibrium polytropes with Maclaurin–like rotation laws, so the threshold for dynamical instability might be far different from 0.27. Another explanation might be that a post–collapse inner core is not isolated, and coupling to the outer core material can drive the bar instability. If this is correct, then growth of the bar mode is a secular process, as discussed for example by Schutz [13]. Note that for isolated, equilibrium polytropes with Maclaurin–like rotation laws, the threshold for growth of the secular instability is about 0.14. The inner core stability parameters for $\Omega24$ and RMR are well above this threshold. On the other hand, the stability parameter of the inner core for model $\Omega20$ is very close to the threshold.

ACKNOWLEDGMENTS

I would like to thank John Blondin for helpful discussions and for his work on the numerical codes. This research was supported by NSF grant PHY-0070892. Computer resources were provided by the North Carolina Supercomputing Center.

REFERENCES

1. Yamada, S., and Katsuhiko, S., *Ap. J.* **450**, 245–252 (1995).
2. Zwerger, T., and Müller, E., *Astron. Astrophys.* **320**, 209–227 (1997).
3. Bethe, H.A., *Rev. Mod. Phys.* **62**, 801–866 (1990).
4. Müller, E., and Janka, H.-T., *Astron. Astrophys.* **317**, 140–163 (1997).
5. See for example Tauris, T.M., and van den Heuvel, E.P.J., to appear in the proceedings of the IAU Colloq. 177 "Pulsar Astronomy—2000 and Beyond". (astro-ph/0001015)
6. Burrows, A., and Hayes, J., *Phys. Rev. Lett.* **76**, 352–355 (1996).
7. Rampp, M., Müller, E., and Ruffert, M., *Astron. Astrophys.* **332**, 969–983 (1998).
8. Tassoul, J., *Theory of Rotating Stars*, Princeton: Princeton University Press, 1978.
9. Centrella, J.M., New, K.C.B., Lowe, L.L., and Brown, J.D., submitted to *Ap. J. Lett.* (astro-ph/0010574). See also New, K.C.B., this volume.
10. Colella, P., and Woodward, P.R., *J. Comput. Phys.* **54**, 174–201 (1984).
11. Heger, A., Langer, N., and Woosley, S.E., Ap. J. **528**, 368–396 (2000).
12. Misner, C.W., Thorne, K.S., and Wheeler, J.A., *Gravitation*, San Francisco: W.H. Freeman, 1973.
13. Schutz, B.F., "Problems in Astrophysical Fluid Dynamics", in *Fluid Dynamics in Astrophysics and Geophysics*, edited by N.R. Lebovitz, AMS Lectures in Applied Mathematics **20**, Providence, 1983, pp. 99–140.

Secular Bar-Mode Evolution and Gravitational Waves From Neutron Stars

Dong Lai

Center for Radiophysics and Space Research, Department of Astronomy
Cornell University, Ithaca, NY 14853

Abstract. The secular instability and nonlinear evolution of the $m = 2$ f-mode (bar-mode) driven by gravitational radiation reaction in a rapidly rotating, newly formed neutron star are discussed. There are two types of rotating bars which generate quite different gravitational waveforms: those with large internal rotation relative to the bar figure (Dedekind-like bars) have GW frequency sweeping downward, and those with small internal rotation (Jacobi-like) have GW frequency sweeping upward. Various sources of viscosity in hot nuclear matter are examined, and the possible effect of star–disk coupling on the bar-mode instability is also discussed.

I INTRODUCTION

Nonaxisymmetric instabilities can develop in rapidly rotating fluid bodies when the ratio $\beta \equiv T/|W|$ of the rotational energy T to the gravitational potential energy W is sufficiently large (e.g., Chandrasekhar 1969). In particular, the $l = m = 2$ f-mode (Kelvin mode), or bar-mode, becomes dynamically unstable when $\beta > \beta_{\mathrm{dyn}} \simeq 0.27$. This β_{dyn}, originally derived for incompressible Maclaurin spheroid, is relatively insensitive to the equation of state and differential rotation (Pickett et al. 1996; Toman et al. 1998), although it tends to be reduced by general relativity (Shibata et al. 2000). The consequence of the dynamical bar-mode instability has been extensively studied using numerical simulations: the mode grows to nonlinear amplitude by shedding mass and angular momentum from the ends of the bar in the form of two-armed spiral pattern, and the central star assumes a bar shape that lasts many rotation periods (e.g., Tohline et al. 1985; New et al. 2000; Brown 2000; Shibata et al. 2000). It is important to note, however, that these simulations start out with a stationary, dynamically unstable star and such an initial condition may not be realized in an actual core collapse (e.g., Rampp, Müller and Ruffert 1998).

However, it is very likely that in a rotating core collapse (or neutron star binary merger), after the messy dynamics is completed, the newly formed NS settles down into a dynamically stable configuration, with $\beta < 0.27$, and yet still suffers further secular, rotational instability. "Secularly unstable" means the instability is slow,

CP575, *Astrophysical Sources for Ground-Based Gravitational Wave Detectors*, edited by J. M. Centrella
© 2001 American Institute of Physics 0-7354-0014-8/01/$18.00

and is driven by gravitational radiation reaction. In this paper, we will focus this secular instability and the possible gravitational wave signals from newly formed neutron stars in the first seconds/minutes of their lives.

II   CFS INSTABILITY: F-, G- AND R-MODES

The gravitational radiation driven instability, or Chandrasekhar-Friedman-Schutz (CFS) instability (see e.g., Friedman 1998 for a review), arises for the following simple reason: Suppose in the rotating frame of the star we set up a perturbation (mode) (with frequency ω_r as in $e^{im\phi+i\omega_r t}$) which travels opposite to the rotation. In the inertial frame, the perturbation will be dragged backward by the rotation, and the mode frequency becomes $\omega_i = \omega_r - m\Omega_s$. If the spin Ω_s is sufficiently large, the perturbation will be prograde in the inertial frame. Since the mode has negative angular momentum (because the perturbed fluid does not rotate as fast as it did without the perturbation), as the mode radiates positive J through gravitational radiation, the mode's angular momentum will be more negative; that means the mode is unstable.

The CFS instability mechanism can be applied to different modes of NSs:

1. F-modes: The f-modes are fundamental acoustic waves, corresponding to a global distortion of the star. For the $m = 2$ mode (bar-mode), the instability occurs

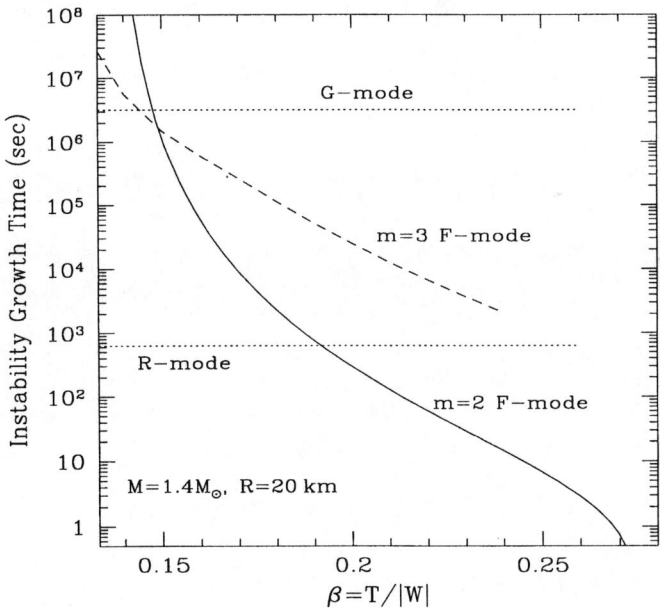

FIGURE 1. Growth times of gravitational radiation driven CFS instability for different modes in a neutron star as a function of β.

when $\beta > \beta_{\rm sec} = 0.14$ — This critical $\beta_{\rm sec}$ is only slightly affected by equation of state, and is somewhat reduced (to as small as 0.1) by strong differential rotation (Imamura et al. 1995) and by general relativity (Stergioulas & Friedman 1998).

2. R-modes: In the last few years it was realized that the so-called r-modes are always secularly unstable (for stars consisting of inviscid fluid) (Andersson 1998) and the growth time of the mode through current quadrupole gravitational radiation can be interestingly short (Lindblom et al. 1998). A lot of works are currently being done on r-modes (see http://online.itp.ucsb.edu/online/neustars00/si-rmode-sched.html for a recent meeting devoted to r-modes), but they are beyond the scope of this paper (see Ushomirsky's paper in this proceedings).

3. G-modes: G-modes are also unstable when Ω_s is (approximately) greater than the mode frequency (of nonrotating stars), which typically is ~ 100 Hz, or $0.1\Omega_{\rm max}$. However, the growth time (despite the finite quadrupole moment) is quite long, an thus not interesting (Lai 1999).

Figure 1 shows a comparison of the CFS instability growth times for different modes in a $M = 1.4 M_\odot$, $R_0 = 20$ km neutron star. We see that for sufficiently large β, the $m = 2$ f-mode (the bar-mode) is the most unstable mode, with the shortest growth time (of order seconds to minutes). So the bar-mode instability is most robust if a neutron star is formed with β in the range of 0.2 to 0.27 — numerical simulations indicate that this is indeed possible (Zwerger & Müller 1997; Rampp et al. 1998; see contributions by New and Brown). In the remainder of the paper we will be concerned with the bar-mode only.

The growth time shown in Fig. 1 is due to gravitational radiation only. There are several complications that can affect the net growth time. We will discuss two of them in the next two sections. Other issues, including the effect of magnetic field, can be found in Ho & Lai (2000).

III   VISCOSITY IN PROTO-NEUTRON STARS

The first complication concerns the viscous dissipation, which tends to suppress the GR driven instability. For a proto-NS with $T \gtrsim 1$ MeV, the shear (kinematic) viscosity due to neutron-neutron scattering, $\nu = \eta/\rho \simeq 14\,\rho_{15}^{5/4} T_{\rm MeV}^{-2}$ cm^{2}s^{-1} (Flowers & Itoh 1976), is negligible (since $t_{\rm visc} \sim R^2/\nu$ is much greater than the growth time of the mode, $t_{\rm GR}$). The shear viscosity due to neutrino-nucleon scattering is $\eta = n_\nu p_\nu l_\nu/5$ (Goodwin & Pethick 1982), where n_ν, p_ν, l_ν are the neutrino number density, momentum and scattering mean free path. Thus

$$\nu = \frac{\eta}{\rho} \sim c\, l_\nu \left(\frac{E_\nu n_\nu}{\rho c^2} \right) \sim 3 \times 10^6 \rho_{15}^{-4/3} T_{\rm MeV} \ {\rm cm}^2 s^{-1}, \tag{1}$$

where we have used $l_\nu \simeq 2000(E_\nu/30\,{\rm MeV})^{-3}$ cm. Contrary to some earlier claims (e.g., Lindblom & Detweiler 1979), the neutrino shear viscosity is also negligible.

The most relevant viscosity in a proto-NS is the bulk viscosity. This arises from the fact that in an oscillation which involves compression, the matter will be temporarily out of chemical equilibrium (in this case, β-equilibrium between n,p,e) and it will try to relax back to the equilibrium by the weak processes

$$e + p \to n + \nu_e, \quad n \to p + e + \bar{\nu}_e, \quad \text{(URCA)} \tag{2}$$

or

$$e + p + N \to n + N + \nu_e, \quad n + N \to p + N + e + \bar{\nu}_e, \quad \text{(Modified URCA)} \tag{3}$$

and therefore emitting neutrinos; this neutrino emission serves as damping of the oscillation.

The standard bulk viscosity widely used in the last decade has been the one derived by Sawyer (1989) (correcting a factor of 100 typographic error)

$$\zeta = 1.5 \times 10^{32} \rho_{15}^2 \, T_{\mathrm{MeV}}^6 \, \omega^{-2} \text{ g/(cm s)}, \quad \text{(Modified URCA)} \tag{4}$$

where T_{MeV} is the temperature in MeV, $\rho = 10^{15} \rho_{15}$ g/cm^3 is the density, and ω is the mode (angular) frequency in s^{-1}. This is quite large, with the corresponding damping time $R^2 \rho/\zeta \sim 10 \, T_{\mathrm{MeV}}^{-6}$ s for $\omega \sim 10^3$ s^{-1}. Thus one may conclude that the bar-mode instability is suppressed for $T_{\mathrm{MeV}} \gtrsim 1$ (see Ipser & Lindblom 1991 for more detailed computations). Similarly, if direct URCA process operates, equation (4) should be replaced by

$$\zeta \simeq 1.5 \times 10^{36} \rho_{15}^2 \, T_{\mathrm{MeV}}^4 \, \omega^{-2} \text{ g/(cm s)} \quad \text{(URCA)} \tag{5}$$

where we have used the free npe gas relation $n_e/n_0 = 0.0765 \, \rho_{15}^2$ ($n_0 = 0.16$ fm^{-3}) to evaluate the equilibrium electron density (this is not consistent since in a free npe gas URCA precesses are suppressed; but it is adequate for estimate). Equation (5) is valid for $\omega \gg 118 \, x^{-2/3} T_{\mathrm{MeV}}^4$ s^{-1} (where the electron fraction $x = n_e/n = 0.02\rho_{15}$).

However, one should be careful when using eqs. (4)-(5) at high temperatures. In fact, when T becomes sufficiently large, the bulk viscosity must go down. The reason is that as T increases, the timescale to relax back to β-equilibrium becomes shorter than the oscillation period. So the matter will stay very close to β-equilibrium during the oscillation and there is very little extra neutrino emission associated with the oscillation. Another effect that needs to be included at high temperatures is neutrino absorption (such as $\nu_e + n \to e + p$) which also helps to speed up relaxation to β-equilibrium.

We now outline the derivation of the neutrino bulk viscosity in the regime where matter is opaque to neutrinos (see Sawyer 1980; Lai 2001). Consider the emission and absorption of ν_e's. The ν_e distribution function f satisfies the Boltzmann equation

$$\frac{\partial f}{\partial t} = j - \kappa f, \tag{6}$$

where j is the emissivity, and $\kappa = j \left(1 + e^{\frac{E - \delta\mu}{T}} \right)$ is the absorption cross section per unit volume (corrected for the effect of stimulated absorption), $\delta\mu \equiv \mu_e + \mu_p - \mu_n$.

In equilibrium ($\delta\mu = 0$; we assume $\mu_\nu = 0$), we have $f = f_0 = f_{eq} = \left(1 + e^{\frac{E}{T}}\right)^{-1}$, and $j_0 = \kappa_0 f_0$. In a perturbation with $\delta f \propto e^{i\omega t}$, we find

$$(i\omega + \kappa_0)\delta f = \frac{e^{E/T}}{(1 + e^{E/T})^2}\left(\frac{\delta\mu}{T}\right)\kappa_0. \tag{7}$$

Similar consideration for $\bar{\nu}_e$ gives $(i\omega+\kappa_0)\delta\bar{f} = -e^{E/T}(1 + e^{E/T})^{-2}(\delta\mu/T)\kappa_0$, where we have used $\bar{j}_0 = j_0$ and $\bar{\kappa}_0 = \kappa_0$. The variation of electron fraction $x = n_e/n$ is

$$\delta x = -\frac{1}{n}\int\frac{d^3p}{(2\pi)^3}\left(\delta f - \delta\bar{f}\right) \simeq -\frac{\delta\mu}{(i\omega + \kappa_0)n}\lambda_{\text{eff}}, \tag{8}$$

where $\lambda_{\text{eff}} = T^2\kappa_0/6$. Since the matter is degenerate to a good approximation, $\delta\mu$ depends only on ρ, x. Thus

$$\delta x = -\frac{\lambda_{\text{eff}}/\rho}{i\omega + \kappa_{\text{eff}}}\left(\frac{\partial\delta\mu}{\delta n}\right)_x\delta\rho, \tag{9}$$

where $\kappa_{\text{eff}} = \kappa_0 + (\lambda_{\text{eff}}/n)(\partial\delta\mu/\delta x)_n$. The energy dissipation rate per unit volume (averaged over oscillation period) can be calculated by

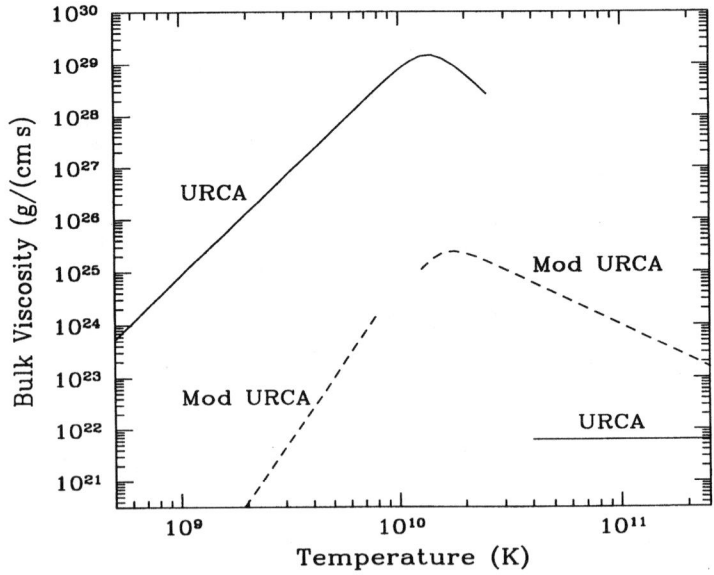

FIGURE 2. Bulk viscosity ζ as a function of temperature. The lower temperature segments correspond to (neutrino) optically thin regime, and the high temperature segments to the opaque regime. The density is $\rho_{15} = 1$, and the mode frequency $\omega/(2\pi) = 10^3$ Hz.

$$\langle \dot{E} \rangle = \frac{1}{\rho} \left\langle \delta P \frac{d\delta\rho}{dt} \right\rangle = \frac{1}{2\rho} (\delta\rho)^2 \text{Re} \left[(-i\omega) \frac{\delta P}{\delta\rho} \right] \equiv \frac{1}{2} \zeta \omega^2 \left(\frac{\delta\rho}{\rho} \right)^2. \qquad (10)$$

Using eq. (9) we then find the bulk viscosity

$$\zeta = \frac{\lambda_{\text{eff}}}{\omega^2 + \kappa_{\text{eff}}^2} \left(\frac{\partial \delta\mu}{\partial n} \right)_x \left(\frac{\partial P}{\partial x} \right)_\rho = \frac{\lambda_{\text{eff}}}{\omega^2 + \kappa_{\text{eff}}^2} \left(\frac{\mu_n}{3} \right)^2, \qquad (11)$$

where the second equality applies for free npe gas ($\mu_n = 140\, \rho_{15}^{2/3}$ MeV is the neutron Fermi energy), for which $(\partial \delta\mu/\partial x)_n = 3(n/n_e)\mu_n$ and thus $\kappa_{\text{eff}} \simeq \kappa_0$.

For modified URCA process, we have

$$\kappa_0 = 1.034 \times 10^3\, \rho_{15}^{2/3} T_{\text{MeV}}^4 \text{ s}^{-1}, \qquad (12)$$

$$\zeta = 7.9 \times 10^{31}\, \rho_{15}^2\, T_{\text{MeV}}^6\, (\omega^2 + \kappa_0^2)^{-1} \text{ g/(cm s)}. \qquad \text{(Modified URCA)} \qquad (13)$$

If direct URCA process operates (which requires sufficiently large x, a condition necessarily satisfied at early times of the proto-neutron star), we have

$$\kappa_0 = 1.13 \times 10^7\, \rho_{15}^{2/3} T_{\text{MeV}}^2 \text{ s}^{-1}, \qquad (14)$$

$$\zeta = 6.7 \times 10^{21}\, \rho_{15}^{2/3} \text{ g/(cm s)}, \qquad \text{(URCA)} \qquad (15)$$

where the ζ expression is evaluated in the $\kappa_0 \gg \omega$ limit. We see from Fig. 2 that for $T \gtrsim$ MeV, the viscosity is less than 10^{25} g/(cm s) (depending on whether URCA or modified URCA processes operates), and thus the bulk viscosity cannot suppress the CFS instability of the bar-mode (see Lai 2001 for more details).

IV   EFFECT OF STAR–DISK COUPLING

A newly formed rotating neutron star is often surrounded by a disk. As the bar grows in the NS, it will excite density waves in the disk, and therefore transfers angular momentum to the disk or from the disk. This will either enhance or suppress the bar-mode instability — this is simply another form of CFS instability.

The angular momentum transfer is mainly through the so-called Lindblad resonances, where $2(\Omega_p - \Omega_k) = \pm \Omega_k$ ($\Omega_p = \omega_i/2$ is the pattern rotation of the bar and Ω_k is the rotation of the disk, assumed to be Keplerian). The torques can be calculated using the formalism of Goldreich & Tremaine (1979). At the inner Lindblad resonance (ILR), $\Omega_k = 2\Omega_p$, the driving rate of the bar-mode due to star–disk coupling is (Lai 2001)

$$\gamma_{\text{ILR}} \simeq -\frac{6\pi^2}{5} \frac{\Omega_p^2}{(\Omega_s - 2\Omega_p)} \frac{\Sigma(r_{\text{ILR}})}{(M/R_e^2)}, \qquad (16)$$

where the negative sign in the front implies that the torque tends to damp the mode, Ω_s is the rotation rate of the star, R_e is the equatorial radius, and $\Sigma(r_{\text{ILR}})$

is the surface density of the disk evaluated at ILR. Similarly, at the outer Lindblad resonance (OLR), we have

$$\gamma_{\rm OLR} \simeq \frac{98\pi^2}{45} \frac{\Omega_p^2}{(\Omega_s - 2\Omega_p)} \frac{\Sigma(r_{\rm OLR})}{(M/R_e^2)}, \qquad (17)$$

where the positive front sign means that the torque drives the CFS instability. Clearly, the net effect of star-disk coupling on the mode depends on the relative importance of ILR and OLR. Compared with driving rate of the mode due to gravitational radiation, we find that star-disk coupling is important when $\Sigma \gtrsim 10^{-6} M/R_e^2$. So even a small amount of material outside the proto-NS may potentially affect the CFS instability of the bar-mode.

V NONLINEAR EVOLUTION OF BAR-MODE AND GRAVITATIONAL WAVEFORMS

In general, to determine the nonlinear evolution of the bar-mode requires 3d hydrodynamical simulations including gravitational radiation reaction, and one needs to follow the system for a time much longer than the dynamical time of the star. This possesses a significant technical challenge (see recent attempt by Lindblom et al. 2000 on the r-mode evolution where an approximate ansatz for the radiation reaction is adopted).

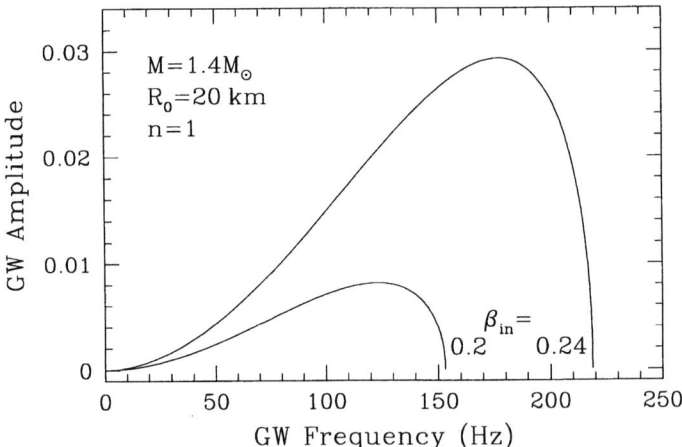

FIGURE 3. The amplitudes of the GWs emitted by a secularly unstable NS, evolving from a Maclaurin spheroid toward a Dedekind ellipsoid. The two curves correspond to initial $\beta = 0.24$ and 0.2 respectively ($n = 1$ is the polytropic index). The GW frequency sweeps downward.

A   Dedekind-like Bar

One of the advantages of the $m = 2$ f-mode (as opposed the higher m f-modes and or r-modes) is that under certain idealized condition, namely for impressible fluid, there exists an exact solution for the nonlinear development of the bar-mode instability driven by gravitational radiation (Miller 1974; see Lai & Shapiro 1995 and references therein; the latter also includes an approximate, compressible generalization).

The evolutionary sequence is as follows: We start with an axisymmetric Maclaurin spheroid with $\beta > 0.14$ (secularly unstable). The bar grows and has a pattern angular frequency Ω_p (which is related to the mode frequency in the inertial frame by $\Omega_p = \omega_i/2$). Relative to the bar, there is also an internal rotation Ω_{in} which is larger than Ω_p. The important point to note is that although the mean rotation of the star $\Omega_s \simeq \Omega_p + \Omega_{in}$ is near breakup, Ω_p can be much smaller (in fact, at the bifurcation point $\beta = 0.14$, we have $\Omega_p = 0$). As the amplitude of the bar continues to grow, the bar also gradually slows down. Eventually we reach a configuration with zero pattern speed. This is the Dedekind ellipsoid, basically stationary "football" with a fixed figure in space but with a lot internal rotation Ω_{in}.

The gravitational wave (GW) emitted during such a quasi-equilibrium secular evolution is quite interesting. Figure 3 shows the GW amplitude as a function of GW frequency. The GW is quasi-periodic. Initially, we have an axisymmetric star, so $h = 0$. Then the GW amplitude increases as the bar grows. In the meantime, the bar slows down and the GW frequency ($= 2\Omega_p$) decreases. So eventually h decreases. Thus we have a non-monotonic GW amplitude evolution, with the GW frequency sweeps downward from a few hundred Hertz toward zero. The timescale

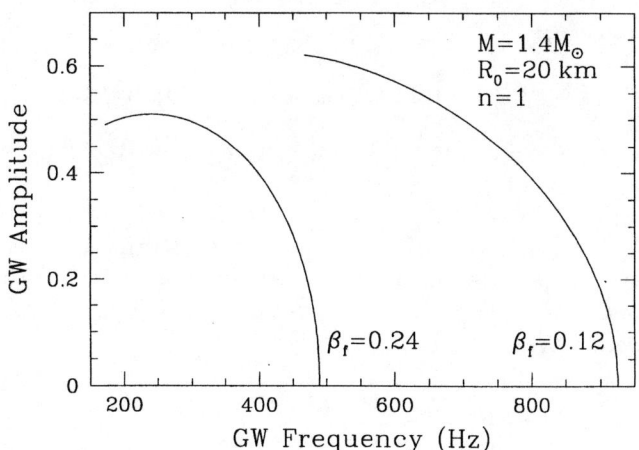

FIGURE 4. The amplitudes of the GWs emitted by Jacobi-like bars. The two curves correspond to final $\beta = 0.24$ and 0.12 respectively. The GW frequency sweeps upward.

of the evolution is of order seconds to minutes, and the characteristic number of cycles of GWs is of order 10^4.

So far we have assumed the star begins its evolution from an axisymmetric state. Recent numerical simulations (e.g., New et al. 2000,; Brown 2000; Shibata et al. 2000), however, indicate that at the end of the dynamical evolution, the star may be already elongated rather than axisymmetric. So one can ask about the long-term evolution of the bar and the emitted GWs. Here the ellipsoid model can also provide qualitative answers. There are two possibilities (see Lai & Shapiro 1995 for details): (1) If the bar has $\Omega_{in} > \Omega_p$, namely if the bar is *Dedekind-like*, then the evolution and waveform discussed above should also apply except that we need to cut off the initial growth phase of the bar. (2) Another possibility occurs when $\Omega_p > \Omega_{in}$, which we discuss in the next subsection.

B   Jacobi-like Bar

If the bar has internal rotation (relative to the bar figure) less than the pattern rotation, i.e., $\Omega_{in} < \Omega_p$, the bar is called *Jacobi-like*, and the evolution is quite different. In this case, gravitational radiation reaction tends to decrease the amplitude of the bar, making the star less elongated. In the meantime, Ω_p increases because, even though J decreases, the moment of inertia decreases faster. So the GW frequency increases and eventually the star becomes axisymmetric. Here again we have a quasi-periodic GW signal (see Fig. 4), except that the frequency sweeps upwards. The timescale foe the evolution is of order a second, and the number of cycles is of order a few hundred.

Note that at the end of the Jacobi-like evolution, the axisymmetric star may still be secularly unstable. If so the star will continue to evolve in the way described in Sec. V.A.

C   Characteristic GW Amplitude

To recapitulate, there are two types of rotating bars, and their evolution and emitted GWs are qualitatively different: (1) A Dedekind-like bar has large internal rotation, and the resulting waveform sweeps downward in frequency; (2) A Jacobi-like bar has relatively small internal rotation, and the GW frequency sweeps upwards. Figure 5 shows the characteristic GW amplitudes for the two types of bars compared with the sensitivity curves ($h_{\rm rms}$) of LIGO I,II and III. The characteristic GW amplitude is given by (Lai & Shapiro 1995)

$$h_c = h \left| \frac{dN}{d \ln f} \right|^{1/2} = \frac{M}{D} \left(\frac{R_0}{M} \right)^{1/4} \left(\frac{5}{2\pi} \left| \frac{d\bar{E}}{d\bar{\Omega}_p} \right| \right)^{1/2} , \tag{18}$$

where D is the distance, $|dN/d \ln f| = |f^2/\dot{f}|$ is the number of cycles of GW spent near frequency f, $E = \bar{E}(M^2/R_0)$ is the energy of the star, and $\Omega_p = \bar{\Omega}_p(M/R_0^3)^{1/2}$

is its angular pattern speed. For a broad band detector such as LIGO, the best signal-to-noise ratio will be obtained by matched filtering of the data, with $S/N \simeq h_c/h_{\rm rms}$. Note that unlike coalescing compact binaries, the phase evolution of the GW from the evolving bars can not be determined with the accuracy needed for matched filtering, so such a high S/N may not be achieved in practice. A new fast chirp transform technique (Jenet & Prince 2000) is promising for detecting such signals.

The event rate of Type II SNe at distance of 30 Mpc is about 100 per year. So even if a small fraction of the NSs are formed rotating rapidly, the GW signals discussed here are promising for lIGO. Of course, it should be emphasized that the waveforms discussed in this section are based on the exact solution for the idealized situation (incompressible fluid). Whether this idealized solution has any resemblance to reality remains to be seen by future studies.

VI PARAMETRIZED WAVEFORMS

Finally we give fitting formulae for the gravitational waveforms generated by a rapidly rotating NS as it evolves from an initial axisymmetric configuration toward a triaxial ellipsoid (Maclaurin spheroid $\Rightarrow$ Dedekind ellipsoid) as discussed in

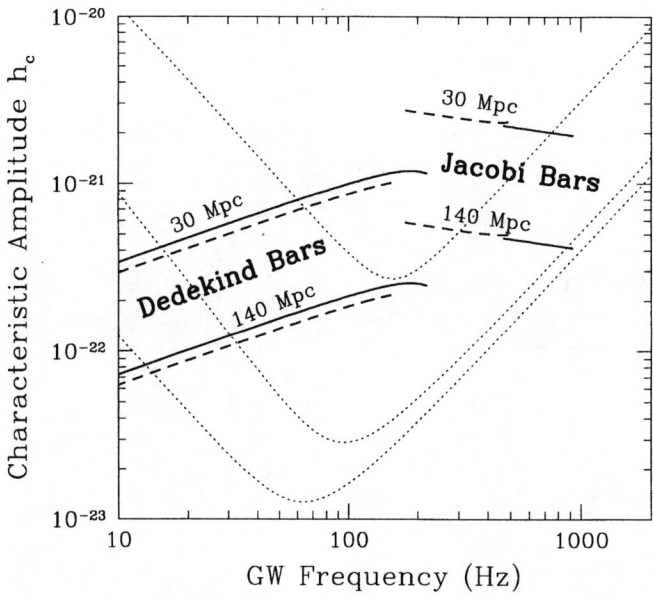

FIGURE 5. Comparison between the characteristic GW amplitudes h_c emitted during the secular evolution of a nonaxisymmetric neutron star and the rms noise $h_{\rm rms}$ of LIGO I,II and III. The solid and dashed lines correspond to the two different cases in Fig. 3 and Fig. 4.

Sec. V.A. We use units such that $G = c = 1$.

The waveform (including the polarization) is given by eq. (3.6) of Lai & Shapiro (1995) (hereafter LS). Since the waveform is quasi-periodic, we will give fitting formulae for the wave amplitude h (Eq. [3.7] of LS) and the quantity $(dN/d\ln f)$ (eq. [3.8] of LS; related to the frequency sweeping rate), from which the waveform $h_+(t)$ and $h_\times(t)$ can be easily generated in a straightforward manner.

1. Wave Amplitude: The waveform is parametrized by three numbers: f_{max} is the maximum wave frequency in Hertz, $M_{1.4} = M/(1.4 M_\odot)$ is the NS mass in units of $1.4 M_\odot$, $R_{10} = R/(10\,\mathrm{km})$ is the NS radius in units of 10 km. (Of course, the distance D enters the expression trivially.) It is convenient to express the dependence of h on t through f (the wave frequency), with $f(t)$ to be determined later. A good fitting formula is

$$h[f(t); f_{max}, M, R] = \frac{M^2}{DR} A \left(\frac{f}{f_{max}}\right)^{2.1} \left(1 - \frac{f}{f_{max}}\right)^{0.5}, \qquad (19)$$

where

$$A = \begin{cases} (\bar{f}_{max}/1756)^{2.7}, & \text{for } \bar{f}_{max} \leq 400\ \mathrm{Hz}; \\ (\bar{f}_{max}/1525)^{3}, & \text{for } \bar{f}_{max} \geq 400\ \mathrm{Hz}; \end{cases} \quad \text{with } \bar{f}_{max} \equiv f_{max} R_{10}^{3/2} M_{1.4}^{-1/2}. \quad (20)$$

Note that if we want real numbers, we have

$$\frac{M^2}{DR} = 4.619 \times 10^{-22} M_{1.4}^2 R_{10}^{-1} \left(\frac{30\,\mathrm{Mpc}}{D}\right). \qquad (21)$$

2. Number of GW cycles: The fitting formula is

$$\left|\frac{dN}{d\ln f}\right| = \left(\frac{R}{M}\right)^{5/2} \frac{0.016^2 (R_{10}^{3/2} M_{1.4}^{-1/2} f/1\,\mathrm{Hz})}{A^2 (f/f_{max})^{4.2}[1 - (f/f_{max})]}. \qquad (22)$$

Notes to the fitting formulae:

Note (i): Using the above equations, we obtain the characteristic amplitude:

$$h_c = h \left|\frac{dN}{d\ln f}\right|^{1/2} = 0.016 \frac{M^{3/4} R^{1/4}}{D} \left(\frac{R_{10}^{3/2} M_{1.4}^{-1/2} f}{1\,\mathrm{Hz}}\right)^{1/2}$$

$$= 5.3 \times 10^{-23} \left(\frac{30\,\mathrm{Mpc}}{D}\right) M_{1.4}^{3/4} R_{10}^{1/4} \left(\frac{R_{10}^{3/2} M_{1.4}^{-1/2} f}{1\,\mathrm{Hz}}\right)^{1/2}, \qquad (23)$$

which agrees with Eq. (3.12) of LS to within 10% [Note that in Eq. (3.12) of LS, the factor $f^{1/2}$ should be replaced by $(R_{10}^{3/2} M_{1.4}^{-1/2} f)^{1/2}$, similar to the above expression.]

Note (ii): The accuracy of these fitting formulae (as compared to the numerical results shown in LS) is typically within 10%. When f is very close to f_{max}, the error in the fitting can be as large as 30%.

Note (iii): The frequency evolution $f(t)$ is obtained by integrating the equation $f^2/\dot{f} = -|dN/d\ln f|$ (note that the frequency sweeps from f_{max} to zero). For example, we can choose $t = 0$ at $f = 0.9f_{max}$. (Note that one should not choose $t = 0$ at $f = f_{max}$ as the time would diverge — the actual evolution near f_{max} depends on the initial perturbations). Once $f(t)$ is obtained, the waveform can be calculated as (cf. Eq. [3.6] of LS):

$$h_+ = h[f(t); f_{max}, M, R]\cos\Phi(t)\frac{1+\cos^2\theta}{2}, \tag{24}$$

$$h_\times = h[f(t); f_{max}, M, R]\sin\Phi(t)\cos\theta, \tag{25}$$

where θ is the angle between the rotation axis of the star and the line of sight from the earth, and $\Phi(t) = 2\pi \int f(t)dt$ is the phase of the gravitational wave.

Note (iv): f_{max} typically ranges from 100 Hz to 1000 Hz (see Fig. 5 of LS); $M_{1.4}$ and R_{10} are of order unity for realistic neutron stars.

REFERENCES

1. Andersson, N. 1998, ApJ, 502, 708
2. Brown, J.D. 2000, gr-qc/0004002.
3. Chandrasekhar, S. 1969, "Ellipsoidal Figures of Equilibrium" (Yale Univ. Press).
4. Flowers, E., & Itoh, N. 1976, ApJ, 206, 218.
5. Friedman, J.L. 1998, in "Black Holes and Relativistic Stars", ed. R.M. Wald (Univ. Chicago Press).
6. Goldreich, P., and Tremaine, S. 1979, ApJ, 233, 857.
7. Goodwin, and Pethick, C.J. 1982, ApJ, 253, 816.
8. Ho, W.C.G., and Lai, D. 2000, ApJ, 543, 386.
9. Imamura, J., et al. 1995, ApJ, 444, 363.
10. Jenet, F.A., and Prince, T.A. 2000, gr-qc/0012029.
11. Lai, D. 1999, MNRAS, 307, 1001.
12. Lai, D., and Shapiro, S.L. 1995, ApJ, 442, 259.
13. Lindblom, L., and Detweiler, S. 1979, ApJ, 232, L101.
14. Lindblom, L., Owen, B.J., & Morsink, S.M. 1998, Phys. Rev. Lett., 80, 4843.
15. Lindblom, L., Tohline, J.E., and Vallisneri, M. 2000, astro-ph/0010653.
16. Miller, B.D. 1974, ApJ, 187, 609.
17. New, K.C.B., Centrella, J.M., and Tohline, J.E. 2000, Phys. Rev. D62, 064019.
18. Pickett, B.K., Durisen, R.H., and Davis, G.A. 1996, ApJ, 458, 714.
19. Rampp, M., Müller, E., and Ruffert, M. 1998, A&A, 332, 969.
20. Saijo, M., Shibata, M., Baumgarte, T.W., and Shapiro, S.L. 2000, astro-ph/0010201
21. Sawyer, R.F. 1989, Phys. Rev. D, 39, 3804.
22. Shibata, M., Baumgarte, T.W., and Shapiro, S.L. 2000, astro-ph/0005378
23. Stergioulas, N., and Friedman, J.L. 1998, ApJ, 492, 301.
24. Tohline, J., Durisen, R., and McCollough, M. 1985, ApJ, 298, 220.
25. Toman, J., Imamura, J.N., Prickett, B.J., and Durisen, R.H. 1998, ApJ, 497, 370.
26. Zwerger, T., & Müller, E. 1997, A&A, 320, 209.

Rotating Neutron Stars, R-Modes, and LXMBs

X-Ray Observations of Low-Mass X-Ray Binaries: Accretion Instabilities on Long and Short Time-Scales

Jean H. Swank

Laboratory for High Energy Astrophysics
NASA/GSFC Greenbelt, MD 20771

Abstract. X-rays trace accretion onto compact objects in binaries with low mass companions at rates ranging up to near Eddington. Accretion at high rates onto neutron stars goes through cycles with time-scales of days to months. At lower average rates the sources are recurrent transients; after months to years of quiescence, during a few weeks some part of a disk dumps onto the neutron star. Quasiperiodic oscillations near 1 kHz in the persistent X-ray flux attest to circular motion close to the surface of the neutron star. The neutron stars are probably inside their innermost stable circular orbits and the x-ray oscillations reflect the structure of that region. The long term variations show us the phenomena for a range of accretion rates. For black hole compact objects in the binary, the disk flow tends to be in the transient regime. Again, at high rates of flow from the disk to the black hole there are quasiperiodic oscillations in the frequency range expected for the innermost part of an accretion disk. There are differences between the neutron star and black hole systems, such as two oscillation frequencies versus one. For both types of compact object there are strong oscillations below 100 Hz. Interpretations differ on the role of the nature of the compact object.

INTRODUCTION

Low-mass X-ray binaries (LMXB) are the binaries of a low-mass "normal" star and a compact star. The compact star could be a white dwarf, a neutron star, or a black hole. The Rossi X-Ray Timing Explorer (*RXTE*) has been observing since the beginning of 1996 and has obtained qualitatively new information about the neutron star and black hole systems. In this paper I review the new results briefly in the context of what we know about these sources. The brightest, Sco X–1, was one of the first non-solar X-ray sources detected, but only with *RXTE* have sensitive measurements with high time resolution been made that could detect dynamical time-scales in the region of strong gravity. *RXTE* also has a sky monitor with a time-scale of hours that keeps track of the long term instabilities and enables in depth observations targeted to particular states of the sources.

CP575, *Astrophysical Sources for Ground-Based Gravitational Wave Detectors*, edited by J. M. Centrella
2001 American Institute of Physics 0-7354-0014-8

The LMXB have a galactic bulge or Galactic Population II distribution. The mass donor generally fills its Roche lobe, is less than a solar mass, and is optically faint, in contrast to the early type companions of pulsars like Cen X–3 or the black hole candidate Cyg X–1. In many cases the optical emission is dominated by emission from the accretion disk, and that is dominated by reprocessing of the X-ray flux from the compact object [1]. The known orbital periods of these binaries range from 16 days (Cir X–1) to 11 minutes (4U 1820–30). The very short period systems (< 1 hr) are expected to have degenerate dwarf mass donors and probably the mass transfer is being driven by gravitational radiation. The different properties of the sources indicate several populations. The longer period systems with more massive companions are probably slightly evolved from the main sequence.

There are about 50 persistent neutron star LMXB [2]. Distances can be estimated in a variety of ways. The hydrogen column density indicated by the X-ray spectrum should include a minimum amount due to the interstellar medium. Many of the sources emit X-ray bursts associated with thermonuclear flashes that reach the hydrogen or helium Eddington limits. In some cases the optical source provides clues. The resulting luminosity distribution appears to range from several times the Eddington limit for a neutron star down below the luminosity of about 10^{35} ergs s^{-1}, corresponding to $\approx 10^{-11}$ $M_\odot$ yr^{-1} [3]. The lower limit has come from instrument sensitivity, but it may also reflect the luminosity below which the accretion flow is not steady, so that the source must be a transient.

"X-Ray Novae" that are among the brightest X-ray sources for a month to a year are sufficiently frequent that they were seen in rocket flights in the beginning of X-ray astronomy. The X-ray missions that monitored parts of the sky during the last three decades found that on average there are 1–2 very bright transient sources each year (e.g. [4]) with durations of a month to a year. In 5 years of *RXTE* operations, we know of 20 transient neutron star sources and and an equal number of transient black hole sources. If they have a 20 yr recurrence time we have seen only a quarter of them and if we have only been watching a third of the region in the sky, 20 observed sources implies more than 240 sources exist. In reality there is a distribution of the recurrence times, some as short as months, others longer than 50 years, if optical records are good. On the basis of such estimates, the number of potential black hole transients is estimated to be on the order of thousands [5].

The separation of sources into persistent and transient sources is a very gross simplification. One of the discoveries of recent missions, and especially of the All Sky Monitor (ASM) [6] has been that the persistent sources have cycles of variations with time-scales ranging from many months to days. If the transient outbursts originate in accretion instabilities, perhaps these variations are related. In the next section I show some of the kinds of behavior being observed.

At radii close to the compact objects the dynamical time-scale gets shorter, till it is the milliseconds of the neutron star or black hole. RXTE's large area detectors detect oscillations on these time-scales which must reflect the dynamics at the innermost stable circular orbit (ISCO) of these neutron stars and black holes.

The neutron stars of this sample are expected to have magnetic dipole moments and surface fields about $10^8 - 10^9$ gauss. Of course the neutron stars have a surface such that matter falling from the accretion disk to the neutron star crashes into the surface and generates X-ray emission. In the case of the black holes matter could fall through the event horizon and disappear with no further emission of energy. Thus the X-rays produced and the dynamics that dominates in the two cases (neutron star versus black hole) could be different. However, a number of similarities appear in the signals we receive.

LONG TIME-SCALE VARIABILITIES

High Accretion Rate - Persistent Sources

Among the persistent LMXB there are characteristic variations on time-scales of months in some sources and days in others [6]. Quasiperiodic modulations were pointed out at 37 days for Sco X–1 (IAUC 6524), 24.7 days for GX 13+1 (IAUC 6508), 77.1 days for Cyg X–2 (IAUC 6452), 37 days for X 2127+119 in M15 (IAUC 6632). The obviously important, but not strictly periodic modulations in 4U 1820–30 and 4U 1705–44 at time-scales of 100–200 days are shown in Figure 1. For Sco X–1, the changes in activity level occur in a day and the activity time-scale is hours. The hardness is often correlated with the rate, although this measure does not bring out more subtle spectral changes.

These time-scales are less regular than the 34 day cycle time of Her X-1, and similar modulations in LMC X-4 and SMC X-1, which are thought to be due to the precession of a tilted accretion disk. The latter sources are high magnetic field pulsars in which the disk is larger than in the LMXB, and is truncated by the magnetosphere at a radius as large as 10^8 cm. The LMXB spectral changes are also different than those of the pulsars. In the LMXB case the changes are thought to be real changes in the accretion onto the neutron star, at least the production of X-rays, rather than a change in an obscuration of the X-rays that we see.

The spectral changes are captured in the color-color diagrams that give rise to the names "Z" and "Atoll" for subsets of the LMXB. These were identified with EXOSAT observations by Hasinger and van der Klis [7]. Characteristics of the bursts from 4U 1636–53 depended on the place of the persistent flux in the atoll color-color diagram [8]. This implied that the real mass accretion rate was correlated with the position on the diagram (although other possibilities such as the distribution of accreted material on the surface of the neutron star may play a role). That the position in the diagram in not uniquely correlated to the flux is as yet not understood. Transients atoll sources like Aql X–1 and 4U 1608–52 go around the atoll diagram during the progress of the outburst.

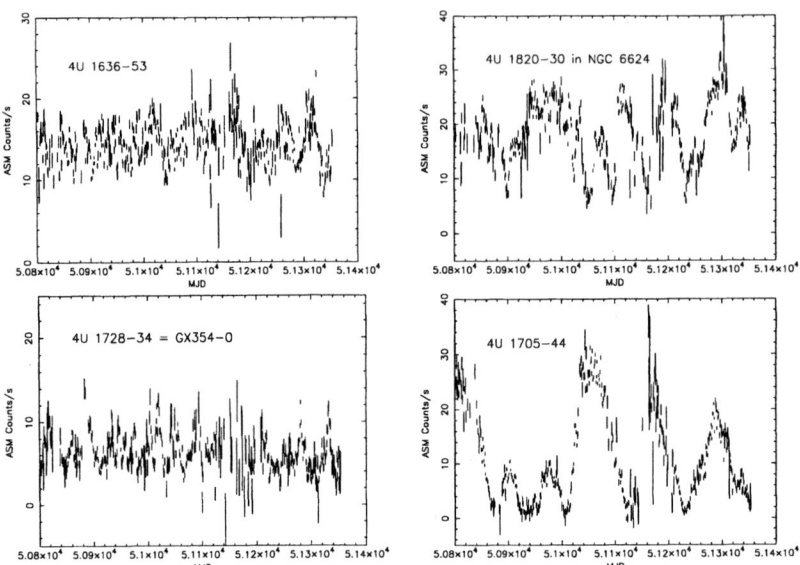

FIGURE 1. *RXTE* ASM Rates from Four Atoll Bursters. Modulations are typically a factor of two, although sometimes more. Many properties vary with these modulations.

Low Average Accretion Rate - Transients

There are only a few persistent LMXB in which the compact object is a black hole. Black hole binaries are for some reason more likely to be transients. Perhaps the binaries harboring them are not being driven to have as much mass exchange, so that it happens that these systems are in the range of mass flow through the disk that makes them transient. There are also neutron star transients with low average mass exchange rates. Figure 2 shows on the left two neutron star transients, a well known atoll burster Aql X–1 and the pulsar GRO J1744–28, which had two outbursts a year apart, but has otherwise not been seen. On the right are two black hole candidates, 4U 1630–47, which recurs approximately every two years, and XTE J1550–564, which like GRO J1744–28, had a dramatic outburst, with a weaker recurrence after a year's hiatus. Black hole candidates can get brighter than the transient bursters, consistent with the Eddington limit for more massive compact objects and they probably go through more different spectral and timing "states", but there are also similarities in the kinds of behavior that are exhibited.

From both BeppoSAX and RXTE results it is clear that there is a population of systems which have transient episodes, but which are an order of magnitude less luminous at peak. BeppoSax has seen bursts from a number of sources for which the persistent flux is below their sensitivity limit. RXTE has seen a dozen sources which may not rise above 10^{36} ergs s^{-1} during transient episodes. Several of these are believed to be neutron stars because Type I (cooling) bursts were observed. They include the source SAX J1808.4-3658, unique to date, that both pulses (2.5

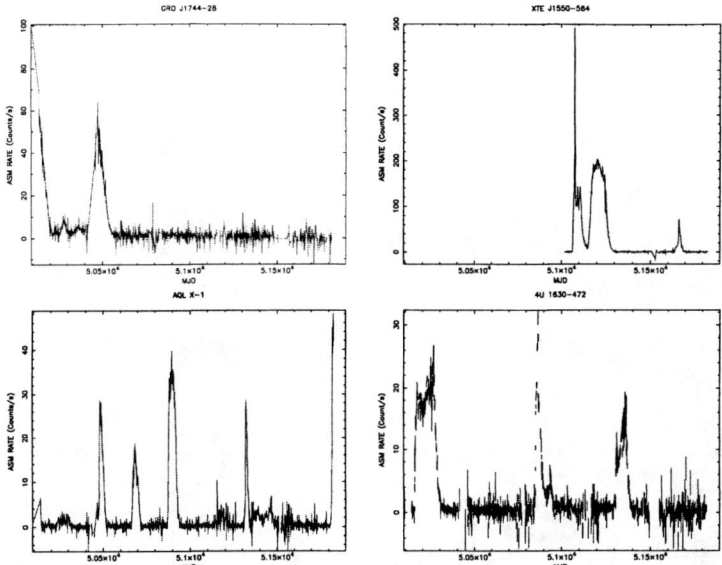

FIGURE 2. *RXTE* ASM Rates for Four Transients. (left) Neutron Star Transients GRO 1744-28 (top) and Aql X-1 (bottom). GRO J1744-28 was bright at the launch of *RXTE*. (right) Black hole candidates XTE J1550-564 (top) and 4U 1630-472 (bottom). Note that the figures have different scales for the ASM counts s^{-1}. (1 Crab = 75 ASM counts s^{-1}.)

msec) and has Type I bursts.

Some sources have spectral and timing properties consistent with black hole candidates which go into the black hole "low hard" state, with strong white noise variability below 10 Hz and hard spectra. One of these was V4641 Sgr, which went into much brighter outburst, with a radio jet, before disappearing.

INSTABILITIES CLOSE TO THE COMPACT OBJECT

Kilohertz Oscillations for Neutron Stars - near the ISCO

More than 22 LMXB have now exhibited a signal at kilohertz frequencies in the power spectra of the x-ray flux (See [9]). Figure 3 (thanks to T. Strohmayer) shows results for samples of data from an atoll and a Z source. Usually this signal is two peaks at 1–15 % power. They indicate quasi-periodic oscillations with coherence (mean frequency/frequency width) as much as 100. The centroid frequencies are not constant for a source, but vary. Over a few hours the frequency is correlated with the X-ray flux, increasing with the flux. The flux variations of a factor of two are correlated with changes of frequency between 500 Hz and 1000 Hz, approximately [10]. The highest reported is 1330 Hz, from 4U 0614+09. Considering that for a circular orbit at the Kepler radius r_K, the observed frequency is

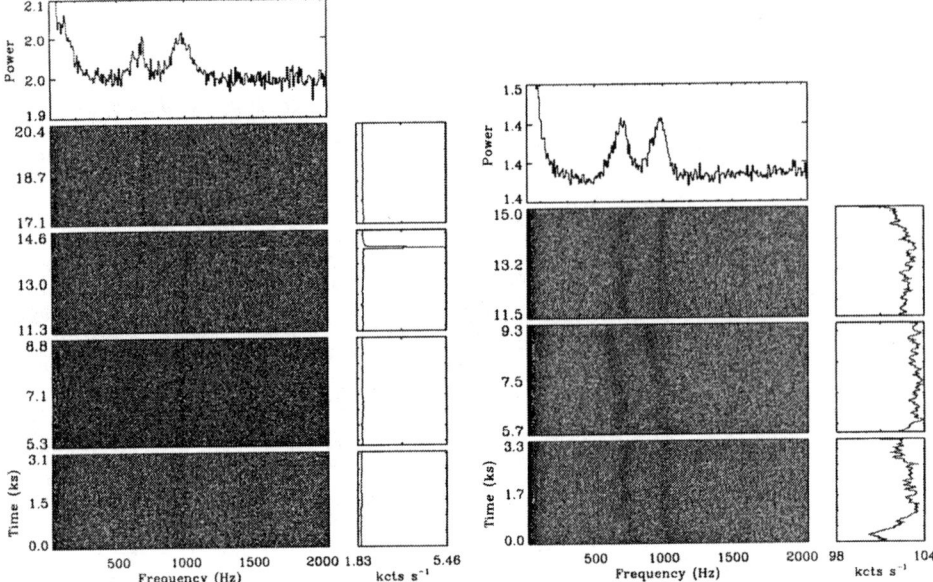

FIGURE 3. *RXTE* Power Density Spectra. (left) Atoll Source 4U1728-34. (right) Z Source Sco X-1. For each the grey scale plot of power as a function of frequency and time is shown for sequential observing intervals. The average PDS is shown above and the count rate during the observations on the right. One burst occurred during the 4U 1728-34 observation. The power spectra used are for 32 s data intervals.

$(2183/M_1)(r_{ISCO}/r_K)^{3/2}$, where $r_{ISCO} = 6GM/c^2$ is the innermost stable circular orbit for a spherical mass $M = M_1 M_\odot$ of smaller radius, neutron stars of masses $M_1 = 1.6$–2.0 would have Kepler frequencies at the ISCO of just such maximum frequencies as are observed.

While the luminosities of the sources exhibiting these QPO range from 10^{36} ergs s^{-1} to above 10^{38} ergs s^{-1}, the maximum values of the upper frequency range only between 820 Hz and 1330 Hz. This suggests [11,12] that it represents a characteristic of the neutron stars fairly independently of the accretion rate. The ISCO and the neutron star radius are candidates. For lower fluxes, the frequencies, at least locally in the light curve, decline, as if the Keplex orbit were further out. Which is more likely, that the inner radius is then at the ISCO or at the radius of the neutron star? In the latter case the neutron star is outside the innermost stable circular orbit. Understanding the boundary requires consideration of the radiation pressure, the magnetic fields, and the optical depth of the inner disk. For sources with flux near the Eddington limit, the optical depth of the material near the surface should be much larger than the optical depth of the material accreting at rates 100 times less. For the inner disk being at the ISCO, and fairly compact neutron stars, this plausibly does not matter. For the inner disk at the surface or a large neutron star, it seems hard to explain the similarity of appearance between luminous Z sources and fainter atoll sources. There are in fact differences in the appearance of the QPOs; one is that the amplitude of the QPOs is larger for the atoll sources than for the Z sources. So the situation is not completely clear.

If a disk is truncated at an inner radius which moves in toward the neutron star as the mass flow through the disk increases and a QPO is generated at near this inner edge, the frequency would be likely to increase with the luminosity. The frequency would not be able to increase beyond the value corresponding to the minimum orbit in which the disk could persist. Miller, Psaltis, and Lamb [13] argued that if radiation drag was responsible for the termination of the disk, optical depth effects would lead to the sonic point radius moving in as the accretion rate increases. There would be a highest frequency corresponding to the minimum possible sonic point radius. In the cycles of 4U 1820-30 the frequency approached a maximum which it maintained as the flux increased further before the feature became too broad to detect. This kind of behavior would arise from a sonic point explanation.

From Figure 4, it can be seen that if the equation of state (EOS) of the nuclear matter at the center of a neutron star is very stiff, near the L equation of state, for $1.4 - 2M_{\odot}$ neutron stars the radius of the star is close to its own ISCO; whether it is inside or outside it is depends sensitively on the mass. If the equation of state is softer, closer to the FPS EOS, interpretation of the maximum frequencies observed as a Kepler frequency *at the surface* would imply a mass significantly less than the $1.4M_{\odot}$ with which many neutron stars are probably formed. In either case, moderately stiff EOS and maximum frequency at the ISCO, or stiff equation of state and maximum frequency either at the ISCO or the surface, the frequency would be from near the ISCO, if not just outside it. Accurate considerations require the rotation rate of the neutron star to be taken into account.

A characteristic of the twin kilohertz peaks is that when the frequency changes, the two frequencies approximately move together, with the difference approximately constant, at least until near the maximum frequencies (and luminosities) for which they are observed in a given source. This suggests a beat frequency and the relation between the difference frequency and the frequencies seen during bursts (See Strohmayer, this volume) suggest the neutron star spin as the origin of the beats. Miller, Lamb and Psaltis [13] explored how the two frequencies could be generated and Lamb and Miller refined the model in agreement with the 5 % changes in the frequency separation, that are observed [14]. However, this varying separation between the two QPO also suggested identification as the radial epicyclic frequency of a particle moving in an eccentric orbit in the field of the neutron star. The lower of the two frequencies is then identified, not with a beat frequency, but with the precession of the periastron [15], although efforts to fit the predictions of this model in terms of particle dynamics produce implausibly large eccentricities, neutron star masses and spins [16]. Psaltis and Norman proposed that similar frequencies could be resonant in a hydrodynamic disk [17]. In these models, at least in their current forms, the difference between the two QPO peaks is not related to the spin, but to something like the radial epicycle frequency.

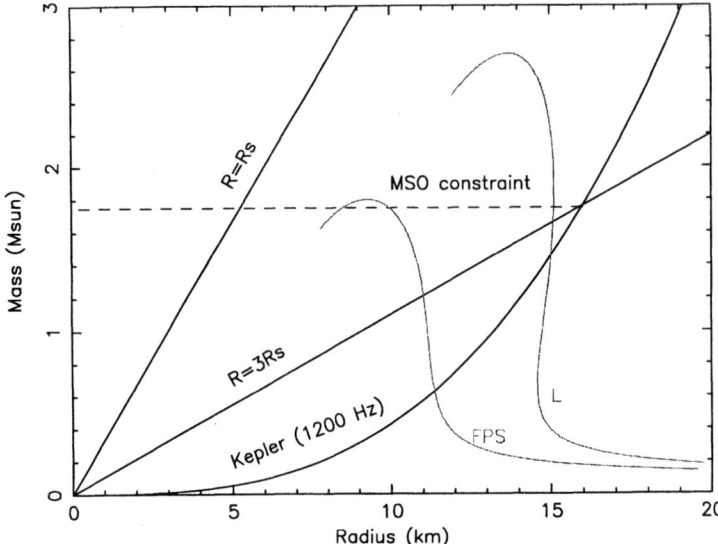

FIGURE 4. Constraints on the Neutron Star Equation of State from the kilohertz QPO. If the frequency of 1200 Hz is a Kepler frequency, the Mass of the star and the radius of the Kepler orbit are constrained. The orbit cannot be inside the ISCO. If it is at the ISCO the mass is determined and any EOS inside that radius which allow a big enough mass would support it.

A quite different class of models are those in which the disk has a boundary layer with the neutron star and the plasma is excited by the magnetic field of the neutron star [18]. The magnetic pole makes a small angle with the neutron star rotation axis. In this case the lower kilohertz QPO frequency is the Kepler frequency, while both the upper frequency and the low frequency oscillation (corresponding to the Horizontal Branch Oscillations in Z sourses) are related to oscillations of plasma interacting with the rotating magnetic field.

Hectohertz oscillations for Black Holes

Although accreting neutron stars and black holes should have important differences, they both presumably have an accretion disk with an inner radius, when the mass flow is high enough. Possible signals from the ISCO of black holes were discussed when accretion onto black holes was first considered [19] and anticipation of *RXTE* inspired detailed calculations [20]. The *RXTE* PCA has detected QPO in 5 black hole candidates at frequencies that are suitable to be signals from the ISCO of black holes in the range of $5 - 30 M_\odot$. They have been observed only in selected observations and are generally of lower amplitude (a few %) than the neutron star kilohertz QPO. For GRS 1915+105, the frequency has always been 67 Hz [21] . For

GRO 1655–40, Remillard identified 300 Hz [22]. For XTE J1550–564, at different times it has been between 185 and 205 Hz [23]. For XTE J1859–262, a broad signal at 200 Hz is observed in the bright phases near the peak of the outburst [24]. For 4U 1630–47 as well, which has had 3 outbursts during the *RXTE* era, Remillard has reported 185 Hz. The black hole candidates have appeared to differ from the neutron stars in having one QPO rather than two. An obvious question is whether the second QPO is associated with the presence of a neutron star with a surface and a rotating magnetic dipole. Recent work by Strohmayer [25] casts doubt on it.

There were other black hole candidates observed with *RXTE*, which did not exhibit high frequency oscillations and the properties of the high frequency signal are not very well defined. Interpretation in terms of Kepler frequency at the ISCO, non-radial g-mode oscillations in the relativistic region of the accretion disk, and Lense-Thirring precession have been discussed. GRO J1655-40 is very interesting because the radial velocities of absorption lines of the secondary have given rather precise measurement of the mass. (The best estimates are so far $5.5 - 7.9 M_\odot$ [26].) In this case the mass well known and the black hole's angular momentum can be the goal. The 300 Hz frequency is high enough that for a g-mode the black hole would have near maximal angular momentum, but if it represents a Kepler velocity, a Schwarzschild black hole would still be possible [27]. The question has been asked whether the microquasars GRS 1915+105 and GRO J1655-40 have powerful radio jets associated with outbursts because they have fast rotation [28].

Decahertz Oscillations for Neutron Stars and Black Holes

In the Z source LMXBs the first QPOs discovered were the Horizontal Branch Oscillations (HBO), first seen by EXOSAT, but then by Ginga. They occur in the range 15–50 Hz, have amplitudes as high as 30 %, increase in frequency with the luminosity, and have strong harmonic structure. With *RXTE* observations the atoll LMXB have also been seen to have these signals, although often the coherence is less and there are other signals (See [29]). These QPO tend to be near in frequency to the break frequency of band-limited white noise at low frequencies.

The black hole transients had already exhibited very similar features in Nova Muscae and GX 339–4 in the range 1-15 Hz. They have very similar properties to the HBO. *RXTE* PCA observations have found these QPO in the power spectra of most black hole candidates [30]. Different origins have been discussed for the neutron star and black hole QPOs, but their similarity is noted. Figure 5 shows examples from a Z source and a black hole candidate (See [31,32]).

The HBO were originally ascribed to a magnetic beat frequency model, assuming the Kepler frequency and the spin were both not seen. Stella and Vietri identified them with the Lense-Thirring precession (See [15]). They appear to have the correct quadratic relation to the high frequency kilohertz QPO. But the magnitude was too high, by even a factor of about four. Assigning them to twice the nodal frequency, a reasonable possibility for the x-ray modulation, relieves the problem in some

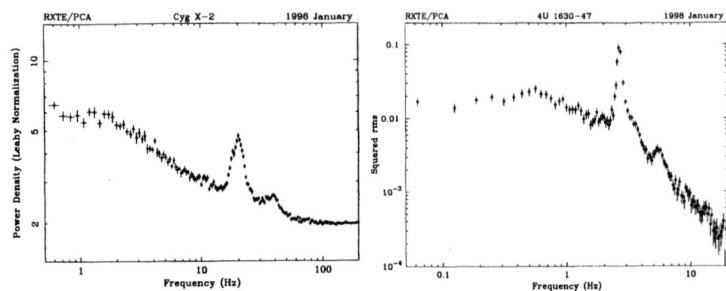

FIGURE 5. Low Frequency Oscillations. (left) Horizontal Branch Oscillations in Cygnus X-2.(right) Decahertz Oscillations in the Black Hole Candidate 4U 1630-47. These are cases with similar harmonic structure, but a different relation to the break frequency of the lower frequency noise. States occur with very different harmonic structure.

cases, but still leaves a factor of two in many. Psaltis argues that a magnitude discrepancy of a factor of two can be accommodated in situations where there is actually complex hydrodynamic flow rather than single particle orbits [33].

In the case of the black holes, the energy spectra seem to distinguish contributions of an optically thick disk and non-thermal, that is "power-law" emission, attributed to scattering of low energy photons off more energetic electrons. This division of components is not observationally so clear in the neutron star LMXB (There are many plausible reasons for this: lower central mass and smaller inner disk, X-rays generated on infall to the surface, possible spinning magnetic dipole.) For the black hole transients, this low frequency QPO is clearly a modulation of the power-law photons. However, there appear to be a variety of correlations with the disk behavior, so that the two components are clearly coupled.

In the case of the neutron stars Psaltis, Belloni, and van der Klis [34] have noted that the HBO and the lower kilohertz oscillation are correlated over a broad range of frequency (1–1000 Hz). Wijnands and van der Klis [35] showed that *both* the noise break and the low frequency QPO are correlated in the same way for certain neutron stars and black hole candidates. Psaltis *et al.* went on to point out that if some broad peaks in the power spectra of some black holes were taken to correspond to the lower kilohertz frequency in the neutron star sources, these points also fell approximately on the same relation.

While the degree to which this relation was meaningful, given the scatter in the points, selection effects, and distinctions of more than one branch of behavior, recent work is suggestive that in some way three characteristic frequencies of the disk in a strong gravitational field are significant, where these correspond to Kepler motion, precession of the perihelion and nodal precession. There remain difficulties however with specific assignments.

It has often been noted that different interpretations implied weaker features in the spectrum, for example modulation of frequencies by the Lense-Thirring precession [16] or excitation of higher modes in the case of g-modes [27]. In the case of the neutron stars, adding together large amounts of data to build up the statistical signal, while Sco X-1 did not show sidebands [36], Jonker et al. [37] found evidence of sidebands at about 60 Hz to the lower kilohertz frequency in three sources. The frequency separation is not the same as the low frequency QPO in those sources although it is in the same range and Psaltis argues is close enough that second order effects can be responsible for the difference. It is not clear yet whether the sidebands imply a modulation of the amplitude or whether they represent a beat phenomenon and are one-sided.

CONCLUSIONS

While it has not yet been possible to fit all the properties of LMXB neatly into a model, it is hard to imagine alternatives for some important results. One of these is that in accordance with the theory of General Relativity, there is an innermost stable orbit, such that quasistatic disk flow does not persist inside it. Nuclear matter at high densities does not meet such a stiff equation of state that the neutron star extends beyond the ISCO. Instead the results suggest the neutron star lies inside the ISCO for its mass.

The accretion flows for both neutron stars and black holes have resonances which, from the observations, are apparently successfully coupled to X-ray flux. QPO are observed with high coherence. They can already be compared to assignments of various frequencies, but they do not match exactly with the identifications that have been made. However before it is possible to use it as diagnostic of gravity, it is necessary to sort out further the physics of the situations.

Extending the measurements to signals an order of magnitude fainter taxes even the abilities of *RXTE*. Continued observations are pushing the limits lower by reducing statistical errors, but must deal with intrinsic source variability on longer time-scales. Observations are also being sought of especially diagnostic combinations of flux and other properties.

REFERENCES

1. van Paradijs, J., and McClintock, J. E., "Optical and Ultraviolet Observations of X-Ray Binaries", in *X-Ray Binaries*, edited by W. H. G. Lewin, J. van Paradijs, and E. P. J. van den Heuvel, Cambridge Univ. Press, New York, 1995, pp. 58-125.
2. van Paradijs, J., "A Catalogue of X-Ray Binaries", in *X-Ray Binaries*, edited by W. H. G. Lewin, J. van Paradijs, and E. P. J. van den Heuvel, Cambridge Univ. Press, New York, 1995, pp. 536-577.
3. Christian, D. J., and Swank, J. H., *ApJS*, **109**, 177-224 (1997).
4. Chen, W., Shrader, C. R., and Livio, M., *ApJ*, **491**, 312-338 (1997).
5. Tanaka, Y., and Shibazaki, N., *Annu. Rev. A&A*, **34**, 607-644 (1996).
6. Bradt, H., Levine, A. M., Remillard, R. E., Smith, D. A., "Transient X–Ray Sources Observed with the Rxte All-Sky Monitor after 3.5 Years", in *Multifrequency Be-

haviour of High Energy Cosmic Sources: III, edited by F. Giovannelli and Lola Sabau-Graziati, *Mem SAIt*, **71**, (2000), in press.

7. Hasinger, G. and van der Klis, M., *A&A*, **225**, 79-96 (1989).

8. van der Klis, M., Hasinger, G., Damen, E., Penninx, W., van Paradijs, J., and Lewin, W. H. G., *ApJ*, **360**, L19-L22 (1990).

9. van der Klis, M., *Annu. Rev. A&A*, **38**, 717-760 (2000).

10. Strohmayer, T. E., Swank, J. H., and Zhang, W., "The periods discovered by *RXTE* in thermonuclear flash bursts", in *The Active X-Ray Sky*, edited by L. Scarsi, H. Bradt, P. Giommi, and F. Fiore, Elsevier, New York, 1998, pp. 129-134.

11. Zhang, W., Smale, A. P., Strohmayer, T. E., and Swank, J. H., *ApJ*, **500**, L171-L174 (1998).

12. Kaaret, P., Piraino, Bloser, P. F., Ford, E. C., Grindlay, J. E., Santangelo, A., Smale, A. P., and Zhang, W., *ApJ*, **520**, L37-L40 (1999).

13. Miller, M. C., Lamb, F. K., and Psaltis, D., *ApJ*, **508**, 791-830 (1998).

14. Lamb, F. K., Miller, M. C., *ApJ*, submitted (2000) (astro-ph/0007460).

15. Stella, L., Vietri, M., and Morsink, S. M., *ApJ*, **524**, L63–L66 (1999).

16. Markovic, D., and Lamb, F. K., *MNRAS*, submitted (2000) (astro-ph/0009169).

17. Psaltis, D., and Norman, C., *ApJ*, submitted (1999) (astro-ph/0001391).

18. Titarchuk, L., Osherovich, V., and Kuznetsov, S., *ApJ*, **525**, L129-L132 (1999).

19. Sunyaev, R., *Sov. Astronom. AJ* **16**, 941– 946 (1973).

20. Nowak, M. A., and Wagoner, R. V., *ApJ*, **418**, 187-201 (1993).

21. Morgan, E. H., Remillard,T.E., Greiner, J., *ApJ*, bf, 482, 993-1009 (1990).

22. Remillard, R. E., Morgan, E. H., McClintock, J. E., Bailyn, C. D., and Orosz, J. A., *ApJ*, **522**, 397-412 (1999).

23. Remillard, R. E., MClintock, J. E., Sobczak, G. J., Bailyn, C. D., Orosz, J. A., Morgan, E. H., and Levine, A. M., *ApJ*, **517**, L127-L130 (1999).

24. Cui, W. E., Shrader, C. R., GHaswell, C. A., and Hynes, R. I., *ApJ*, **535**, L123-L127 (2000).

25. Strohmayer, T. E., *ApJ*, submitted (2001).

26. Shahbaz, T., van der Hooft, F., Casares, J., Charles, P. A., and van Paradijs, J., *MNRAS*, **306**, 89-94 (1999).

27. Wagoner, R., *Phys. Rep.*, **311**, 259-269 (1998) (astro-ph/9805028).

28. Mirabel, I. F., and Rodriguez, L. F., *Ann. Rev. A&A*, **37**, 409-443 (1999).

29. Wijnands, R., *Adv. Space Res.*, submitted (2000) (astro-ph/0002074).

30. Swank, J. H., "Disk Corona Oscillations", in *The Third Microquasar Workshop*, editors A. Castrado and J. Greiner, in press (2000) (astro-ph/0011494).

31. Focke, W., *ApJ*, **470**, L127-L130 (1996).

32. Dieters, S. *et al.*, *ApJ*, **538**, 307-314 (2000).

33. Psaltis, D., *ApJ*, submitted (2000) (astro-ph/0101118).

34. Psaltis, D., Belloni, T., and van der Klis, M., *ApJ*, **520**, 262-270 (1999).

35. Wijnands, R., and van der Klis, M., *ApJ*, **514**, 939-944 (1999).

36. Mendez, M. , and van der Klis, M., *MNRAS*, **318**, 938-942 (2000).

37. Jonker, P. G., Mendez, M., and van der Klis, M., *ApJ*, **540**, L29-L32 (2000).

X-ray Observations of Neutron Star Binaries: Evidence for Millisecond Spins

Tod E. Strohmayer

Laboratory for High Energy Astrophysics
NASA's Goddard Space Flight Center
Mail Code 662, Greenbelt, MD 20771

Abstract.

High amplitude X-ray brightness oscillations during thermonuclear X-ray bursts were discovered with the *Rossi X-ray Timing Explorer* (RXTE) in early 1996. Spectral and timing evidence strongly supports the conclusion that these oscillations are caused by rotational modulation of the burst emission and that they reveal the spin frequency of neutron stars in low mass X-ray binaries (LMXB), a long sought goal of X-ray astronomy. I will briefly review the status of our knowledge of these oscillations. So far 10 neutron star systems have been observed to produce burst oscillations, interestingly, the observed frequencies cluster in a fairly narrow range from ∼ 300 − 600 Hz, well below the break-up frequency for most modern neutron star equations of state (EOS). This has led to suggestions that their spin frequencies may be limited by the loss of angular momentum due to gravitational wave emission. Connections with gravity wave rotational instabilities will be briefly described.

INTRODUCTION

X-ray binaries are potentially among the most interesting sources of gravitational wave emission which current and future gravity wave detectors will attempt to study. The high frequency gravity wave signal produced during binary inspiral and ring down of black hole and neutron star binaries contains detailed information on the properties of the compact object as well as the structure of spacetime in its vicinity [1,2]. These objects will be prime targets for ground based detectors such as LIGO which because of seismic noise are only sensitive in the high frequency range above ∼ 100 Hz [3].

Neutron stars are compelling targets of investigation because of the extreme physical conditions which exist in their interiors and immediate environs. For example, the gravity wave signals produced by inspiral of a neutron star depend on the equation of state (EOS) at supranuclear density, a quantity which is still not precisely constrained by currently available astrophysics and nuclear physics data (see for example Heiselberg & Hjorth-Jensen [4]). Moreover, fundamental

CP575, *Astrophysical Sources for Ground-Based Gravitational Wave Detectors,* edited by J. M. Centrella
2001 American Institute of Physics 0-7354-0014-8

properties of the star, such as its mass, can be extracted if the gravity wave signal can be measured. Thus gravity wave astronomy can in principle provide new probes of fundamental physics as well as advancing neutron star astrophysics.

Radio observations provided the first indications that some neutron stars are spinning with periods approaching 1.5 ms [5]. These rapidly rotating neutron stars are observed as either isolated or binary radio pulsars. Binary evolution models indicate that neutron stars accreting mass from a companion can be spun up, or 'recycled', to millisecond periods (see for example Webbink, Rappaport & Savonije [6]). This formation mechanism likely accounts for a substantial fraction of the observed population of millisecond radio pulsars, however, other formation scenarios have also been proposed [7,8]. In recent years direct evidence linking the formation of rapidly rotating neutron stars to accreting X-ray binaries has been provided by data from the *Rossi X-ray Timing Explorer* (RXTE). The first evidence came from the discovery of high amplitude, nearly coherent X-ray brightness oscillations (so called 'burst oscillations') during thermonuclear flashes from several neutron star binaries (see Strohmayer [9] for a recent review). These oscillations likely result from spin modulation of either one or a pair of antipodal 'hot spots' generated as a result of the thermonuclear burning of matter accreted on the neutron star surface. Indisputable evidence that neutron stars in X-ray binaries can indeed be rotating rapidly then came with the discovery of the first, and so far only, accreting millisecond X-ray pulsar SAX J1808-369 [10,11], which is spinning at 401 Hz.

The observed distribution of burst oscillation frequencies, including the 401 Hz pulsar, is very similar to the observed distribution of millisecond radio pulsars. The RXTE observations suggest that the spin frequencies of neutron stars in accreting binaries span a relatively narrow range from $\sim 300 - 600$ Hz. Moreover, these observed frequencies are significantly less than the maximum neutron star spin rates for almost all but the stiffest neutron star equations of state [12]. This has led to the suggestion that some mechanism may limit the spin periods of these accreting neutron stars. Bildsten [13] recently proposed that the angular momentum gain from accretion could be offset by gravitational radiation losses if a misaligned quadrupole moment of order $10^{-8} - 10^{-7} I_{NS}$ could be sustained in the neutron star crust. Here I_{NS} is the moment of inertia of the neutron star. In this scenario the strong spin frequency dependence of the gravitational radiation losses sets the limit on the observed spin frequencies. Recent theoretical work has also shown that an r-mode instability in rotating neutron stars may also be important in limiting the spins of neutron stars via gravity wave emission [13,14,15,16].

In the remainder of this contribution I will present an overview of the RXTE data and lay out the evidence for the conclusion that the observed burst oscillation frequencies are a manifestation of the spin frequencies of neutron stars (or perhaps twice the spin frquency in a few cases). I will summarize the status of current X-ray measurements of neutron star spin periods and the implications for gravity wave emission.

BURST OSCILLATIONS: OBSERVATIONAL OVERVIEW

Burst oscillations with a frequency of 363 Hz were first discovered from the LMXB 4U 1728-34 by Strohmayer et al. [17]. Oscillations in an additional nine sources have since been reported, with four of these only appearing in the last few months. The sources and their observed frequencies are given in Table 1. In the remainder of this section I will briefly review the important observational properties of these oscillations and summarize the evidence supporting spin modulation as the mechanism. Because of the rapid pace of developments this will no doubt be an incomplete review.

Oscillation Amplitudes near Burst Onset

Some bursts show large amplitude oscillations during the $\approx 1-2$ s rises commonly seen in thermonuclear bursts. An example of this behavior in a burst from 4U 1636-53 is shown in figure 1. Strohmayer, Zhang & Swank [18] showed that some bursts from 4U 1728-34 have oscillation amplitudes as large as 43% within 0.1 s of the observed onset of the burst. They also showed that the oscillation amplitude decreased monotonically as the burst flux increased during the rising portion of the burst lightcurve. Strohmayer et al. [19] reported on strong pulsations in 4U 1636-53 at 580 Hz with an amplitude of $\approx 75\%$ only ~ 0.1 s after detection of burst onset. The presence of modulations of the thermal burst flux which can approach nearly 100% right at burst onset fits nicely with the idea that early in the burst there exists a localized hot spot which is then modulated by the spin of the neutron star. In this scenario the largest modulation amplitudes should be produced when the spot is smallest, and as the spot grows to encompass more of the neutron star surface, the amplitude decreases. This behavior is consistent with the observations. X-ray spectroscopy during burst rise also indicates that the X-ray emission is localized on the neutron star near the onset of bursts. Strohmayer, Zhang, & Swank [18] also found that during burst rise the flux is *underluminous* compared with intervals later in the burst which have the same observed black body temperature, and suggested that during the rise a localized, but growing segment of the surface of the neutron star is producing the X-ray emission. As the burst progresses the burning area increases in size until the entire surface is involved.

Coherence and Stability of Burst Oscillations

The observed oscillation frequency during a burst is usually not constant. Often the frequency is observed to increase by $\approx 1-3$ Hz in the cooling tail, reaching a plateau or asymptotic limit [20]. This behavior is common to all the burst oscillation sources, and it would appear that the same physical mechanism is involved, however, there have been reports of decreases in the oscillation frequency with time.

For example, Strohmayer [21] and Miller [22] identified a burst from 4U 1636-53 with a spin down of the oscillations in the decaying tail. This burst also had an unusually long decaying tail which may have been related to the spin down episode. Muno et al. [23] also reported an episode of spin down in a burst from KS 1731-260.

Strohmayer et al. [24] have suggested that the time evolution of the burst oscillation frequency results from angular momentum conservation of the thermonuclear shell. The burst expands the shell, increasing its rotational moment of inertia and slowing its spin rate. Near burst onset the shell is thickest and thus the observed frequency lowest. The shell spins back up as it cools and recouples to the underlying neutron star. Calculations indicate that the ~ 10 m thick pre-burst shell expands to $\sim 30 - 40$ m during the flash [25,26,27], which gives a frequency shift of $\approx 2\ \nu_{spin}(20\text{ m}/R)$, where ν_{spin} and R are the stellar spin frequency and radius, respectively. For the several hundred Hz spin frequencies inferred from burst oscillations this gives a shift of ~ 2 Hz, similar to that observed. Recently, Galloway et al. [28] reported the discovery of a 270 Hz oscillation in a burst from 4U 1916-053. They measured a frequency drift of ~ 3.6 Hz during this burst and suggested that thermal expansion of the burning layer may not be sufficient to explain the shift. We note that the current theoretical estimates do not include the rotational lowering of the effective gravity and are also hydrostatic calculations. Dynamic motions of the layer may also contribute to changes in the height of the burning layer. These effects could increase the height of the burning layer and allow for greater frequency drifts than have been currently calculated.

Burst oscillations have a much higher coherence than is typical for other

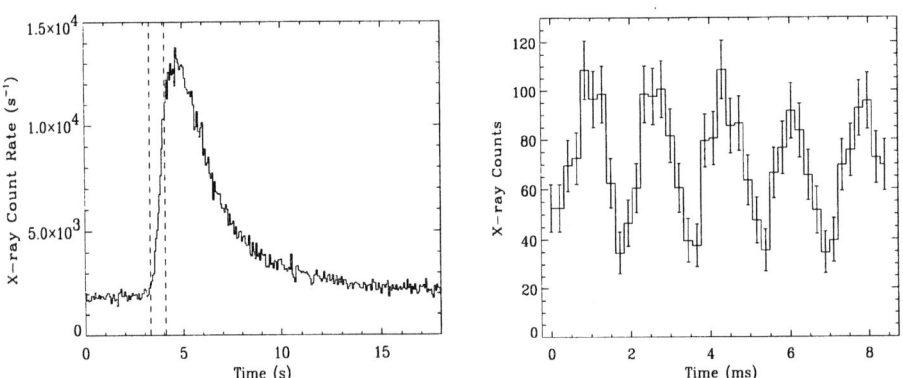

FIGURE 1. Burst oscillations during the rising phase of a burst from 4U 1636-53 at 580 Hz. The left panel shows the RXTE/PCA (2-60 keV) X-ray countrate during a burst recorded on August 20th, 1998 at 05:14:09 UTC. The right panel shows the pulsations during the rising interval denoted by vertical dashed lines in the top panel. The date were folded at intervals of $5 \times P_{spin}$ where $P_{spin} = 1.725$ ms. Note the large amplitude of the pulsations. The preburst countrate intensity of about 35 cts/sec has been subtracted.

Table 1. Burst Oscillation Sources and Properties

Sources	Frequency (Hz)	$\Delta\nu$ (kHz QPO, in Hz)	References[1]
4U 1728-34	363	363 - 280	1, 2, 3, 4, 5, 13, 14
4U 1636-53	290, 580	251	6, 7
4U 1702-429	330	315 - 344	4, 9
KS 1731-260	524	260	10, 11, 12
Galactic Center	589	Unknown	15
Aql X-1	549	Unknown	16, 17
X1658-298	567	Unknown	18
4U 1916-053	270	290 - 348	19, 20
4U 1608-52	619	225 - 325	8, 21
SAX J1808-369	401	Unknown	22,23

[1]References: (1) Strohmayer et al. [17]; (2) Strohmayer, Zhang, & Swank [18]; (3) Mendez & van der Klis [49]; (4) Strohmayer & Markwardt [29]; (5) Strohmayer et al. [19]; (6) Strohmayer et al. [20]; (7) Miller [50]; (8) Mendez et al. [51]; (9) Markwardt, Strohmayer & Swank [52]; (10) Smith, Morgan, & Bradt [53]; (11) Wijnands & van der Klis [54]; (12) Muno et al. [23]; (13) van Straaten et al. [30]; (14) Franco [32]; (15) Strohmayer et al. [24]; (16) Zhang et al. [55]; (17) Ford [56]; (18) Wijnands, Strohmayer & Franco [57]; (19) Boirin et al. [58]; (20) Galloway et al. [28]; (21) Chakrabarty [59]; (22) Heise [60]; (23) Ford [61]

quasiperiodic X-ray variations observed from neutron star systems. For example, the kHz QPO seen in many LMXBs have maximum coherence values, $Q \equiv \nu/\Delta\nu \sim 100$. Strohmayer & Markwardt [29] showed that the frequency evolution of burst oscillations in 4U 1728-34 and 4U 1702-429 is highly phase coherent. They modelled the frequency drift and showed that a simple exponential "chirp" model of the form $\nu(t) = \nu_0(1 - \delta_\nu \exp(-t/\tau))$, works remarkably well, producing quality factors $Q \equiv \nu_0/\Delta\nu_{FWHM} \sim 4,000$. Muno et al. [23] performed a similar analysis on bursts from KS 1731-26 and concluded that the burst oscillations from this source were also phase coherent. These results argue strongly that the mechanism which produces the modulations is intrinsically a highly coherent one.

The accretion-induced rate of change of the neutron star spin frequency in a LMXB is approximately 1.8×10^{-6} Hz yr^{-1} for typical neutron star and LMXB parameters. The Doppler shift due to orbital motion of the binary can produce a frequency shift of magnitude $\Delta\nu/\nu = v\sin i/c \approx 2.05 \times 10^{-3}$, again for representative LMXB system parameters. This Doppler shift easily dominates over any possible accretion-induced spin change on orbital to several year timescales. Therefore the extent to which the observed burst oscillation frequencies are consistent with possible orbital Doppler shifts, but otherwise stable over $\approx$ year timescales, provides strong support for a highly coherent mechanism which sets the observed frequency. At present, the best source available to study the long term stability of burst oscillations is 4U 1728-34. Strohmayer et al. [20] compared the observed asymptotic frequencies in the decaying tails of bursts separated in time by ≈ 1.6 years. They found the burst frequency to be highly stable, with an estimated time scale to change the oscillation period of about 23,000 year. We illustrate this behavior in figure 2 which compares the observed burst oscillation frequency of two bursts from 4U 1728-34 which ocurred ~ 3 years apart. van Straaten et al. [30] showed evidence that the frequency track made by a given burst from 4U 1728-34 is dependent on the position in the X-ray color - color diagram (a surrogate for mass accretion rate). The two bursts shown in figure 2 have similar frequency tracks and there closeness in frequency over a span of 3 years argues for a highly stable process, such as rotation, as the mechanism which sets the frequency.

Strohmayer et al. [20] also suggested that the stability of the asymptotic periods might be used to infer the X-ray mass function of LMXB by comparing the observed asymptotic period distribution of many bursts and searching for an orbital Doppler shift. The source 4U 1636-53 is a good candidate for such an effort because its orbital period is known (3.8 hrs). Strohmayer et al. [20] compared the highest observed frequencies in three different bursts from 4U 1636-53. The frequencies in these bursts alone were consistent with a typical orbital velocity for the neutron star. However, study of additional bursts reveals a greater range of highest frequencies than can likely be accounted for by orbital motion alone [31]. A possible explanation of this within the context of the spin modulation scenario is that not every burst has relaxed to the asymptotic value before the oscillations fade below the detection level. Nevertheless, the observed distribution does suggest the existence of an upper limit, which can naturally be associated with the spin frequency [31].

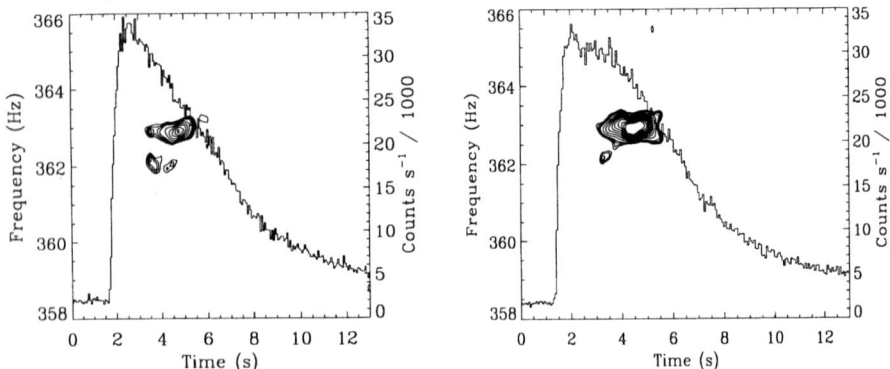

FIGURE 2. Burst oscillations in two bursts from 4U 1728-34 separated in time by ~ 3 yr. The frequency tracks are almost identical and suggests that the mechanism which sets the frequency is highly stable. The burst on the left was observed in Feb. 1996 while that on the right was seen in Feb. 1999.

Burst Oscillations and Mass Accretion Rate

Recent studies have focused on how the presence and properties of burst oscillations correlate with other properties of these sources, for example, their spectral state and inferred mass accretion rates. Muno et al. [23] were the first to conduct such a study and found that bursts from KS 1731-26 with oscillations appear to only occur when the source is on the banana branch in the X-ray color-color diagram. They also found that these bursts were all photospheric radius expansion bursts. Cumming & Bildsten [27] suggested that such bursts were likely pure Helium flashes and that it would be more likely for these to show oscillations because the radiative diffusion time is short compared to the inferred shearing time of the thermonuclear burning layer, making it more likely that a modulation would survive. Franco [32] and van Straaten et al. [30] showed that bursts from 4U 1728-34 with oscillations also occur preferentially on the banana branch, but they did not find a similar relationship with radius expansion as for KS 1731-26. They also found that the portion of a full frequency track which is present in a burst appears to also depend on mass accretion rate. Franco [32] also found that for 4U 1728-34 the strength of oscillations was correlated with position in the color-color diagram. Since other burst properties, such as peak flux, fluence and durations, are also known to correlate with mass accretion rate it is not surprising, given the fact that the physics of thermonuclear burning is dependent on mass accretion rate, that the properties of burst oscillations appear also to be strongly dependent on mass accretion rate.

Theoretical Expectations

The notion that X-ray bursts are caused by thermonuclear instabilities in the accreted layer on the surface of a neutron star is now universally accepted. There is no doubt that interesting puzzles remain and our detailed understanding is incomplete, but the basic model is firmly established. The thermonuclear instability which triggers an X-ray burst burns in a few seconds the fuel which has been accumulated on the surface over several hours. This makes it very unlikely that the conditions required to trigger the instability will be achieved simultaneously over the entire stellar surface. This notion, first emphasized by Joss [25], led to the study of lateral propagation of the burning front over the neutron star surface [26,33,34]. The short risetimes of thermonuclear X-ray bursts suggest that convection plays an important role in the physics of the burning front propagation, especially in the low accretion rate regime which leads to large ignition columns (see Bildsten [35] for a recent review of thermonuclear burning on neutron stars). These studies emphasized that the physics of thermonuclear burning is necessarily a multi-dimensional problem and that *localized* burning is to be expected, especially at the onset of bursts. The properties of oscillations near burst onset described above fit nicely with this picture of thermonuclear burning on neutron stars.

IMPLICATIONS FOR GRAVITY WAVE EMISSION

The strong ν_s^6 frequency dependence of the energy radiated by gravitational waves means that rapidly rotating neutron stars with misaligned quadrupole moments might have observationally interesting gravitational wave amplitudes. The spin periods of neutron stars inferred from burst oscillations cluster rather tightly in the range from $\sim 300 - 600$ Hz. As pointed out earlier, these frequencies are well below the maximum break-up frequencies for most modern neutron star equations of state [12]. White & Zhang [36] suggested that the observed range of spin frequencies could be produced if these neutron stars were spinning in magnetic equilibrium. However, in order for the observed frequencies to be similar would require that M and the magnetic moment μ_b be correlated. It is not presently known if such a correlation is to be naturally expected based on theoretical grounds.

Crustal Deformation Quadrupole Moments

An alternative model has been proposed by Bildsten [13]. He suggested that the spins of these neutron stars may be limited by the emission of gravity waves. The spin down torque due to gravitational wave emission is proportional to ν_s^5 so that one would expect a critical spin frequency above which accretion torques are cancelled out by gravity wave losses. By equating the characteristic accretion torque with the gravity wave torque one can determine the average quadrupole moment required to maintain the critical frequency. For the mass accretion rates characteristic of LMXBs and a critical spin frequency of 300 Hz one obtains a quadrupole $Q \sim 4.5 \times 10^{37}$ g cm^2, or about 5×10^{-8} I_{NS} [13]. The question that remains is whether or not such a quadrupole moment can be routinely generated in a neutron star.

The idea that the spin frequencies of neutron stars might be limited by gravitational radiation losses was initially proposed by Papaloizou & Pringle [37] and Wagoner [38]. Wagoner [38] argued that the Chandrasekhar - Friedman - Schutz instability would excite non-axisymmetric modes which would radiate gravity waves and limit the spin frequency. However, Lindblom [39] and Lindblom & Mendell [40] showed that this instability will only set in near the break-up frequency, which is at much higher frequency than the observed burst oscillations for most modern equations of state. Bildsten [13] has suggested that electron captures on heavy nuclei in the neutron star crust might be able to produce a quadrupole of the required amount. The basic idea is that the electron capture process produces density jumps in the crust. Since the electron capture rate is temperature sensitive in the crust a lateral temperature gradient could lead to density jumps which as a function of lateral position occur at slightly different depths in the crust. Ushomirsky, Cutler, & Bildsten [41] have investigated this process in more detail and concluded that electron captures could produce the observed quadrupole if there are $\sim 5\%$ lateral temperature variations at crustal depths where the density is in excess of 10^{12} g cm^{-3}. They also computed the dimensionless strain σ which gives a quadrupole sufficient to balance the accretion torque and found that $\sigma \sim 10^{-2}$ at near Eddington accretion rates. Although promising, more theoretical work will be required to convincingly establish whether this mechanism can produce a sufficient quadrupole moment to balance the accretion torque.

Regardless of the mechanism, if the accretion torque is indeed balanced by gravity wave losses then the amplitude of the gravitational radiation can be calculated [13,38]. The dimensionless strain h is in the range from $h \sim 10^{-27}-10^{-26}$. Although this strain is significantly less than the estimated sensitivity for LIGO I, one can greatly improve the sensitivity by pulse folding if the rotational ephemeris of the neutron star is known [14,42]. Current estimates indicate that a narrow band configuration for LIGO-II will reach interesting search limits for these neutron stars, especially for the brightest of the LMXBs, for example, Sco X-1 [14]. This also provides strong motivation for additional deep X-ray timing searches in order to detect coherent pulsations in more LMXBs and to measure the pulse ephemerides so as to improve searches for gravity wave emission. It also illustrates the strong synergism between X-ray and gravity wave astronomy in the context of neutron star binaries.

R-Mode Instability

Andersson [43] recently discovered that the r-modes of rotating relativistic stars are excited by gravitational radiation at all rotation frequencies (see also Friedman & Morsink [44]). Shear and bulk viscosity damps these modes, so whether or not they can attain significant amplitudes depends on the competition between gravitational radiation excitation and viscous damping. This discovery has led to a flurry of theoretical activity to try and understand the implications of the r-mode instability for rapidly rotating neutron stars. Here I will only give a brief summary. See the contribution by Ushomirsky in these proceedings [45] for all the details.

For accreting neutron stars the basic idea is that a star will be spun up by accretion to some critical frequency at which the r-mode instability sets in. Work by Andersson, Kokkotas & Stergioulas [16] suggested that the star would spin up to this critical frequency and then remain in equilibrium at this frequency with the accretion torque balanced by angular momentum losses from the excited r-modes. However, Levin [15] and Spruit [46] showed that this evolution is not possible because the r-modes heat the stellar interior which reduces the viscosity and increases the growth rate. Thus the r-modes would grow rapidly and spin the star down on a very short timescale ($\sim$ a few months), much shorter in fact than the timescale to spin the star up via accretion. During this time the star would be a powerful source of gravity waves, however, the timescale is so short that the effective event rate is very low [15,47].

Recent efforts have focused on studying the sources of viscous damping, most importantly the influence of the solid crust of the neutron star. Bildsten &

Ushormirsky [48] estimated the effect the solid crust would have on the damping and concluded the crust disspation would greatly exceed that from standard viscosity in the core. More recently, Andersson et al. [47] have revised these estimates and argue that the damping is not as strong as suggested by Bildsten & Ushormirky [48]. They conclude that the r-mode instability could explain the observations of most millisecond pulsars with periods between about 1.5 and 6 ms as well as the lack of any pulsars spinning faster than ~ 1.5 ms.

ACKNOWLEDGEMENTS

I would like to thank the organizers of the Astrophysical Sources of Gravitational Radiation Meeting for their gracious hospitality and for producing such a stimulating scientific program. I thank Jean Swank, Greg Ushomirsky, and Lars Bildsten for many helpful discussions.

REFERENCES

1. Lee, W. 2001, these proceedings
2. Faber, J. & Rasio, F. 2001, these proceedings
3. Barish, B. 2001, these proceedings
4. Heiselberg, H. & Hjorth-Jensen, M. 1999, ApJ, 525, L45
5. Backer, D. C., Kulkarni, S. R., Heiles, C., Davis, M. M. & Goss, W. M. 1982, Nature, 300, 615
6. Webbink, R. F., Rappaport, S. A. & Savonije, G. J. 1983, ApJ, 270, 678
7. van den Heuvel, E. P. J. & Bitzaraki, O. 1995, A& A, 297, L41
8. van Paradijs, J., van den Heuvel, E. P. J., Kouveliotou, C., Fishman, G. J., Finger, M. H. & Lewin, W. H. G. 1997, A& A, 317, L9
9. Strohmayer, T. E. 2000, in Proceedings of X-ray Astronomy '99, Stellar Endpoints, AGN and the Diffuse X-ray Background. Bologna, Italy, (astro-ph/9911338)
10. Wijnands, R. & van der Klis, M. 1998, Nature, 394, 344
11. Chakrabarty, D. & Morgan, E. H. 1998, Nature, 394, 346
12. Cook, G. B., Shapiro, S. L. & Teukolsky, S. A. 1994, ApJ, 424, 823
13. Bildsten, L. 1998, ApJ, 501, L89
14. Ushomirsky, G., Bildsten, L. & Cutler, C. 2000, in 3rd Edoardo Amaldi Conference on Gravitational Waves, (astro-ph/0001129)
15. Levin, Y. 1999, ApJ, 517, 328
16. Andersson, N. Kokkotas, K. D. & Stergioulas, N. 1999, ApJ, 516, 307
17. Strohmayer, T. E. et al. 1996, ApJ, 469, L9
18. Strohmayer, T. E., Zhang, W. & Swank, J. H. 1997, ApJ, 487, L77
19. Strohmayer, T. E., Zhang, W., Swank, J. H., White, N. E. & Lapidus, I. 1998a, ApJ, 498, L135
20. Strohmayer, T. E., Zhang, W., Swank, J. H. & Lapidus, I. 1998b, ApJ, 503, L147
21. Strohmayer, T. E. 1999, ApJ, 523, L51

22. Miller, M. C. 2000, ApJ, 531, 458

23. Muno, M. P., Fox, D. W., Morgan, E. H. & Bildsten, L. 2000, ApJ, 542, 1016

24. Strohmayer, T. E., Jahoda, K., Giles, A. B. & Lee, U. 1997, ApJ, 486, 355

25. Joss, P. C. 1978, ApJ, 225, L123

26. Bildsten, L. 1995, ApJ, 438, 852

27. Cumming, A. & Bildsten, L. 2000, ApJ, in press (astro-ph/0004347)

28. Galloway et al. 2000, ApJ, submitted (astro-ph/0010072)

29. Strohmayer, T. E. & Markwardt, C. B. 1999, ApJ, 516, L81

30. van Straaten, S. et al. 2000, ApJ, submitted, (astro-ph/0009194)

31. Giles, A. B. & Strohmayer, T. E. 2001, in preparation

32. Franco, L. 2000, ApJ, submitted (astro-ph/0009189)

33. Fryxell, B. A., & Woosley, S. E. 1982, ApJ, 261, 332

34. Nozukura, T., Ikeuchi, S., & Fujimoto, M. Y. 1984, ApJ, 286, 221

35. Bildsten, L. 1998, in "The Many Faces of Neutron Stars", ed. R. Buccheri, A. Alpar & J. van Paradijs (Dordrecht: Kluwer), p. 419

36. White, N. E. & Zhang, W. 1997, ApJ, 490, L87

37. Papaloizou, J. & Pringle, J. E. 1978, MNRAS, 184, 501

38. Wagoner, R. V. 1984, ApJ, 278, 345

39. Lindblom, L. 1995, ApJ, 438, 265

40. Lindblom, L. & Mendell, G. 1995, ApJ, 444, 804

41. Ushomirsky, G., Cutler, C. & Bildsten, L. 2000, MNRAS, submitted, (astro-ph/0001136)

42. Brady, P. & Creighton, T. 1999, Phys. Rev. D, in press (gr-qc/9812014)

43. Andersson, N. 1998, ApJ, 502, 708

44. Friedman, J. L. & Morsink, S. 1998, ApJ, 502, 714

45. Ushomirsky, G. 2001, these proceedings

46. Spruit, H. C. 1999, A& A, 341, L1

47. Andersson, N., Jones, D. I., Kokkotas, K. D. & Stergioulas, N. 2000, ApJ, L75

48. Bildsten, L. & Ushomirsky, G. 2000, ApJ, 529, L33

49. Mendez, M. & van der Klis, M. 1999, ApJ, 517, L51

50. Miller, M. C. 1999, ApJ, 515, L77

51. Mendez, M., van der Klis, M. & van Paradijs, J. 1998, ApJ, 506, L117

52. Markwardt, C. B., Strohmayer, T. E., & Swank, J. H. 1999, ApJ, 512, L125

53. Smith, D., Morgan, E. H. & Bradt, H. V. 1997, ApJ, 479, L137

54. Wijnands, R., & van der Klis, M. 1997, ApJ, 482, L65

55. Zhang, W. et al. 1998, ApJ, 495, L9-12

56. Ford, E. C. 1999, ApJ, 519, L73

57. Wijnands, R. Strohmayer, T. E. & Franco, L. M. 2000, ApJ, in press (astro-ph/0008526)

58. Boirin, L., et al. 2000, A& A, 361,121

59. Chakrabarty, D. 2000, Talk presented at AAS HEAD meeting, Honolulu, HI

60. Heise, J. et al. 2000, Talk presented at AAS HEAD meeting, Honolulu, HI

61. Ford, E. C. 2000, ApJ, 535, L119

R-modes in Accreting and Young Neutron Stars

Greg Ushomirsky

Theoretical Astrophysics, M/C 130-33, California Institute of Technology, Pasadena, CA 91125

Abstract. Recent work has raised the exciting possibility that r-mode pulsations (Rossby waves) in rotating neutron star cores may be strong gravitational wave sources. Rapidly rotating young neutron stars born in supernovae enter the r-mode instability region within the first minutes of their lives and may spin down by substantial amounts due to gravitational radiation from r-modes. Accreting neutron stars in low-mass X-ray binaries (LMXBs) are spun up by accretion to such short rotation periods that they may be unstable to r-mode pulsations as well. Gravitational waves from these neutron stars are strong enough to be detectable by second-generation, "enhanced" gravitational wave interferometers. I review the recent progress in understanding the r-mode instability in young and accreting neutron stars, with the focus on the issues of the coupling of the pulsations to the crust and nonlinear saturation amplitudes.

I INTRODUCTION

Since the work of Chandrasekhar [1], it has been known that the loss of energy and angular momentum due to emission of gravitational waves (GW) can make certain pulsation modes of rotating stars grow, rather than decay. This counter-intuitive situation is easy to understand. Imagine a star that rotates with angular frequency Ω and undergoes a pulsation of frequency $\omega_r > 0$ (in the rotating frame), so that the azimuthal dependence of the perturbation is $e^{im\phi + i\omega_r t}$. Such a wave propagates in the direction opposite to the rotation (retrograde), and carries negative angular momentum. As seen from the inertial frame, this wave has a frequency $\omega_i = \omega_r - m\Omega$. The spin of the star or the m of the mode may be large enough that $\omega_i < 0$, i.e., the mode appears prograde in the inertial frame. Gravitational radiation, since it lives in the inertial frame, removes positive angular momentum from such mode. This means that the mode's *negative* angular momentum (in the rotating frame) must become more negative, i.e., the mode amplitude must grow. This instability is generic: in any rotating perfect-fluid star one can find a mode with sufficiently high m that it is unstable to gravitational radiation reaction [2].

Gravitational radiation therefore offers an exciting possibility to set an upper limit on the spin frequencies of rotating neutrons stars (NS), both young pulsars

CP575, *Astrophysical Sources for Ground-Based Gravitational Wave Detectors*, edited by J. M. Centrella
© 2001 American Institute of Physics 0-7354-0014-8/01/$18.00

born in supernovae and old millisecond pulsars thought to be spun up by accretion in close binaries. However, the generic nature of the CFS (Chandrasekhar-Friedman-Schutz) instability does not guarantee that it is applicable to real NSs. Indeed, up until a few years ago, research in this field concentrated on stability of modes that couple to gravitational radiation via mass multipoles, mainly f-modes ("fundamental" modes, or surface waves). Frequencies of such modes are comparable to the breakup frequency[1], Ω_b, and do not change significantly with rotation. In order to satisfy the CFS instability criterion, $\omega_r(\omega_r - m\Omega) < 0$, the NS has to be rotating close to Ω_b. Moreover, real NSs have viscosity, which damps oscillations. The net result [4] (see also Lai's contribution in this volume) is that the instability in normal-fluid NSs is completely suppressed unless the temperature is between 10^7 and 10^{10} K. Even then, the critical frequency Ω_c (the frequency such that NSs with $\Omega < \Omega_c$ are stable and ones with $\Omega > \Omega_c$ are unstable) is $\gtrsim 0.9\Omega_b$. Thus, the parameter space for the CFS instability of f-modes is rather small.

This field underwent a renaissance in 1998, when Andersson [5] and Friedman & Morsink [6] realized that a special class of fluid modes, called r-modes, is CFS-unstable at *any* Ω. R-modes are global equivalents of what geophysicists call Rossby waves, which have been studied extensively since 1930s. They are predominantly horizontal fluid motions for which the restoring force is the Coriolis force. For example, sinusoidal distortions of the jet stream are caused by a long-wavelength Rossby wave propagating through the atmosphere. Rossby waves have also been detected on the Sun [7]. In the astrophysical context, they were first considered by Papaloizou & Pringle [8], who coined the term "r-modes".

For slow rotation, the r-mode velocity field is mostly transverse:

$$\delta\vec{v} = \alpha\Omega R \left(\frac{r}{R}\right)^m e^{i\omega_r t} \vec{r} \times \vec{\nabla} Y_{mm}(\theta, \phi) + \mathcal{O}(\Omega^3), \tag{1}$$

where $l = m$. A snapshot of this velocity field is shown by arrows in Figure 1a. Individual fluid elements oscillate around their equilibrium positions, while the entire pattern propagates counter-clockwise in the corotating frame. Radial motion in an r-mode is generally negligible (radial displacements at the surface are $\sim 0.1\alpha R(\Omega/\Omega_b)^2$), but is quite important for pulsations in NSs with crusts (Sec. II).

The dispersion relation for r-modes (for small Ω) is

$$\omega_r = \frac{2m\Omega}{l(l+1)} + \mathcal{O}(\Omega^3), \tag{2}$$

so they indeed satisfy the CFS instability criterion, $\omega_r(\omega_r - m\Omega) < 0$ at any spin. However, unlike the case with f-modes, various damping processes are not nearly as effective in suppressing the instability of r-modes. Remarkably, even when viscous damping is included [9,10], the critical frequency for the onset of the r-mode instability, Ω_c, can be as small as $0.1\Omega_b$! The solid line in Figure 1b shows Ω_c

---

[1] Almost independently of the equation of state, NS breakup frequency is $\Omega_b \approx \frac{2}{3}\Omega_o$, where $\Omega_o = \sqrt{\pi G \bar{\rho}}$ [3].

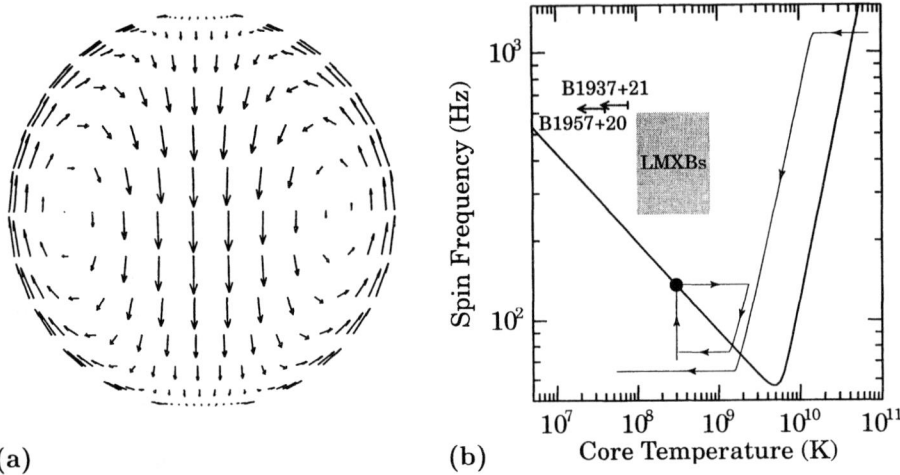

(a)   (b)   Core Temperature (K)

FIGURE 1. (a) Velocity field of an $l = m = 2$ r-mode. (b) Critical stability curve (thick solid line) for $l = m = 2$ r-modes in fluid NSs. The curve and the loop marked with arrows denote the evolution scenarios for newborn NSs and NSs in LMXBs, respectively.

for $l = m = 2$ r-mode of a particular NS model [9–11]. The low-temperature ($T \lesssim 5 \times 10^9$ K) part of the stability curve is determined by the competition between GW growth, with e-folding time $\tau_{\mathrm{gw}} \propto \Omega^{-6}$, and damping due to shear viscosity, with $\tau_{\mathrm{v}} \propto T^2$. At high temperatures ($T \gtrsim 5 \times 10^9$ K) bulk viscosity (neutrino emission which arises because compression and rarefaction of matter during oscillations drive the fluid out of β equilibrium) is the dominant dissipation mechanism, with $\tau_{\mathrm{b}} \propto \Omega^{-2}T^{-6}$.

Clearly, the parameter space for the r-mode instability to operate is quite large (for comparison, Ω_c for f-modes would be off the vertical axis in Figure 1b). A purely fluid NS can stably exist only if its Ω and T place it below the instability line. In any NS that somehow found itself above this instability line an r-mode would quickly grow, and angular momentum loss due to GW emission would quickly spin the star down and out of the instability region.

This discovery dramatically widened the applicability of the CFS instability in astrophysical situations. Two major scenarios have so far been proposed. Lindblom et al. [9], Owen et al. [12], and Andersson et al. [10] considered the effect of this instability on newborn NSs, which may rotate near breakup at birth. These NSs rapidly cool from $T_i \sim 10^{11}$ K by neutrino emission, and reach $T \sim 10^9$ K within a year from birth. However, seconds after the start of cooling they enter the r-mode instability region (follow the curve marked with arrows in Figure 1b). Any initial velocity perturbations get rapidly amplified by the CFS instability. If the amplitude of the unstable r-mode saturates near $\alpha \sim 1$, the GW torque is so large that it can spin down the NS from near-breakup to roughly $0.1\Omega_b$. In this scenario, a NS would lose as much as 99% of its rotational energy to GWs, which would be detectable by LIGO-II from as far as the Virgo cluster, where the event rate could be several per year [12]. The major uncertainty in this scenario is the nonlinear saturation of r-modes, discussed in Sec. III.

Bildsten [13] and Andersson et al. [14] pointed out that the r-mode instability may also explain the spin frequencies of NSs in LMXBs. There is now a mounting body of evidence (see contributions by Strohmayer and Bildsten in this volume) that accreting NSs in LMXBs have a rather narrow range of spin frequencies, $\approx 300 - 600$ Hz. Remarkably, despite the fact that the spinup time for a NS in an LMXB is much smaller than the lifetime of such a system, none of the observed NSs are spinning anywhere near the breakup. Bildsten [13] conjectured that accretional spinup in these systems is halted by GW emission. The steep frequency dependence of the GW torque can then account for the small range of observed frequencies, in spite of the large range of accretion rates in LMXBs. GWs could be due to either "mountains" supported by NS crusts [13,15] (reviewed in Bildsten's article in this volume) or r-modes in their cores.

Accretion of high angular momentum gas spins up the NS in an LMXB (follow the loop marked with arrows in Figure 1b) at roughly a constant temperature until it reaches the r-mode instability line. Since the r-mode amplitude needed to balance the accretion torque is rather small (corresponding to fluid displacements of tens of centimeters), it was originally believed [16,13,14] that the NS can just hover at the instability line. At this equilibrium point (thick dot in Figure 1b) the accretion torque would be balanced by GW emission from an unstable r-mode. The narrow range of spin frequencies stems from the similarity between core temperatures of accreting NSs. If such equilibrium obtains, then Sco X-1, with $h_c \approx 2 \times 10^{-26}$, would be detectable by LIGO-II, as would a handful of other bright LMXBs [13].

However, this cannot be the whole story. First, Levin [17] and Spruit [18] showed that steady-state equilibrium between accretion and r-modes is thermally unstable. Dissipation from shear viscosity in the r-mode quickly heats the star. Shear viscosity of a normal fluid scales as T^{-2}, so the increase in the core temperature decreases the viscosity, thereby increasing the r-mode's growth rate. The ensuing runaway cycle [17] is depicted by a loop in Figure 1b. Instead of just hovering near the instability line, the r-mode rapidly grows to saturation, heats up the star, and spins it down and out of the instability region in less than 1 yr. The implications of this scenario for detectability of GWs from r-modes in LMXBs is discussed in Sec. II. Second, at $T = $ few $\times 10^8$ K, typical of LMXBs (shaded box in Figure 1b), the equilibrium spin frequency would be ≈ 150 Hz, rather than $300 - 600$ Hz. Moreover, the existence of two 1.6 ms recycled radio pulsars (spins and upper limits on temperatures of which are shown by arrows in Figure 1b) means that rapidly rotating NSs are formed in spite of the r-mode instability. While it is not clear whether their *current* core temperatures place them inside the r-mode instability region, normal-fluid r-mode theory says that they were certainly unstable during spinup. It was speculated that including superfluid dissipation ("mutual friction") would move the instability curve to higher frequencies, and perhaps into agreement with the observed spins [13,14]. But tour-de-force calculations of Lindblom & Mendell [19] showed that, except for $\approx 2\%$ of the allowed parameter space, mutual friction gives only a modest increase in the damping rate, insufficient to account for the observed high spins of NSs. It now appears likely that the presence of a solid crust plays a crucial role in this mechanism, and, as argued by Bildsten & Ushomirsky [20], dissipation in the boundary layer between the solid crust and the core is indeed sufficiently strong to reconcile this discrepancy.

In the remainder of this contribution, I will describe recent work on the role of the crust in the r-mode instability (Sec. II), as well as the current understanding of nonlinear saturation of r-modes in young neutron stars. Several in-depth reviews have recently been published [21,22], and I refer the reader to them for the many issues not covered here.

II CRUSTS AND R-MODES IN LMXBS

The instability line in Figure 1b is computed [9–11] using a very simple NS model: a fluid ball with a polytropic equation of state. As discussed in Sec. I, the low-temperature part of the instability curve is determined by energy loss due to the shearing motions of the r-mode, and the damping timescale is $\tau_v = 3 \times 10^6$ s T_8^2 for a particular NS model [9]. For comparison, the gravitational wave growth timescale for the same model is $\tau_{gw} = 20$ s $(1 \text{ kHz}/f_{spin})^6$. Clearly, shear viscosity in the bulk of the star is quite feeble compared to radiation reaction. This is because the energy loss rate due to shear viscosity depends on the strength of the velocity shear, i.e., the length scale for the velocity gradient, and, for an r-mode in a fluid star, the length scale for the shear is of order the stellar radius R.

However, all but the hottest NSs have crusts that occupy the outer $\sim 1 - 2$ km of the star. The fluid motions of the r-mode are mostly transverse, indicated schematically by arrows in Figure 2a. This means that the fluid rubs against the solid crust. Neglecting viscosity is an excellent approximation in the bulk of the star, away from the crust-core boundary, and, in the absence of viscosity, there is no extra dissipation due to the rubbing as well. However, for a viscous fluid, there can be no relative motion at the crust-core boundary, leading to a boundary layer where the relative transverse velocity drops from a large value to zero [20], as indicated schematically in Figure 2a.

The width d of this boundary layer is set by the balance between the viscous term in Navier-Stokes equations, $\sim \nu \delta v / d^2$ and either the acceleration term, $\sim \omega \delta v$, or the Coriolis force term, $\sim \Omega \delta v$. The former is referred to as a viscous boundary layer, while the latter is called an Ekman layer. For r-modes in rotating stars, Ekman layer treatment is applicable [23,24], however, the widths of the layers and the damping rates are comparable since $\omega \sim \Omega$ (i.e., Coriolis force is responsible for the acceleration of the fluid elements). It turns out that the boundary layer is very thin,

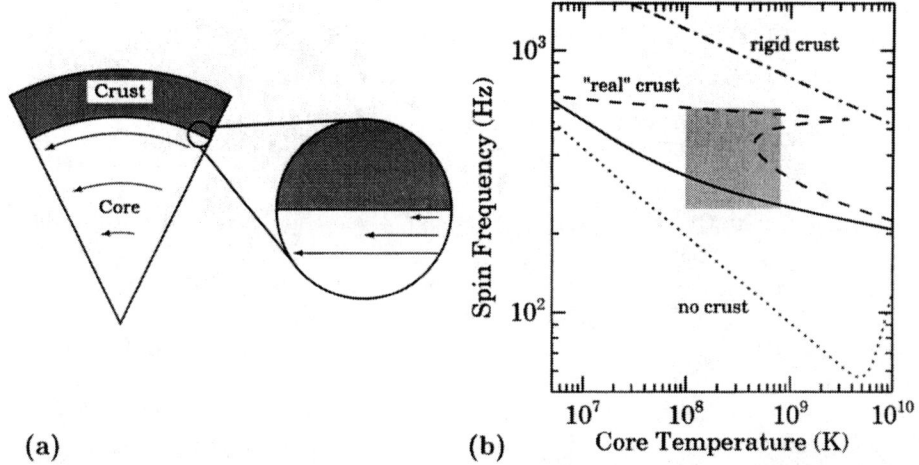

(a)   (b)

FIGURE 2. (a) Schematic depiction of r-mode fluid motion in the bulk of the star and in the boundary layer at the crust-core interface. (b) R-mode instability curves for fluid NSs (dotted line), NSs with perfectly rigid crusts (dot-dashed line), and realistic, elastic crusts (solid and dashed lines).

$$d = \left(\frac{\nu}{2\Omega}\right)^{1/2} = 1 \text{ cm } \frac{1}{T_8}\left(\frac{1 \text{ kHz}}{f_{\text{spin}}}\right)^{1/2} \tag{3}$$

Thus, the rate of viscous dissipation in the boundary layer exceeds the damping rate in the interior of the star by a factor $\approx R/d = 10^6$. Lindblom et al. [24] performed an exhaustive survey for a variety of equations of state, and found that the dissipation timescale is in the range

$$\tau_{\text{rub}} \approx 30 - 60 \text{ s } T_8 \left(\frac{1\text{kHz}}{f_{\text{spin}}}\right)^{1/2}. \tag{4}$$

The new instability line for NSs with crusts is shown by the dot-dashed line in Figure 2b. For comparison, $\Omega_c(T)$ for fluid NSs is shown by the dotted line (same as the corresponding line in Figure 1b). The critical frequency for the r-mode instability is a factor of $\sim (R/d)^{2/11}$ higher for neutron stars with crusts. Clearly, boundary layer damping plays an important role in the r-mode instability.

The dissipation rate, Eq. (4), depends on the square of the relative velocity between the bulk of the liquid and the crust. In the above discussion, it was implicitly assumed that the crust is stationary in the rotating frame, and does not participate in the oscillations. However, Levin & Ushomirsky [25] and Yoshida & Lee [26] demonstrated that this picture is not accurate for crusts of real NSs. The crust does participate in r-mode oscillations to some degree, so the amplitude of the relative velocity $\Delta v/v$ ("slippage") between the crust and the core is not 1, as is implicit in Eq. (4). The damping rate, taking into account the motion of the crust, is then [25]

$$\tau_{\text{bl}} = \tau_{\text{rub}} \left(\frac{v}{\Delta v}\right)^2.$$

(5)

The crux of the matter is that the NS crust is not very rigid. NS matter crystallizes when the ratio of the Coulomb energy of the ion lattice, $Z^2 e^2/a$ to the thermal energy, $k_B T$, exceeds ≈ 172 [27], i.e., at densities $\gtrsim 10^8$ g cm^{-3}. The ratio of the shear modulus to the bulk modulus (i.e., pressure) is rather small, $\mu/p \approx 10^{-2} - 10^{-3}$, throughout the crust. At best, the crust is more like Jell-o rather than a rigid solid. The small $(\mathcal{O}(\Omega/\Omega_b)^2)$ radial motions of the r-mode push on the crust, and are quite effective in making the crust move almost in unison with the core. Another way to look at this situation is as follows. The $l = m = 2$ torsional mode in a non-rotating crust, which has the same angular displacement pattern as the corresponding r-mode, has $f_{\text{cr}} \simeq 50$ Hz [28]. This frequency is a few times lower than the r-mode frequency in rapidly rotating stars, indicating that the elastic restoring force is quite weak compared to the Coriolis force. Therefore, at a sufficiently high spin, one would expect the crust to oscillate more or less like a liquid, with elasticity only slightly modifying the mode's structure. If the shear modulus μ of the crust were exactly zero, there would be no slippage Δv between the crust and the core. Since μ is non-zero but small, one would expect the slippage to be proportional to the ratio of the elastic restoring force to the Coriolis force, $\Delta v/v \sim (f_{\text{cr}}/f_{\text{rmode}})^2$.

Quantitatively [25], the behavior of $\Delta v/v$ with frequency is somewhat more complicated. The crust possesses a spectrum of torsional and spheroidal modes, so the whole (core+crust) system has many modes that look like simple r-modes in the core, but have non-trivial behavior in the crust. The restoring force in the crust then depends on which particular crustal mode is closest in frequency to the core's preferred frequency. At some Ω the crust may be able to effectively expel r-mode pulsations, while at other, nearby frequencies, pulsations may easily penetrate the crust. This phenomenon of avoided crossings (see Figure 1 of [25]) is well-known in asteroseismology. The net result is that the slippage $\Delta v/v$ is not a monotonically decreasing function of frequency. Instead, while typically $\Delta v/v \approx 10^{-2} - 10^{-1}$, it rises sharply to ≈ 1 at the frequencies of avoided crossings. The resulting r-mode instability lines are shown by the solid and dashed lines in Figure 2b, computed for a rather simple NS model, in which the crust is elastic, but has a constant density [25]. The solid line corresponds to the crust occupying the outer $0.1R$, while the dashed line is for a thicker crust, $0.2R$ (this is approximately the range of crustal thicknesses in realistic NS models). The instability line for a thick crust model has a peculiar double-valued shape due to an avoided crossing between the r-mode and a crustal mode at $f_{\text{spin}} \approx 550$ Hz. Qualitative behavior of the instability lines for more realistic NSs is expected to remain the same.

The new stability curves cut right through the observed range of spins of both NSs in LMXBs and of millisecond pulsars, for temperatures of interest for accreting NSs (shaded box in Figure 2). An accreting NS can continue its spinup so long as it is located to the left of the stability curve in Figure 2. However, once it crosses the critical stability curve and is located to the right of it, the r-mode will grow and halt the spinup, probably forcing the NS into a thermal runaway cycle [17,18]. The details of what happens during this runaway and how violent it is are a topic of current research. For example, the displacements induced in the crust may cause it break when the strain (which is of order the r-mode amplitude α) exceeds a critical value ($\lesssim 10^{-2}$, see § 6.1 of [15] and references therein for a summary) or melt if the heating due in the boundary layer raises its temperature above $\approx 10^{10}$ K (see [24] and Sec. III). The evolution of the spin frequency and the temperature depends on the saturation of r-mode's growth (see below). Despite these uncertainties, it is clear that, if the r-mode instability operates in accreting NSs, the non-trivial shape of the stability curve and its sensitivity to the crustal structure will be reflected in a peculiar, non-uniform spin distribution of accreting NSs and millisecond pulsars (cf. Strohmayer; Bildsten; this volume). At present, the role of r-modes in setting the spins of NSs in LMXBs cannot be ruled out, and appears to be as plausible as the crustal quadrupole GW equilibrium [13] or magnetospheric equilibrium [29] scenarios. However, many issues still need to be explored to ascertain the location of the r-mode instability curve, such as (1) the presence of the magnetic field that could couple the crust to the core, reducing the effectiveness of boundary layer damping (2) possible pinning of core superfluid to the crust (3) presence of superfluid neutrons in the crust, etc.

The presence of the crust also leads to saturation of r-mode growth. As noted by Wu et al. [30], the Reynolds number in the boundary layer, defined appropriately for oscillatory flow as $Re = \delta v^2 / \omega_r \nu$ exceeds the (experimentally measured) critical value of 2×10^5 at a rather small α. Turbulent energy loss rate scales as α^3, while energy input rate from radiation reaction is $\propto \alpha^2$. Turbulence in the boundary layer can halt r-mode growth at $\alpha_{\text{sat}} = 3.5 \times 10^{-3} (v/\Delta v)^3 (f_{\text{spin}}/1 \text{ kHz})^5$ [30], where the dependence of the prefactor on the temperature is only logarithmic. It therefore seems certain that, for most of the parameter space, the saturation amplitude in the presence of the crust is expected to be less than 1, but is not exceedingly small. This gives us the first indication that the r-mode instability may indeed be directly relevant in astrophysical situations. However, the actual value of α_{sat} is still quite uncertain. The drag coefficient used in deriving it [30] is extrapolated from experimental data for flow of water over solid surfaces, and it is not clear how well they would apply to (possibly superfluid) flow at the crust-core boundary. Convective motions in the turbulent boundary layer may lead to powerful energy losses due to bulk viscosity (neutrino emission), which would lower α_{sat}. It is also not clear what role an equipartition magnetic field that may be set up by the turbulence in the boundary layer may play. Despite the uncertainties, the work by Wu et al. [30] is clearly a step in the right direction.

Are GWs from r-modes in LMXBs detectable by ground-based interferometers?

If $\alpha_{\text{sat}} \lesssim 3.5 \times 10^{-5} (\dot{M}/\dot{M}_{\text{Edd}})^{1/2} (300 \text{ Hz}/f_{\text{spin}})^{7/2}$ (i.e., the value needed to balance the accretion torque by GW emission), and is not temperature-sensitive, then accretion-GW equilibrium [16,13,14] is still possible and stable. In this case, GW emission is persistent, and the signal strength h_c is roughly the same as in the crustal quadrupole emission scenario (see Sec. I), but would be distinguishable from it since the radiation would be emitted at $4f_{\text{spin}}/3$, rather than $2f_{\text{spin}}$. Such stable equilibrium can be ruled out on the basis of existing X-ray observations of low-luminosity, transiently accreting LMXBs [31], since the heat deposited by viscosity into the NS core would make these systems (Aql X-1 in particular) appear about 5-10 times brighter between accretion outbursts than is actually observed. However, current observations cannot rule out this possibility in bright X-ray sources, such as Sco X-1, where core neutrino emission can easily get rid of the extra heat.

If α_{sat} exceeds the above value, then LMXBs are caught in a runaway cycle [17,18]. For $\alpha_{\text{sat}} \approx 1$, the spindown phase, during which GWs are emitted, lasts only a fraction of a year. GW signal in this case is similar to that from spindown of newborn NSs, so such LMXBs should be detectable by LIGO-II from distances of tens of Mpc. However, even for a star accreting at $\dot{M}_{\text{Edd}}$, the "inactive" spinup phase lasts $\sim 10^7$ yr. The duty cycle for $\alpha_{\text{sat}} \approx 1$ is therefore quite low, and Levin [17] concluded that, in order to observe one event per year, the detector must reach a volume encompassing $\gtrsim 10^6$ galaxies (assuming $10 - 100$ LMXBs per galaxy), beyond the capabilities of LIGO-II.

Two things that have changed since the original analysis [17] may bring some renewed optimism for detecting GWs from LMXBs. First, X-ray timing observations (see Swank's article in this volume) have revealed that, in addition to 50 or so persistent LMXBs, there are perhaps as many as several hundred transient LMXBs in the Galaxy. Moreover, recent *Chandra* observations of nearby galaxies (see [32,33] and many others) have resolved most of their diffuse X-ray backgrounds into similar numbers of individual LMXBs. It is now possible to obtain arcsecond-accurate positions of these extragalactic accreting NSs. Second, since the spindown time during the runaway cycle scales as $1/\alpha_{\text{sat}}^2$, a small saturation amplitude would lead to a much more favorable duty cycle for GW emission. Using the results [30] for the turbulent saturation amplitude, the spindown phase lasts ≈ 1500 yr $(\Delta v/v)^{0.4}$, leading to a duty cycle of $\approx 2 \times 10^{-4} (\dot{M}/\dot{M}_{\text{Edd}}) (\Delta v/v)^{0.4}$. Therefore, the volume one needs to sample is much smaller. Moreover, since the spindown time much exceeds the typical observation time, GW signal is nearly monochromatic. Therefore, a viable strategy is to search for GWs from extragalactic LMXBs, using existing GW pulsar search techniques and codes, with precise positions obtained by *Chandra*.

III SPINDOWN OF YOUNG NEUTRON STARS

About a minute after its birth, a young NS can cool to $T_m \approx 10^{10}$ K, the temperature at which a crust begins to form at $\rho \approx 1.5 \times 10^{14}$ g cm^{-3}. The r-mode instability curves applicable to this situation are then those shown in Figure 2b, with Ω_c a factor of 2–3 higher than the pure-fluid value. Therefore, if a solid crust does form, the final frequency to which a young NS can spin down would be a factor of several higher than if the crust did not form. Since most of the GW signal-to-noise (S/N) during spindown is accumulated at the lowest spin frequencies, where the source spends most of the time, raising the frequency cutoff by a factor of few would correspondingly reduce the S/N.

However, the crust has to form in the presence of the shearing motion of the r-mode, which can substantially delay this process.[2] If a solid crust manages to form, viscous dissipation in the boundary layer will heat the fluid at the crust-core boundary. An r-mode with amplitude exceeding $\alpha_c = 4.7 \times 10^{-3}(T_m/10^{10} \text{ K})^{5/2}(\Omega_b/\Omega)^{5/4}$ (obtained simply by balancing boundary-layer heating with local neutrino emission and heat conduction into the core, [24]) will heat the crust-core boundary to T_m and melt the crust. The would-be crust is constantly losing energy to neutrino emission and wants to solidify. However, as soon as it does, it can be re-melted by the r-mode if its amplitude exceeds α_c. Clearly, neither a completely solid, nor a completely liquid state is possible in this case. Lindblom et al. [24] argued that the outer layers of a nascent NS will be in a mixed liquid-solid state: a NS ice flow. The viscosity in such granular flow is set by the sizes of individual ice chunks and the distances between them, and much exceeds ordinary molecular viscosity. This flow is self-regulating: chunk sizes adjust (typical size ~ 1 cm) to keep the dissipation at the level needed to keep the temperature at the melting point. The net result is that the star can spin down to a frequency much lower than if a solid crust were present. The spindown can continue as long as the r-mode has enough energy to melt the crust. Calculations of Lindblom et al. [24] show that the final frequency is $\Omega \approx 0.14\Omega_b\alpha_{\text{sat}}^{-1/4}$ (where α_{sat} is the *pure-fluid* saturation amplitude), i.e., almost the same as if the crust did not form at all.

This brings us to the central question for r-modes in young NSs: what is the saturation amplitude in a fluid star? So far, we have assumed that the r-mode would be able to grow to a "large" amplitude ($\alpha = 0.01 - 1$). Do the full equations of hydrodynamics allow this, or does the nonlinearity present in them limit the growth at small values of α? For example, radial pulsations in Cepheids easily reach amplitudes ~ 1, while g-modes in white dwarfs saturate at rather small amplitudes, so either possibility is not excluded a priori. One can approach this problem either by direct numerical simulations (i.e., studying the fully nonlinear regime), or by retaining, in addition to the linear terms, the terms of $\mathcal{O}(\alpha^2)$ or higher in the equations of motion (i.e., weakly nonlinear regime).

What magnitude of saturation amplitude may one expect? In the linearized equations of motion, employed to compute r-mode frequencies and damping times, all modes are independent and do not couple to each other. If the mode amplitude is small but finite, we expect that the effect of the nonlinear terms is to couple the linear modes and allow energy transfer among them. Suppose that lowest order, quadratic coupling to daughter modes is responsible for saturation of the

---

[2] Differential rotation present in the nascent NS may also delay crust formation.

parent r-mode. This means that the rate at which gravitational radiation reaction is depositing energy into the parent r-mode, $\sim \alpha E_{\mathrm{mode}}/\tau_{\mathrm{gw}}$, is the same as the rate at which quadratic coupling drains the energy from it, $\sim \alpha^2 C E_{\mathrm{mode}}\omega_r$, where C is a dimensionless number signifying the efficiency of coupling (Phinney, private communication). If the modes are well-coupled ($C \approx 1$), then $\alpha_{\mathrm{sat}} \sim 10^{-3}$ for stars rotating near breakup, and even smaller at slower Ω. However, there is no easy way to estimate C. It has to be computed, and there are no compelling theoretical arguments to prefer large or small values of C. The main problem becomes to identify the modes that couple most strongly to the parent r-mode, and to compute the coupling coefficients C for them. Very preliminary indications (Morsink, private communication) are that, if only coupling to other r-modes is considered, no saturation is observed until the parent mode amplitude grows to unrealistically large values. Coupling to other classes of modes (e.g., g-modes) may turn out to play a significant role.

So far, direct numerical simulations are hinting at roughly the same conclusion. Stergioulas & Font [34] imposed a large-amplitude ($\alpha \approx 1$) r-mode on a rotating star, and evolved the system for 26 rotation periods without much change in the mode amplitude. Lindblom et al. [35] evolved a small-amplitude r-mode under the influence of radiation reaction (artificially increased by a factor of 4500). They observed exponential growth, as predicted by perturbation theory, until $\alpha \approx 2$, while at $\alpha \approx 3.4$ shocks developed and quickly damped out the mode. These results seem to imply that even an r-mode with $\alpha \approx 1$ is dynamically stable, and that nonlinear hydrodynamical coupling, if it occurs, happens on timescales much longer than the dynamical time. However one needs to exercise caution in interpreting the results: these pioneering simulations were carried out for barotropic stars that do not have g-modes, and their resolution was likely insufficient to resolve the length scales ($l \gtrsim 50$) on which microscopic viscosity is able to dissipate the energy input due to gravitational radiation reaction. It is also uncertain what role the differential rotation, observed in simulations of Lindblom et al. [35] (see also [36,37]), or the magnetic field [38] will play.

If shock formation does turn out to be the mechanism that limits the growth of r-modes, then it has some interesting implications for the young NS spindown scenario [35]. Since the spindown timescale is $\propto 1/\alpha^2$, the spindown will occur in $\approx 1/10$ the time originally expected [12]. GWs will then be emitted in a much narrower band, $\Delta f \approx 0.05 f$, which somewhat simplifies the data analysis problem. Finally, after the shocks dissipate the r-mode, the star is still left rather rapidly rotating, so subsequent bursts of r-mode spindown are possible.

REFERENCES

1. S. Chandrasekhar, Phys.Rev.Lett, **24**, 611 (1970).
2. J. L. Friedman and B. F. Schutz, ApJ, **222**, 281 (1978).
3. S. Bonazzola, J. Frieben, and E. Gourgoulhon, ApJ, **460**, 379 (1996).

4. L. Lindblom, ApJ, **438**, 265 (1995).

5. N. Andersson, ApJ, **502**, 708 (1998).

6. J. L. Friedman and S. M. Morsink, ApJ, **502**, 714 (1998).

7. J. R. Kuhn, J. D. Armstrong, R. I. Bush, and P. Scherrer, Nature, **405**, 544 (2000).

8. J. Papaloizou and J. E. Pringle, MNRAS, **182**, 423 (1978).

9. L. Lindblom, B. J. Owen, and S. M. Morsink, Phys.Rev.Lett, **80**, 4843 (1998).

10. N. Andersson, K. D. Kokkotas, and B. F. Schutz, ApJ, **510**, 846 (1999).

11. L. Lindblom, G. Mendell, and B. J. Owen, Phys.Rev.D, **60**, 064006 (1999).

12. B. J. Owen *et al.*, Phys.Rev.D, **58**, 084020 (1998).

13. L. Bildsten, ApJ, **501**, L89 (1998).

14. N. Andersson, K. D. Kokkotas, and N. Stergioulas, ApJ, **516**, 307 (1999).

15. G. Ushomirsky, C. Cutler, and L. Bildsten, MNRAS, **319**, 902 (2000).

16. R. V. Wagoner, ApJ, **278**, 345 (1984).

17. Y. Levin, ApJ, **517**, 328 (1999).

18. H. C. Spruit, A&A, **341**, L1 (1999).

19. L. Lindblom and G. Mendell, Phys.Rev.D, **61**, 104003 (1999).

20. L. Bildsten and G. Ushomirsky, ApJ, **529**, L33 (2000).

21. J. L. Friedman and K. H. Lockitch, Prog. Theor. Phys. Suppl. **136**, 121 (1999).

22. N. Andersson and K. Kokkotas, Intl. J. Mod. Phys. D (2001); astro-ph/0010102.

23. M. Rieutord, ApJ, (2000), in press; astro-ph/0003731.

24. L. Lindblom, B. J. Owen, and G. Ushomirsky, Phys.Rev.D, **60**, 084030 (2000).

25. Y. Levin and G. Ushomirsky, MNRAS, (2000), in press; astro-ph/0006028.

26. S. Yoshida and U. Lee, ApJ, (2000), submitted; astro-ph/0006107.

27. R. Farouki and S. Hamaguchi, Phys. Rev. E, **47**, 4330 (1993).

28. P. N. McDermott, H. M. Van Horn, and C. J. Hansen, ApJ, **325**, 725 (1988).

29. N. E. White and W. Zhang, ApJ, **490**, L87 (1997).

30. Y. Wu, C. D. Matzner, and P. Arras, ApJ, (2001), in press; astro-ph/0006123.

31. E. F. Brown and G. Ushomirsky, ApJ, **536**, 915 (2000).

32. C. L. Sarazin, J. A. Irwin, and J. N. Bregman, ApJ, **544**, L101 (2000).

33. E. L. Blanton, J. A. Sarazin, and J. A. Irwin, ApJ, (2001), in press; astro-ph/0012481.

34. N. Stergioulas and J. A. Font, Phys.Rev.Lett, (2000), in press; gr-qc/0007086.

35. L. Lindblom, J. E. Tohline, and M. Vallisneri, Phys.Rev.Lett, (2001), in press; gr-qc/0010653.

36. Y. Levin and G. Ushomirsky, MNRAS, (2000), in press; astro-ph/9911295.

37. L. Rezzolla, F. K. Lamb, and S. L. Shapiro, ApJ, **531**, L139 (2000).

38. W. C. G. Ho and D. Lai, ApJ, **543**, 386 (2000).

Conference Rapporteurs

The Future Looks Bright[1]

Joel E. Tohline*, Peter Saulson†, and Rainer Weiss††

*Department of Physics & Astronomy, Louisiana State University , Baton Rouge, Louisiana 70803
†Department of Physics, Syracuse University, Syracuse, New York 13244
††Department of Physics, Massachusetts Institute of Technology, Cambridge, Massachusetts 02139

Abstract. As we have attempted to digest the broad range of material that has been presented at this workshop over the past couple of days, one common theme has shown through more brightly than any other: *This first decade of the new millennium is certain to be a tremendously exciting period of discovery for the fields of gravitational physics and astrophysics.* Indeed, it appears destined to be the decade in which these two fields finally blend into one. This workshop has been very timely, therefore, as we have gathered together at the beginning of the decade to assess where we stand in terms of the development of experimental and theoretical tools that will be drawn upon to carry us along this path of discovery. From our point of view, the future looks bright!

THE OBSERVATIONAL AND DATA-ANALYSIS PERSPECTIVE

Ground-Based Gravitational Wave Detectors

It is fair to say that the most exciting news came from Barish during the opening session as he reported on the status of LIGO I. Construction of the primary infrastructure for both detectors (LHO and LLO) is complete, and LHO has reached a fundamental milestone by successfully shining light down both of its 2 km-arm segments and achieving "first lock" as an interferometer. That is to say, for the first time all required servo loops functioned together to hold all mirrors at the appropriate operating point to allow full functionality as a gravity wave detector. (Toward the middle of 2001, the LLO interferometer is expected to acquire first lock in its 4 km-arm configuration.) The LIGO project development is progressing on schedule, with both observatory sites scheduled to be fully operational in 2002. Barish also reported that significant progress is being made in the GEO, VIRGO, and TAMA groups, with TAMA being the closest to completion. Reportedly, the TAMA research team expects to be in a position to run its detector in an operational mode (although not yet with optimal sensitivity) for a month or more, early in the 2001 calendar year.

In two separate talks, Fritschel and Thorne reviewed for us the technical challenges that lie ahead during the development of instrumentation for LIGO II. In an effort to build a "quantum-limited" interferometer: The laser power will be increased by more than an order of magnitude; signal recycling will be employed; the test masses will increase from 11 kg to 30-40 kg and are likely to be made from sapphire, which has a

---

[1] Supported in part by grants PHY-9900775 and AST-9987040 from the National Science Foundation.

CP575, *Astrophysical Sources for Ground-Based Gravitational Wave Detectors,* edited by J. M. Centrella
© 2001 American Institute of Physics 0-7354-0014-8/01/$18.00

much higher Q than does silica; and a quadruple suspension system (along the lines of what is now being developed by GEO) will be coupled with a system for active seismic isolation. Completion of LIGO II should bring us to a level of sensitivity that is sufficient to see the one *guaranteed* source (a Hulse-Taylor type neutron star-neutron star binary inspiral) out to a distance of 200 Mpc. Over this volume of space, the expectation is that LIGO II will detect roughly one event per month. (See further discussion, below, of Kalogera's event rate estimates.) The number of events detected by LIGO II during its first few hours of operation during the latter half of this decade is expected to equal the number of events seen during an entire year of LIGO I operation.

Thorne emphasized that, at a fundamental level, vacuum fluctuations are responsible for both the shot noise and the radiation pressure noise in a quantum-limited interferometer. Hence he encouraged everyone to stop thinking of these as separate sources of noise. With this realization, it becomes possible to envision a detector system that can beat the standard quantum limit (by perhaps a factor of five or so) in a given frequency band. Issues of this nature are being carefully considered in the design of an advanced LIGO. Signal recycling will be absolutely necessary; and an optimum design is likely to incorporate cryogenics.

Hamilton reviewed the status of resonant detectors around the world. Five detectors are operational (ALLEGRO in Louisiana, USA; AURIGA in Legnaro, Italy; Explorer at CERN in Switzerland; Nautilus in Frascati, Italy; and Niobé in Perth, Australia) at varying degrees of sensitivity in a relatively narrow frequency band near 1 kHz. As a single instrument, to date ALLEGRO has demonstrated the highest sustained level of sensitivity, but the duty cycle of other instruments has been steadily increasing. In their data analysis efforts, the groups now work closely together through an International Gravitational Event Collaboration (IGEC). To date, no significant coincident events have been seen, but by focusing on *possible* coincidences between two or more detectors, they have been able to establish significantly improved upper limits to the event rate of burst sources around 1 kHz.

Astronomy and Astrophysics in the 21st Century

As we learned in separate talks from Gaisser and White, the astronomical community — and NASA, in particular — has developed aggressive plans for advancing detector technologies and observational capabilities over the first decade of this new century. Many of the planned ground-based experiments or space-based missions will enable us to study particle as well as photon radiation from astrophysical sources at unprecedented levels of sensitivity, at unprecedented spatial resolution, and – most importantly in the context of this workshop – *from precisely the same types of astrophysical sources that are most likely to produce detectable levels of gravitational radiation.* For example, the final stages of a binary inspiral (NS-NS and NS-BH), or the core collapse associated with a supernova explosion, or a dynamical instability (such as the *r*-mode) that develops to nonlinear amplitude in a newly formed hot neutron star, is likely to generate large photon luminosities at X-ray and γ-ray energies as well as an energetic burst of neutrinos. Dynamical processes (*eg.*, relativistic jets, or shocks arising from blast waves)

300

associated with these types of events are also likely to be the primary source of high-energy cosmic rays. Through projects such as ACCESS and AUGER (cosmic rays), SuperK, KamLAND, and IceCube (neutrinos), Constellation-X, GLAST and EXIST (high energy photons) we are extremely optimistic that over the coming decade the fields of high energy astrophysics and gravitational physics will merge as research groups in both arenas put forth an enormous amount of effort to understand radiation phenomena associated with dynamically evolving "compact objects."

Looking beyond the present decade, Prince presented the plans for LISA – a space-based interferometer that will be sensitive to gravitational radiation in a frequency band that is much lower than the frequency band to which LIGO and other ground-based detectors are sensitive. With the expectation that, before the close of this decade, ground-based instruments will provide the first direct detection of gravitational radiation through our ability to extract a faint signal from the detector noise, LISA promises to take us into a band where gravitational wave signals are no longer buried in the noise of the detector. Indeed, LISA expects to be confusion limited because it will be sensitive to a much larger population of astrophysical sources, such as binary systems that include white dwarf stars (rather than just neutron stars and/or black holes). Gaisser and White also sketched out various ideas that have been proposed for high energy particle and photon detectors in future decades.

Finally in this session, Hanisch described the concept of a "National Virtual Observatory" (NVO). By utilizing high-speed networks, distributed high-performance computers, and efficient data-mining and data-analysis software, a properly constructed NVO would provide astronomers worldwide with immediate access to all archival data in a form that is easily adapted to a multitude of different analysis efforts. The accessible data will cover all regions of the electromagnetic spectrum, as it has been acquired over the years (and will continue to be acquired) through the various national observatories. The concept of a NVO has received strong support from the astronomy community's recent decadal report (the NRC's McKee-Taylor report: http://www.nap.edu/books/0309070317/html/) and promises to usher in an entirely new era of discovery.

LIGO Data and Data Analysis

Lazzarini summarized for us the information technology challenges that will accompany LIGO I in terms of data acquisition, data storage and data analysis. He also described how the LIGO collaboration's involvement with GriPhyN (the Grid Physics Network) is designed to begin to address many of these challenges. LIGO I will collect and store digital data from over 1000 separate channels, with an aggregate acquisition data rate $\sim$ 7 MBytes/s, that is $\sim$ 220 TeraBytes/yr. Less than 2% of the channels will contain strain data; the rest will provide reference data to gauge the instrument's performance and the level of various expected sources of noise. At both sites (LHO and LLO), capabilities will be available to perform a first, near real-time analysis of the data. Much more thorough, follow-on analyses – including searches for correlations between the signals from both sites – will be carried out at the various institutions that participate in the LIGO Science Collaboration (LSC) after the data is shipped to and archived at Caltech.

In separate talks, Brady and Finn outlined how the LSC plans to examine the data that is collected from the strain channels at both LIGO sites. They also estimated how successful they expect the data analyses to be at identifying real astrophysical events during the first decade of LIGO's operation. Matched filtering and FFT techniques will be used to locate the "chirp" signal that should arise from the last minute or so of the inspiral of a NS-NS binary system (as mentioned above, this is the only *guaranteed* source). Other techniques are being designed and developed to search for possible continuous-wave sources (akin, for example, to radio pulsar signals), burst sources, or stochastic signals, and, more generally, to enable the identification of unexpected sources.

Using the chirp signal from a NS-NS binary inspiral as a point of reference, Brady estimates that it should be possible to search through ~ 300 matched filter templates in real time on a single 0.6 GFlops processor. Thirty processors searching through the data in real time should be able to thoroughly examine a sufficiently dense array of templates to identify any system having $M \gtrsim 1 M_{\odot}$, if the signal arrives at the Earth with a sufficiently large signal-to-noise ratio. Fifty times this number of processors will be required to adequately search in real time for systems having masses down to $0.2 M_{\odot}$. This demonstrates why most of the analysis will have to be performed off-line, and why it will be wise to draw upon a very large, distributed network of computers such as will be provided through the GriPhyN project. As Finn has pointed out, only systems closer to us than 14 Mpc will have a signal-to-noise $\gtrsim 8$, and therefore be readily detectable in LIGO I data; at the same signal-to-noise, LIGO II should be able to detect sources out to ~ 200 Mpc. Based on the expected rate of such events in a typical galaxy (see our discussion of Kalogera's presentation, below), it is unlikely that such an event will be detected until LIGO II is operational – unless, of course, we get lucky!

THE THEORETICAL PERSPECTIVE

The second day of the workshop focused on discussions of astrophysical systems that are likely to produce detectable levels of gravitational radiation *and* whose population is large enough to provide more than one event per year that is likely to be detectable by ground-based, gravitational-wave detectors. It is important to appreciate that the astrophysical systems that are most often mentioned in this context — coalescing compact binaries, the collapse of stellar cores, and nonaxisymmetric distortions in rotating neutron stars — are interesting to study for a variety of other reasons as well. As mentioned earlier, such systems are also likely to produce bursts of neutrinos, large fluxes of high-energy photons, and/or cosmic rays. It was not surprising, therefore, to see that the second day of talks raised many interesting issues beyond just ones directly related to the detectability of gravitational radiation. It was also entirely appropriate that the day's presentations begin with a focus on the often mentioned *guaranteed* source, coalescing compact binaries.

Coalescing Compact Binaries

Event Rates

How often will LIGO (or any other gravitational-wave interferometer) detect the last phase of the inspiral of a NS-NS binary system? If we demand a certain signal-to-noise ratio, then for a given level of noise in the detector we know with reasonable confidence the distance out to which an inspiraling system will be detectable. (See, for example, Finn's article in these proceedings.) The frequency of detectable events can then be determined if we know how many short-period NS-NS binary systems reside within a spherical volume out to this distance. Unfortunately, the number density of such systems is very poorly known. At present, we know of only two NS-NS binary systems in our Galaxy; they are both relatively long-period systems (*i.e.*, in an evolutionary sense, neither system is close to its final phase of inspiral), and both have been identified only because they contain a relatively loud radio pulsar. Obviously it is difficult to accurately project from this extremely small sample what the rate of NS-NS binary mergers is in our Galaxy, let alone project what the number density of such systems is in the Universe.

In her talk, Kalogera presented the best estimates, to date, of this number density. She has taken into account models of the formation and evolution of binary systems in which both stars are expected to end up as neutron stars. She also has taken into account such factors as pulsar beaming and pulsar recycling; and an appreciation that the true population of these binary systems is likely to be dominated by the faint end of the pulsar luminosity function, whereas the observed sample is dominated by the bright end. Her estimates lead to an expectation that LIGO II will detect anywhere from $2-100$ NS-NS binary mergers every year. It remains unlikely, however, that any such events will be detectable by LIGO I during its first few years of operation because it will be operating at a significantly higher noise level that LIGO II.

Kalogera also offered updated estimates for the event rates for NS-BH ($\lesssim 1-1000\mathrm{yr}^{-1}$) and BH-BH ($\lesssim 10-1000\mathrm{yr}^{-1}$) mergers, but these estimates are understandably even less reliable than the event rates projected for NS-NS binaries because, to date, no compact binary systems containing a black hole have been found in our Galaxy.

In the context of the formation of compact binary systems in which at least one of the two stars is a black hole, McMillan summarized results from recent precise N-body simulations of the evolution of globular star clusters. The simulations have shown that, because massive stars naturally sink toward the cluster center, and the cores of globular clusters are dense enough so that three-body encounters happen relatively frequently, stellar-mass black holes in globular clusters will preferentially be trapped into tight binary systems. This is good news because it raises previous estimates of the formation rate of BH binary systems and, by inference, universal estimates of the merger event rate for such systems. (A search for tight, BH binary systems in the cores of the globular clusters of our Galaxy is unlikely to be productive, however, because the simulations also indicate that these compact binary systems are very likely to get kicked out of the globular cluster core and may escape the cluster altogether.) On a pessimistic note, however, McMillan pointed out that when this type of binary system forms, it is sufficiently tight (has a sufficiently small separation and orbital period) that it is conceivable all such systems have already coalesced by the present epoch. Hence, when this mechanism for forming BH binaries is incorporated into models that estimate merger event rates, it may not actually produce higher event rates at the present epoch.

Merger Simulations

As Rasio reviewed, more than a dozen groups around the world are now attempting to model the final phase of coalescence of NS-NS and NS-BH binaries. These studies differ from one another in many respects as the various groups are utilizing different numerical simulation tools, they are including relativistic effects with different levels of sophistication, they are assuming varying degrees of compressibility for the neutron star equation of state, and simulations are begun from a wide variety of different initial conditions. This variety reflects, in part, that the objectives of the groups are often different. As described in these proceedings, for example, Ruffert has focused his efforts on showing the possible connection between coalescing compact binaries and the central engine that drives γ-ray bursts. While focusing on the NS-BH merger problem, Lee has investigated in depth how simulation results depend on the system mass ratio, the adopted NS equation of state, and the initial spin rate of the NS. And in collaboration with Rasio, Faber has tried to show how changes in the adopted NS equation of state and changes in the spin rate of the initial stars affects the luminosity and the spectrum of gravitational radiation that will be emitted during the final coalescence of an equal mass NS-NS binary.

Perhaps most importantly in the context of this workshop, Faber has shown that the gravitational wave power spectrum from the NS-NS merger usually has three identifiable features: a characteristic (orbital) frequency f_{dyn} at which the power begins to drop noticeably below the level predicted by the inspiral of two point masses (there may or may not be a slight peak in power just before the drop); an isolated, but relatively broad peak in the power output at a frequency $f_{peak} \gtrsim f_{dyn}$ that seems to signal the main phase of coalescence; and finally a somewhat sharper peak at a frequency $f_{bar} \gtrsim f_{peak}$ associated with the formation of a spinning, triaxial bar. Because the precise location and amplitude of these features in the power spectrum varies from simulation to simulation, it is not yet clear how easy it will be to translate these general features into a useful spectral template that can be incorporated effectively into software that is being developed, for example, by the LSC to search for events in the LIGO data. Overall, though, when initially Newtonian simulations were rerun with key post-Newtonian effects taken into account, all the characteristic features shifted to lower frequencies (*i.e.*, more into the frequency band to which LIGO is sensitive), but all of the features dropped in luminosity. It is worth noting as well that as Lee has investigated NS-BH mergers, he has noticed that neutron stars with a stiffer equation of state do not get totally disrupted during their first plunge toward the BH. As a result, we might expect a single binary to produce two (or more) signals of "coalescence" that are separated in time by roughly one orbital period.

Closing out the session on coalescing compact binaries, Baumgarte presented a simple but very insightful explanation of why an ISCO (innermost stable circular orbit) must arise in any compact binary star system; and Matzner gave a brief update on the progress that is being made toward simulating the off-center collision of two black holes. Suffice it to say that the binary BH problem in its most general form has proven to be an extremely tough problem to solve numerically. It now seems quite likely that LIGO will be collecting useable data well before the numerical relativity community is prepared to predict what features are likely to be found in the signal from a BH-BH merger event.

Cosmology

In separate talks, Kosowsky and Spergel reminded us that gravitational radiation offers us a unique and direct probe of the structure and dynamical evolution of the universe at times well before recombination ($z \gtrsim 1300$), when the universe was opaque to photons. We may ultimately be able to use gravitational radiation to examine the universe back to within a Planck time of its origin.

Kosowsky explained that a stochastic radiation background is expected to have arisen from any of a number of different physical processes: inflation, primordial magnetic fields, cosmological phase transitions, or cosmological turbulence, for example. However, current models of such processes are sufficiently crude or non-unique that it is difficult to predict with a high level of confidence how strong the stochastic background is likely to be. It does not seem likely that LIGO II will have the sensitivity to detect a stochastic signal of cosmological origin, but a first-order phase transition at the electroweak energy scale is potentially detectable with a space-based instrument such as LISA.

Spergel pointed out that indirect evidence for a cosmological source of gravitational radiation may ultimately come from precise measurements of the cosmic microwave background (CMB) radiation because gravity waves will produce background temperature fluctuations that have a unique polarization signature. It is unlikely that the next two space-based CMB detectors (MAP, within the coming year; and Planck in ~ 2008) will have sufficient angular resolution to measure such a polarization signature, however. We look forward, then, to a third generation of such detectors that will have the potential to probe the universe back to Planck scales.

Stellar Collapse and Rotational Instabilities

The process of nuclear fusion proceeds in a relatively uneventful fashion in the core of massive stars until the typical atomic nucleus has grown to the size of iron or nickel, at which point no additional energy can be liberated through nuclear fusion. Then, as the iron-nickel core grows in mass to roughly $1.4M_\odot$, a catastrophic event occurs: the core collapses dynamically, almost in free-fall, under the force of gravity. If the core is not rotating, in approximately one second it shrinks to roughly 1% of its original radius, forming a neutron star (or, perhaps sometimes a black hole). Generally speaking, the neutron star's formation is accompanied by a supernova explosion that ejects the massive envelope of the star out into the interstellar medium and ultimately reveals the compact star remnant. In the absence of rotation, however, this dynamical evolution proceeds in a nearly spherically symmetric fashion and, as a result, is not expected to release much energy in the form of gravitational radiation.

As Fryer emphasized in his presentation, from the perspective of the gravitational wave community, it is much more interesting to consider scenarios in which the core of the massive star is initially rotating and, better yet, scenarios in which the *collapsed* core is rotating sufficiently fast that it becomes susceptible to the development of non-axisymmetric structure with a rapidly varying mass quadrupole moment. In this context, he pointed out that it is now becoming possible to build realistic, axisymmetric stellar models in which the nuclear-burning core is flattened by rotation. These models can then be followed in a self-consistent manner through the phase of core collapse to find out how rotationally flattened the final compact object is likely to be.

With this basic idea in mind, Brown has used 3D hydrodynamical tools to model the collapse of stellar cores having a variety of different initial conditions, for example, differing amounts of rotational kinetic energy. Consistent with earlier but much simpler studies along these lines, Brown's models show that: (1) With only a modest amount of rotation in the initial stellar core, the collapse can be halted by rotation and be "centrifugally hung up" well before reaching neutron star densities. (2) A final neutron star configuration that is significantly flattened by rotation can be produced as a result of core collapse, but the initial stellar core parameters may have to be tuned fairly carefully to generate such an outcome. By following the evolution of the most rapidly rotating collapsed core well beyond its initial formation, Brown also showed that such a core can deform spontaneously into a rotating bar-like configuration. A number of other groups have also recently demonstrated that a relatively long-lived bar-like structure is expected to develop in this manner. (See, for example, part of New's presentation at this workshop.) In addition to the burst of radiation that might be associated with the initial core collapse, this type of configuration also would be expected to generate a brief but strong continuous wave gravitational wave signal.

In her talk, New showed results from one of the first attempts to examine the dynamical stability of a stellar core that becomes centrifugally hung up during its collapse. She examined a "collapsed" model that had a great deal of its angular momentum stored it its outermost layers and, as a result, resembled a torus more than a centrally condensed spheroid. To everyone's surprise, even though this model had a ratio of rotational to gravitational potential energy $T/|W| \sim 0.14$ — that is, significantly less than the often quoted critical value of 0.27 — it spontaneously deformed into a nonlinear-amplitude, $m = 1$ nonaxisymmetric structure. This work, then, increases the theoretical likelihood that stellar core collapse will produce a detectable gravitational wave signal.

Finally, Lai showed us that analytical models can still be used very effectively to illustrate how a newly formed NS might "evolve" through LIGO's frequency band if it is sufficiently rapidly rotating $(0.14 \lesssim T/|W| \lesssim 0.27)$ to be *secularly* but not dynamically unstable to a bar-like deformation. If the NS is Dedekind-like — that is, if the NS "fluid" is on average spinning more rapidly than the natural frequency of the bar-like deformation to which it is unstable — then in $\lesssim 1$ minute (through $\sim 10^4$ cycles), its gravitational-wave signal will shift *downward* in frequency in a non-monotonic fashion from ~ 100 Hz to 0 Hz. If, however, the NS is Jacobi-like — that is, if the NS "fluid" is on average spinning less rapidly than the natural frequency of the bar-like deformation to which it is unstable — then in ~ 1 second (through ~ 300 cycles), its gravitational-wave signal will shift *upward* in frequency in a non-monotonic fashion from ~ 100 Hz to ~ 1000 Hz.

Rotating Neutron Stars, *r*-Modes, and LMXBs

Because we are presently unable to detect gravitational radiation from astrophysical systems, we must rely upon indirect methods to probe the properties of the deep gravitational potential wells associated with compact objects like neutron stars and black holes. In separate talks, Swank and Strohmayer presented exciting evidence that high time-resolution observations of the flux of X-rays from accreting compact binary systems is providing such a probe. Swank explained that quasiperiodic oscillations near 1 kHz appear to be associated with the ISCO (innermost stable circular orbit) in LMXBs (low-mass X-ray binaries) that contain a neutron star as well as LMXBs that contain a BH. Then Strohmayer showed evidence that the high amplitude X-ray brightness oscillations that have been observed by the RXTE in over half a dozen LMXBs reveals the spin frequency of the accreting neutron star. The observed frequencies cluster around $300 - 600$ Hz.

Bildsten argued that the clustering of NS rotation frequencies around $300 - 600$ Hz — which is below the angular frequency required for centrifugal breakup — in systems that are undergoing accretion is an indication that NSs are able to effectively get rid of excess angular momentum that would otherwise raise the rotation frequency above about 600 Hz. He proposed that these stars have a small, but nonzero mass-quadrupole moment and that the excess angular momentum is thereby being radiated away via gravitational radiation. The size of the required quadrupole moment can be calculated. This, in turn, allows one to predict the luminosity of the gravitational radiation that would have to be emitted by each system. His prediction is that such sources will be detectable by LIGO II.

Finally, Ushomirsky reviewed recent work which suggests that very young neutron stars are unstable toward the development of a so-called "*r*-mode" instability. It now appears likely that the *r*-mode will be able to grow to nonlinear amplitude and thereby give rise to a short-lived (less than one minute), but relatively strong flux of gravitational radiation. In addition to the late stages of binary inspiral, and the possible development of a bar-mode instability in rapidly rotating NSs, the *r*-mode now appears to offer a third type of source for gravitational radiation that is likely to be detectable by LIGO II, if not by LIGO I and its sister interferometers that are presently under development worldwide.

A TRIBUTE TO JOSEPH WEBER

Many speakers took time to pay tribute to Professor Joseph Weber, who passed away on 30 September 2000. The field of gravitational wave detection was founded single-handedly by Joe Weber in the 1960s, in an act of intellectual courage and perseverance that has few equals in the history of science.

The existence of this workshop would have been inconceivable without Joe's work. Today's resonant-mass detectors are direct descendants of Webber's first bars. Even interferometric detectors, which look so different, contain technologies pioneered by Weber. Joe and a student of his, Robert Forward, were among the first to think of the interferometric style of detector, and Forward was among the first to develop the technology seriously. Even more important than these crucial ideas about specific hardware was Joe's key idea — that gravitational waves were not just a sterile concept for theorists to think about, but could actually be detected, if only one were clever enough, patient enough, and lucky enough.

The estrangement between Joe and the community he created, caused by disagreement about interpretation of his observations from the 1970s onward, was painful. We will all miss him nevertheless, in part for his sharp observations about our present work and even more for his example of what one person can do.

List of Participants

Barry C. Barish *Caltech - LIGO*
Thomas Baumgarte *University of Illinois*
Krzysztof Belczynski *Harvard-Smithsonian Center for Astrophysics*
Lars Bildsten *Institute for Theoretical Physics*
Erin Bonning *University of Texas*
Patrick Brady *University of Wisconsin - Milwaukee*
David Brown *North Carolina State University*
Joan Centrella *Drexel University*
Philip Charlton *Caltech*
Melvyn Cheslow
Dae-Il (Dale) Choi *Drexel University*
Patrick Conley *ASI*
Orhan Donmez *Drexel University*
Joshua Faber *MIT*
L. Samuel Finn *The Pennsylvania State University*
David Fiske *University of Maryland*
Peter Fritschel *MIT - LIGO*
Chris Fryer *Los Alamos National Laboratory and UC Santa Cruz*
Thomas Gaisser *University of Delaware*
Jimin Gao *Drexel University*
Shantilal Goradia *University of Notre Dame*
James S. Graber
William Hamilton *Louisiana State University*
Robert Hanisch *Space Telescope Science Institute*
Xiaolan Huang *Drexel University*
Richard Isaacson *NSF*
Vassiliki Kalogera *Harvard-Smithsonian Center for Astrophysics*
Arthur Kosowsky *Rutgers University*
Dong Lai *Cornell University*
Albert Lazzarini *Caltech - LIGO*
William Lee *Instituto de Astronomia, UNAM*
Keith H. Lockitch *The Pennsylvania State University*
Lisa L. Lowe *Drexel University*
Andrew Mack *Rutgers University*
Ernest Mamikonyan *Drexel University*
Richard Matzner *University of Texas*

Steve McMillan *Drexel University*
Charles W. Misner *University of Maryland*
Kimberly New *Los Alamos National Laboratory*
Roland Oechslin *University of Basel*
Tim Olson *Salish Kootenai College*
Thomas Prince *Caltech*
Fred Rasio *MIT*
Maximilian Ruffert *University of Edinburgh*
Peter Saulson *Syracuse University and LIGO Livingston Observatory*
Conrad Schiff *University of Maryland*
Hisaaki Shinkai *The Pennsylvania State University*
Deirdre Shoemaker *The Pennsylvania State University*
David Spergel *Institute for Advanced Study & Princeton University*
Tod Strohmayer *NASA Goddard Space Flight Center*
Jean Swank *NASA Goddard Space Flight Center*
Bonnard Teegarden *NASA Goddard Space Flight Center*
Joel Tohline *Louisiana State University*
Kip Thorne *Caltech*
Greg Ushomirsky *Caltech*
Michel Vallieres *Drexel University*
Michael Vogeley *Drexel University*
Rainer Weiss *MIT - LIGO*
Nicholas White *NASA Goddard Space Flight Center*